TECHNEAU: Safe Drinking Water from Source to Tap

State of the Art & Perspectives

TECHNEAU: Safe Drinking Water from Source to Tap
State of the Art & Perspectives

Edited by
Theo van den Hoven
and Christian Kazner

Publishing
London • New York

Published by **IWA Publishing**
Alliance House
12 Caxton Street
London SW1H 0QS, UK
Telephone: +44 (0)20 7654 5500
Fax: +44 (0)20 654 5555
Email: publications@iwap.co.uk
Web: www.iwapublishing.com

First published 2009, Reprinted 2009 and 2010

Typeset in India by Alden Prepress Services Private Limited.

British Library Cataloguing in Publication Data
A CIP catalogue record for this book is available from the British Library

Library of Congress Cataloging-in-Publication Data
A catalog record for this book is available from the Library of Congress

ISBN: 1843392755
ISBN13: 9781843392750

Table of Contents

 TECHNEAU: Safe Drinking Water from Source to Tap. Edited by Theo van den Hoven and Christian Kazner. ISBN: 9781843392750. Published by IWA Publishing, London, UK.

Preface

Throughout the world water utilities are in a transition from a focus on mainly local and isolated issues to challenges of a more global nature. Global challenges include shortage of good quality and readily treatable resources as a result of climate change and urbanisation, aging infrastructures being vulnerable to accidental and deliberate contamination and new emerging chemical and microbiological pollutants. These global challenges greatly increase the complexity of the tasks of the water utilities.

Science can assist the utilities in these complex tasks. However, the current water science system in Europe is very fragmented with hundreds of institutions mainly addressing pieces of the puzzle. To be more effective, more innovative and more supportive to the water utilities, the science system has to scale up. More coordination and more integration are needed.

Through various instruments (such as the Framework Programmes and WSSTP), the European Commission is very much stimulating cooperation in research on European level. TECHNEAU is an example of an European cooperative research effort. The 19 M€ mainly EU-funded Integrated Project (2006–2011) brings together 30 leading research institutions and technology developers across Europe and beyond with the task to develop adaptive strategies, novel technologies and operational practices for safe water supplies.

From the beginning we have strived to establish close links to water utilities across Europe and beyond. Their needs are the basis for our research and in return they are offering their systems and infrastructure for case studies to validate and demonstrate TECHNEAU outcomes. Currently we work with more than 25 end-users in 15 countries. This cooperation is very stimulating for the researchers and it facilitates the uptake of project outcomes by the utilities.

As TECHNEAU consortium we are proud to share with you the project outcomes thus far obtained. We are happy that other research teams from Europe and beyond are present to show their latest findings.

We wish to extend special thanks to Limburg Water Company for hosting the conference, to IWA for co-organising the event and to IWA-Publishing for the support in producing the conference proceedings.

We wish all of you an inspiring conference with good networking and lots of take home messages. And don't forget to enjoy the beautiful city of Maastricht.

On behalf of the TECHNEAU consortium,

Theo van den Hoven
Project coordinator

Edited by Theo van den Hoven and Christian Kazner. ISBN: 9781843392750. Published by IWA Publishing, London, UK.

Challenges of the Water Sector

TECHNEAU – a collective effort towards safe water supply

C. Kazner[a], T. van den Hoven[b] and R. Wielinga[b]

[a]Dept of Chemical Engineering, RWTH Aachen University, Turmstrasse 46, 52056 Aachen, Germany (E-mail: christian.kazner@avt.rwth-aachen.de)
[b]KWR Watercycle Research Institute, Groningenhaven 7, 3433 PE Nieuwegein, The Netherlands (E-mail: theo.van.den.hoven@kwrwater.nl; ronald.wielinga@kwrwater.nl)

Abstract TECHNEAU, an integrated project funded by the European Commission, challenges the ability of traditional system and technology solutions for drinking water supply to cope with present and future global threats and opportunities. This will be initiated through rethinking of current water supply options and by providing researched and demonstrated new and improved technologies for the whole water supply chain. This paper presents an overview of the methodology and key results from the FP6 project TECHNEAU.

Keywords adaptation, drinking water supply, modelling, operation, treatment technologies, water quality monitoring

INTRODUCTION

The water supply sector worldwide is facing tremendous challenges to assure drinking water safety. Climate change, continued population growth and urbanisation, emerging contaminants, aging infrastructure, an increasing shortage of good quality and readily treatable resources as well as more demanding regulators and consumers are just a few examples of these challenges. During the last years, several research endeavours and construction measures have started to identify and implement solutions for adaptation of the water supply chain to global change, especially in these parts of the world which are severely impacted by main drivers such as water scarcity and population growth. The TECHNEAU project running for 5 years and currently the largest EU project on drinking water addresses these challenges in an integrated manner by bringing together more than 100 water supply researchers and technology developers from 30 institutes across Europe and abroad to develop adaptive strategies, novel technologies and operational practices.

Objectives

The project has the following main objectives:

Integrated approach to the design and operation of the water supply chain

- Adaptive strategies including flexible systems for water supply.
- An integrated framework for risk assessment and risk management and risk communication.
- Understand consumer expectations and perception around drinking water issues.

Novel and improved technologies for water treatment and water quality monitoring

- Concepts and technologies for the cost effective production of safe drinking water both for large and small supply systems. Focus will be given to membrane and oxidation based multi-barrier water treatment systems, providing safety against a broad spectrum of chemical and microbiological contaminants. All processes and practices should enable to reduce the amount of disinfectant residuals in distribution systems and should minimise deterioration of water quality in distribution networks and plumbing systems.
- New monitoring technologies for on-line, on site detection and early-warning for water quality threats and for the control of treatment processes. The project will also develop bioassays for lab-scale detection of categories of contaminants.

Modelling and improved operational practices

- Web based, predictive and integrated computer models enabling control and optimisation of existing water supply systems with respect to water quality, reliability, customer service level, environmental impact and costs.

Widespread communication and dissemination

- This includes a web based portal to access outputs generated by the project, newsletters, conferences and regional technology platforms.

This paper presents the global picture of the EU funded project TECHNEAU.

METHODOLOGY

To deliver its objectives, TECHNEAU has been divided into eight integrated Work Areas (WA, cf. Figure 1). The integrated approach to the design and operation of the water supply chain is addressed in WA1 'Rethink the system', WA4 'Risk Assessment and Risk Management' and in WA6 'Consumer Acceptance and Trust'. WA1 Rethink the system shall evaluate the capability of present water systems to cope with ongoing and future challenges, and identify to which degree they are future proof and where they are vulnerable. Adaptive strategies including new supply concepts and multi-scale systems, technologies and management options, will be developed to meet future requirements.

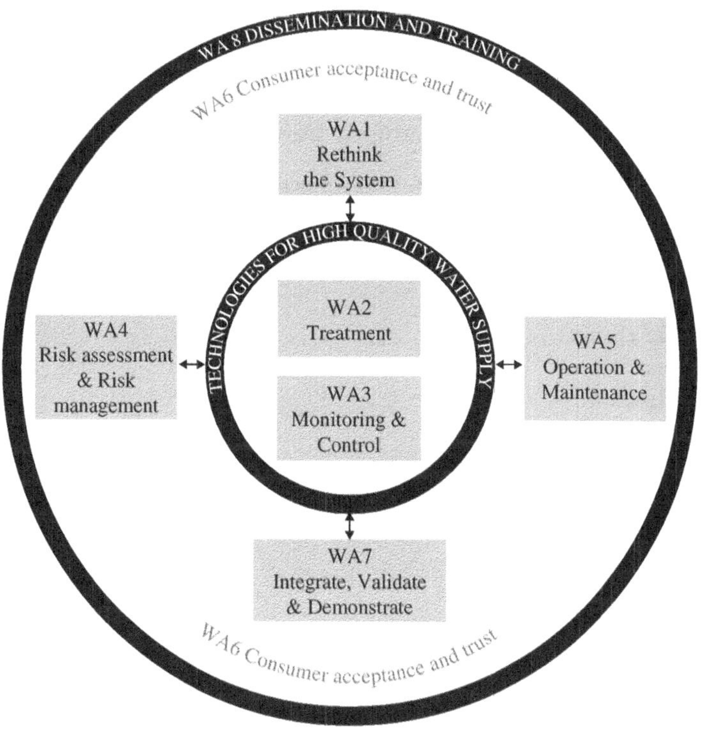

Figure 1. Conceptual approach of TECHNEAU.

WA4 shall develop an integrated framework for risk assessment and risk management for large and small supply systems, taking account of security issues. Consumer acceptance of new technologies and trust in the supplied water is a main driver for TECHNEAU and will be addressed in WA6. Emerging Technologies and monitoring tools will constitute the core of TECHNEAU and will be developed in WA2 'Treatment technologies' and WA3 'Monitoring and control technologies'. These work areas will identify research and further develop promising technologies for treatment, monitoring and control fitting within the future supply systems and the adaptive strategies. Modelling and improved operation of existing supply systems is the subject of WA5 'Operation and maintenance'. An essential part of this work area is the development of integrated computer models that enable simulation and optimisation of supply systems with respect to water quality, costs, customer service level and environmental impact. Supply options, technologies and management practices thus developed will be integrated, demonstrated and validated in real life cases in WA7.

KEY RESULTS

Rethinking the system: challenges, trends and adaptation of water supply

Data were gathered on current and future trends in different regions in Europe and abroad represented in the TECHNEAU project and on basis of these reports, the following main trends were identified (Segrave *et al.*, 2007):

- Climate change
- Urbanisation
- Globalisation
- Emerging pollutants
- Increasing energy demand and costs
- Ageing infrastructure
- Community involvement & consumer intelligence
- Emerging technologies
- Increase bottled water use
- The efficiency driven water sector.

In addition, a major challenge exists in reaching the Millennium Development Goals (MDG): to halve the proportion of the population without access to safe drinking water. This can not so much be regarded as a trend, but is a major aspect which has to be recognized in developing strategies for the future decades in the developing as well as in the industrialized countries.

Some of the trends can be considered to be a threat for the drinking water sector and an anticipation is required (e.g. micropollutants). Other trends can be considered to be an opportunity (e.g. emerging technologies). Consumer involvement can be regarded as a threat as well as an opportunity; for example, increased transparency improves ownership and on the other hand, too large openness may serve to erode consumer trust.

Based on trends, impacts and challenges for the water sector are derived, and actions are defined to cope with these impacts. Taking into account that especially the medium and long term development of trends is highly unsure – e.g. while the trend report expects an ongoing increase of bottled water sales, the US bottled water market decreased perceptibly in the course of the financial crisis – and given that water infrastructure has to be planned for a long time frame, guiding principles have been defined which are generally valid.

The following guiding principles for the development of adaptive strategies have been defined (Pronk and Kazner, 2008):

- Integration
- Flexibility
- Locally optimised solutions.

These guiding principles and the derived adaptive strategies have been applied and tested in a set of case studies representing most relevant boundary conditions: Cyprus (adaptation to water scarcity), Brabant Water, NL (increased flexibility in sources and supply), Riga (decreasing water demand and aging infrastructure), South Africa (safe water supply for rural areas under developing country conditions). The case studies proved that adaptation of our supply infrastructure is feasible and alternative sources are available. Successful adaptation requires, however, appropriate and well acquainted legal, financial and administrative boundary conditions to transform the water infrastructure in a sustainable way with the right timing and economic efficiency (van Schaik *et al.*, 2009).

Treatment technologies for high quality water supply

New and changing boundary conditions require new water treatment technologies and treatment trains based upon the multi-barrier principle to cope for example with emerging pollutants. Among these new technologies the project's research focuses mainly on membrane technologies and oxidation processes with the following fields:

1. Technologies for the exploitation of non-conventional water sources such as sea water desalination.
2. Development of the oxidation-biofiltration-membrane (OBM) process.

3. Optimal integration of membrane filtration in drinking water treatment including treatment of concentrates.
4. Oxidation processes.
5. Compact units for decentralized supply (small scale systems).

Some of these technologies and studies address certain issues related to Southern Europe. Most relevant to this region are technologies that address the issue of water scarcity such as dense membrane processes. Improved technologies can treat non-conventional water sources, sources with high loads of pollutants and/or organic matter, or treat the "wastes" from conventional water treatment files:

- Membrane technologies are an important focus with ceramic membranes showing potential for the direct treatment of water. Another application under investigation is the treatment of backwash water from rapid sand filters. Ceramic membranes are becoming more competitive compared to conventional polymeric membranes allowing a broader application. Another field of investigation is the treatment of concentrates from nanofiltration or reverse osmosis (Schrotter *et al.*, 2009).
- A novel approach, the Oxidation Biodegradation membrane Filtration technology, is able to treat impaired water sources with high levels of colour or high concentrations of contaminants, and produces small amounts of wastes. Because it includes multiple barriers, it can handle a variety of contaminants and proved in this study to be most adapted to low and medium Dissolved Organic Carbon containing water (Azrague *et al.*, 2009). It is also appropriate for decentralised supply.
- The effect of chemical oxidants on organic and inorganic micropollutants is another important aspect of the research work. A database of kinetic of mechanistic information on more than 300 compounds including pharmaceuticals, endocrine disruptors, and pesticides as well as inorganic compounds was built in order to understand the kinetic and by-product formation during chlorine treatment. A software is also being developed to predict the fate of micropollutants during ozone or chlorine treatment, taking into account the reaction mechanisms and the reactor hydraulics. This should constitute a useful tool for the prediction of DBP formation.

Decentralised supply, an issue which is not only relevant in developing countries, but also in Europe, is another main focus of this work area and shall contribute to solutions for reaching the MDGs (Peter-Varbanets *et al.*, 2009).

Sensors and monitoring

Fast, sensitive and reliable monitoring and control technologies are indispensable for the production of safe drinking water. They allow for the surveillance of source water quality and the detection of biological and chemical threats, thus defining the boundary conditions for the subsequent treatment and providing early-warning in case of unexpected contaminations; they are mandatory for the permanent control of the treatment process and the efficacy of each single treatment step, and they safeguard the high quality of the finished water. Furthermore, appropriate analytical techniques are indispensable for the detection of changes in water quality during distribution and for monitoring drinking water quality at consumers' tap. Reliable monitoring technologies contribute to a large extent to the consumers' trust in a high drinking water quality (Sacher and Hambsch, 2009).

Various teams work on novel water quality detection systems, both for field and laboratory applications. The following technologies, previously developed, have reached a status which allows an estimation of their potential in practice.

- Biological detection systems for contaminants in water, based on bioassays and DNA-arrays.
- A generic method for the concentration and purification of pathogens.
- Analytical tools for new algal toxins.
- A novel assay for AOC determination.
- A pathogen growth potential (PGP) assay.
- A toolbox for identifying and quantifying membrane biofouling in drinking water treatment.
- Flow cytometric based toolkit to quantify total and viable bacterial concentrations in water.
- Methods for quantification of indicator bacteria and pathogens by FISH-techniques.
- Electronic nose/tongue technology for monitoring of drinking water quality.
- Fish real time biomonitor for drinking water protection.
- Multi-sensor platform.

Risk assessment and risk management

Risk assessment and risk management builds a main cornerstone of TECHNEAU. The World Health Organisation (WHO) concludes that end-product testing in treatment and distribution systems is not sufficient to guarantee safe drinking water to consumers. Instead, a holistic and proactive risk management approach from catchment to consumer is emphasised. The Framework for Safe Drinking

Water presented by WHO includes the preparation of Water Safety Plans (WSPs) in which risks in source waters, treatment systems and distribution networks should be assessed and managed. WSPs are currently being implemented in several countries and are expected to become an increasingly important part of water management in both developed and developing countries. The Bonn Charter strategy presented by the International Water Association (IWA) further specifies the use of WSP in drinking water management.

In work area 4, a generic framework and specific methods for an integrated risk assessment and risk management in WSP have been developed. Examples of methods are (cf. Figure 2): TECHNEAU *hazard database* (THDB), with numerous generic hazards listed; a *risk reduction option database*, suggesting measures to reduce and/or control the associated risks; and *a decision support tool* for cost-effective selection of risk reduction options. Good examples on integrated risk assessment carried out in six different *case studies* are presented in case study reports intended as guiding examples to water utilities. Finally, *training sessions* for practical training in TECHNEAU methods are arranged as three seminars in 2008 in different parts of the world: Portugal, Norway and South Africa. Documentation and experiences gained from these training seminars as well as all results from the project are accessible through the project website (TECHNEAU, 2009).

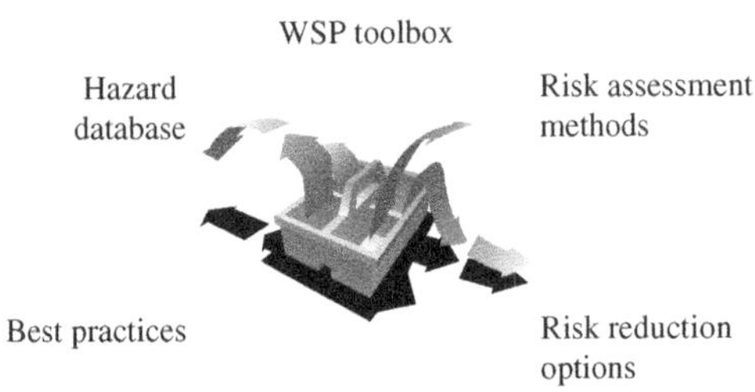

Figure 2. The Water Safety Plan toolbox of TECHNEAU.

Operation and maintenance

As most of the water supply systems in Europe are already built and in operation since decades, optimising operation and maintenance of existing systems are another crucial and previously often neglected field of research (Eikebrokk, 2009). Continued optimisation research on enhanced coagulation (EC) and ozonation-biofiltration (OBF) in the laboratory and at pilot scale was conducted in Riga and Trondheim and yielded more knowledge on optimum performance requirements. The EC optimization procedure was implemented with good results at the VIVA pilot treatment plant in Trondheim.

Computer models offer great opportunities to improve the performance of water supply systems as they enable an easy analysis of complex treatment systems. Hence a key product within this work area is the harmonisation of the models from the different European modelling schools to derive an 'European Water Treatment Simulator' (Ross *et al.*, 2009) including process modules for softening, rapid sand filtration, coagulation, pH adjustments, ozone, aeration, chlorination, activated carbon filtration, biological filtration. A first prototype is meanwhile available and can be accessed through the project website.

Continued pilot and field research on particles, sediment, corrosion and biofilm growth in distribution systems yielded interesting findings on the iron/manganese ratio of deposits in water pipes, which is useful for source tracking and definition of remedial actions (Vreeburg *et al.*, 2009).

Operational procedures for networks, including flushing, valve operation, rehabilitation and disinfection strategies were established and applied to several cases. Within the project a method to analyse particle transport in water networks yielded promising results in Lisbon, Riga and Bergen.

The project consortium completed sampling campaigns and field tests in the Delhi case study on Managed Aquifer Recharge (Grützmacher *et al.*, 2009). The vulnerability of river bank filtration systems to global environmental changes was assessed.

Consumer acceptance and trust

By adopting an end-user based perspective this work area looks at consumer expectations, decisions and behaviours around individual water consumption as well as the way in which water technologies and water management systems are accepted within communities. It is critical that the design, regulation and management of water supply systems occurs in-line with individual and community expectations.

A mixture of qualitative and quantitative methodologies has been applied in order to examine consumer acceptance and trust in the water industry and new technologies. Hence one multi-country survey and one set of focus groups and interviews were conducted within this work area (Fife-Schaw *et al.*, 2009).

Integration, validation and demonstration

As major activity, especially in the second half of the project, outcomes are being tested in case studies with water utilities throughout Europe and beyond (Figure 3), a.o. in Lisbon (Portugal), Bergen (Norway), Riga (Latvia), Windhoek (Namibia), New Delhi (India) and Amsterdam (The Netherlands). This promotes the uptake of the developed technologies and gives the research team feedback for further improvement of the technologies.

Figure 3. Case studies, demonstration sites and cooperating water utilities within TECHNEAU.

CONCLUSIONS

TECHNEAU provided a comprehensive overview on the current state-of-the-art in drinking water practices with a set of reports available on the project website. Through the development of adaptive strategies and novel technologies as well as operational practices for safe water supplies the project contributes to safeguarding drinking water supply while addressing identified current and future challenges. The project also strives to contribute to achieve the MDGs by transferring and adapting European practices, such as river bank filtration or membrane filtration, to the developing countries' context. Comprehensive information on the project can be found on the project website: www.techneau.eu

ACKNOWLEDGEMENTS

The European Commission is acknowledged for co-funding the TECHNEAU Project under contract number 018320 in the Global Change and Eco-system sub-priority of the 6th Framework Programme.

REFERENCES

Azrague, K., Osterhus, S.W. and Leiknes, T. (2009). Development and evaluation of a new concept for drinking water treatment: The OBM process. In: *TECHNEAU: Safe Drinking Water from Source to Tap – State-of-art & Perspectives*, T. van den Hoven and C. Kazner (eds.), IWA Publishing, London, UK.

Eikebrokk, B. (2009). Water treatment: Optimization with respect to what? In: *TECHNEAU: Safe Drinking Water from Source to Tap – State-of-art & Perspectives*, T. van den Hoven and C. Kazner (eds.), IWA Publishing, London, UK.

Fife-Schaw, C., Barnett, J., Morrison, G., Åström, J. and Hartung, J. (2009). Enhancing consumer relations: the role of trust and confidence. In: *TECHNEAU: Safe Drinking Water from Source to Tap – State-of-art & Perspectives*, T. van den Hoven and C. Kazner (eds.), IWA Publishing, London, UK.

Grützmacher, G., Hülshoff, I., Wiese, B., Moreau-Le Golvan, Y., Sprenger, C., Lorenzen, G. and Pekdeger, A. (2009). Function and relevance of aquifer recharge techniques to enable sustainable water resources management in developing or newly industrialized countries. In: *TECHNEAU: Safe Drinking Water from Source to Tap – State-of-art & Perspectives*, T. van den Hoven and C. Kazner (eds.), IWA Publishing, London, UK.

Pronk, W. and Kazner, C. (2008). Adaptive strategies – integrated approach and flexibility under recognition of local conditions. Techneau deliverable D 1.2.1. http://www.techneau.org/fileadmin/files/Publications/Publications/Deliverables/D1.2.1.pdf (accessed on 14 Apr 2009).

Peter-Varbanets, M., Zurbrügg, C., Swartz, C. and Pronk, W. (2009). Decentralized systems for potable water and the potential of membrane technology. *Water Research*, **43**(2), 245–265.

Ross, P.S., van Schagen, K.M., van der Helm, A.W.C., Dudley, J. and Rietveld, L.C. (2009). Modelling with the European drinking water treatment simulator. In: *TECHNEAU: Safe Drinking Water from Source to Tap – State-of-art & Perspectives*, T. van den Hoven and C. Kazner (eds.), IWA Publishing, London, UK.

Sacher, F. and Hambsch, B. (2009). State-of-the art in drinking water monitoring. In: *TECHNEAU: Safe Drinking Water from Source to Tap – State-of-art & Perspectives*, T. van den Hoven and C. Kazner (eds.), IWA Publishing, London, UK.

Schrotter, J.-C., Breda, C., Vince, F., de Roubin, M.-R., Roche, P. and Bréant, P. (2009). Development of advanced treatment trains and decision support tools to meet future drinking water challenges. In: *TECHNEAU: Safe Drinking Water from Source to Tap – State-of-art & Perspectives*, T. van den Hoven and C. Kazner (eds.), IWA Publishing, London, UK.

Segrave, A., Pronk, W., Ramaker, T. and Zulegg, S. (2007). Global trends affecting the water cycle – winds of change in the water world. Techneau deliverable D 1.1.7. http://www.techneau.org/fileadmin/files/Publications/Publications/Deliverables/D1.1.7.pdf (accessed on 14 Apr 2009).

TECHNEAU (2009). Project Website: http://www.techneau.eu.

van Schaik, H., Tyagi, A. and Kasten, T. (2009). Theme 1: Global changes and risk management. Thematic Report. 5th World Water Forum, Istanbul 2009.

Vreeburg, J.H.G., Menaia, J., Branco, L., Benoliel, M., Aprisco, P., Rebola, N. and Cordeiro, B. (2009). Conceptual model for discolouration in drinking water systems: Who's to blame and what to do? In: *TECHNEAU: Safe Drinking Water from Source to Tap – State-of-art & Perspectives*, T. van den Hoven and C. Kazner (eds.), IWA Publishing, London, UK.

Highlights from EU Research

Water management in the city of the future

K. Vairavamoorthy, C. Howe and P. van der Steen

UNESCO-IHE Institute for Water Education, Westvest 7, Delft, 2601 DA, Netherlands (E-mail: k.vairavamoorthy@unesco-ihe.org; c.howe@-unesco-ihe.org; p.vandersteen@unesco-ihe.org)

Abstract With increasing global change pressures (urbanisation, climate change etc.) coupled with existing un-sustainability factors and risks inherent to conventional urban water management, cities of the future will experience difficulties in efficiently managing scarcer and less reliable water resources. SWITCH is an EU 6th Framework integrated research project that aims to develop scientific, technological and socio-economic solutions for the sustainable and effective management of water in the city of the future – 2050. The project involves 32 partners from across the globe, including 17 from the EU and 12 from developing countries. SWITCH calls for a paradigm shift based on several key concepts of urban water management including: resilience of urban water systems to global change pressures; interventions over the entire urban water cycle; reconsideration of the way water is used (and reused); greater application of natural systems for water and wastewater treatment; governance and financial management structures, covering the entire urban water cycle.

Keywords flexibility, global change pressures, paradigm shift, robustness, sustainable development, uncertainties

INTRODUCTION

It is widely accepted that one of the major challenges of the 21st century is to provide safe drinking water and basic sanitation for all. Presently, more than 1 billion people lack access to improved water sources, and over 2.6 billion people lack access to basic sanitation – nearly all of these people live in developing countries (UNICEF/ WHO, 2004; Elimelech, 2006). Unsafe water and poor sanitation are one of the major causes of disease in the world. Every year, unsafe water coupled with a lack of basic sanitation causes death of at least 1.6 million children under the age of five years (WHO/UNICEF, 2006). Consequently, over a billion people are locked in a cycle of poverty and disease (UNICEF/ WHO, 2004).

Providing adequate water supply and sanitation, particularly in urban areas, is a challenging task for governments throughout the world. This task is made even more difficult due to predicted dramatic global changes. For example, climate change is predicted to cause significant changes in precipitation patterns and their variability affecting the availability of water; technological and financial challenges for maintaining and upgrading infrastructure assets to deliver water to

 TECHNEAU: Safe Drinking Water from Source to Tap. Edited by Theo van den Hoven and Christian Kazner. ISBN: 9781843392750. Published by IWA Publishing, London, UK.

all sectors while maintaining the quality of water distributed to the various users; population growth, urbanization, industrial activities are leading to a dramatic increase in water consumption and wastewater discharge.

Under the aforementioned circumstances current models of urban water management (UWM) and their corresponding infrastructure have already failed or are on the verge of collapse from the perspective of cost effectiveness, performance and sustainability. Cities are now faced with difficult future strategic decisions (e.g. the choice between centralised and the decentralised systems; the choice regarding the level of involvement of individual citizens, NGOs and companies; the choice between an institutional framework where separate institutions are responsible for a certain element of the urban water system or moving towards a more integrated institutional set-up).

This paper presents the current challenges and possible future challenges to be faced by cities in relation to urban water due to the global change pressures and proposes innovative approaches for sustainable UWM. Finally, it presents the research process, approaches and research areas undertaken in the SWITCH project as a way forward for a paradigm shift towards the achievement of the goals of sustainable UWM for the city of the future.

FUTURE CHALLENGES

Cities around the world are facing a range of dynamic regional and global pressures (Figure 1), that will impact urban water management.

Climate change

There is little dispute that the earth system is undergoing very rapid changes as a result of increased human activities. Clearly these changes will severely impact the urban water cycle. Components of the urban water cycle, like water supply, wastewater treatment, and urban drainage etc. are generally planned for life-spans over several decades. Hence there is a need for us to pay attention to these changes in the context of how these systems will be designed and operated in the 'city of the future'.

Although the regional distribution is uncertain, the frequency and severity of droughts are likely to increase in some areas as a result of a decrease in total rainfall, more frequent dry spells, and higher evapotranspiration. Flood frequencies are also likely to increase in many areas, although the amount of increase for any given climate scenario is uncertain and impacts will vary among basins. The impact of climate change will be observed throughout the world; only the types and degree of vulnerabilities will be different.

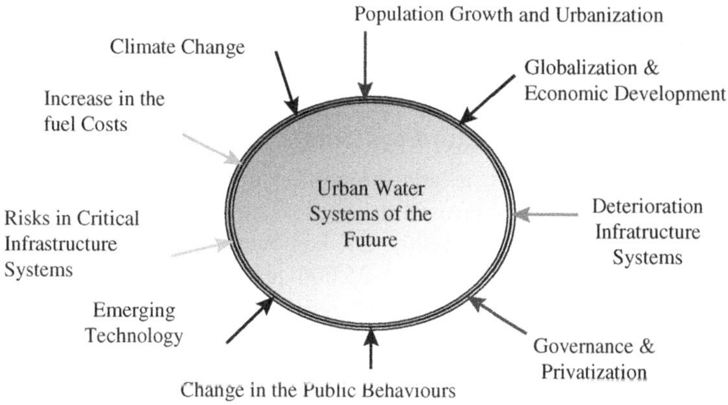

Figure 1. Global pressures affecting urban water (Khatri and Vairavamoorthy, 2007).

In addition to water scarcity and flooding, water quality problems may increase where there is less flow to dilute contaminants introduced from natural and human sources. The potential increase in water temperature could alter the rate of bio-geo-chemical processes and lower dissolved oxygen concentrations. Also, increased runoff coupled with more frequent stormwater overflow activities will increase the load of pollutants into the water bodies. There are other more obvious impacts, such as increased risk of damage to stormwater infrastructure and facilities (e.g. underground drains, levee banks, pump stations etc.) due to higher peak flows, and other less obvious ones such as increased risk of pipe failure and collapse due to drier soil conditions.

Population growth and urbanization

Population growth and urbanization will be one of the world's most important challenges in the next few decades. From 2007 the urban population of the world will exceed the rural population.

Urban settlements in the developing countries are, at present, growing five times as fast as those in the developed countries. In developing countries urban population is predicted to grow from 1.9 billion in 2000 to 3.9 billion in 2030, averaging 2.3% per year. On the other hand, in developed countries, the urban population is expected to increase, from 0.9 billion in 2000 to 1 billion in 2030 overall growth rate 1% (Brockerhoff, 2000).

Population growth and rapid urbanization will create a greater demand for water while simultaneously decreasing the ability of ecosystems to provide more regular and cleaner supplies. Moreover, rapid increase in built-up areas disturbs the local hydrological cycle and environment by reducing the natural infiltration opportunity and producing the rapid peak storm water flow.

Deterioration of infrastructure systems

In most cities of the world, a large proportion of this infrastructure is over 100 years old, placing it at increased risk for leaks, blockages and malfunctions due to deterioration. For example, the UK has over 700,000 km of mains and sewers pipes, and these require over 35,000 maintenance works per month (Vahala, 2004). Consequently the cost of rehabilitation of water infrastructure system is increasing substantially due to their deterioration. In Germany, for example, estimates are that over the next 15 years, €12 billion per annum is required to keep the urban wastewater systems operational (Hiessl *et al.*, 2001). Higher rates of water leakage means higher water losses and higher chances of in-filtration and ex-filtration of water. This will create higher chances of drinking water contamination and outbreak of water-borne diseases (Vairavamoorthy *et al.*, 2007a).

The deterioration processes is more severe for the developing countries due to the poor construction practices, little or no maintenance and rehabilitation activities, and operation at higher capacities than design. This is compounded by the lack of records and data about the location and condition of the infrastructure and the lack of efficient decision support tools for managing the infrastructure.

SWITCH: INNOVATION IN URBAN WATER MANAGEMENT

Currently, providing safe water supply and basic sanitation services across the globe is a major challenge. With increasing global change pressures coupled with existing un-sustainability factors and risks inherent to conventional urban water management, cities of the future will experience difficulties in efficiently managing scarcer and less reliable water resources. Realising the shortcoming of conventional urban water systems, SWITCH calls for a paradigm shift.

SWITCH (Sustainable Water Management Improves Tomorrow's Cities Health) is an EU funded action research program being implemented and co-funded by a cross-disciplinary team of 33 partners from across the globe, including 17 from the EU and 12 from developing countries (see Figure 2). The "consortium" is from the fields of academic, urban planning, water utility and consulting interests. This network of researchers and practitioners are working directly with stakeholders in ten cities around the globe. The overall goal behind this global consortium is to catalyse change towards more sustainable urban water management in the "City of the Future". Demonstrating research and sharing knowledge across a range of different geographical, climatic and socio-cultural settings is expected to lead to global adoption and acceleration of more sustainable solutions.

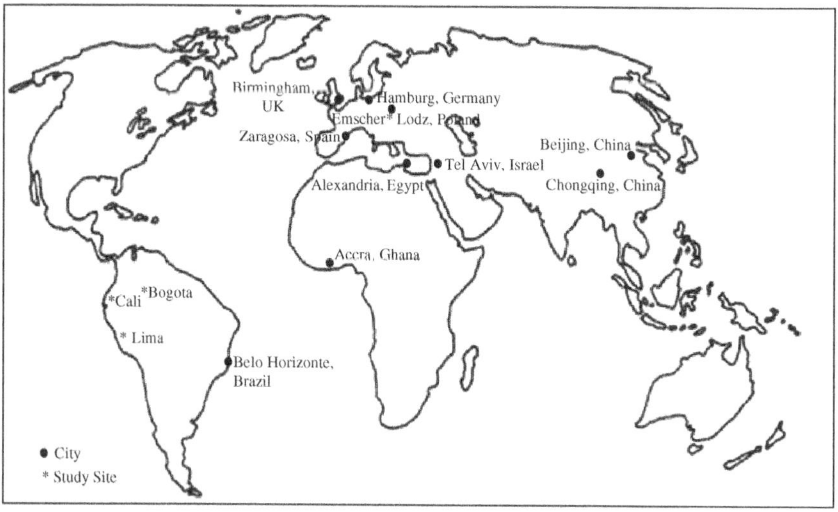

Figure 2. SWITCH cities and study sites.

SWITCH RESEARCH PROCESS

The SWITCH research process is a combination of:

- **Learning alliances** – SWITCH aims to link up a wide range of stakeholders at city level to interact productively and to create win-win solutions along the water chain. They consist of a series of structured platforms, at different institutional levels (national, river basin, city, community etc.), designed to break down barriers to both horizontal and vertical information sharing. This will speed up the process of identification, adaptation, and uptake of new innovations.

- **Action research** – SWITCH aims to carry out more demand-led, action-orientated research in cities with a view to achieving greater integration and wider impact through the Learning Alliances (LAs). Hence, SWITCH will address problems through innovation based upon involvement of users in local demonstrations (that are designed to show application of the new technologies in practical cases).
- **Multiple-way learning** – SWITCH aims to promote multiple-way learning, where European cities learn from each other and from developing countries, and vice versa. These multiple-way learning pathways will be developed by linking experts on urban water from developed and developing countries, by pooling scientific, technological and financial resources from the partners and in the demonstration cities resulting in an integrated, multi-disciplinary research effort.

The three main components of the SWITCH research process described above, will lead to greater impact and more potential for taking innovations to scale through the development of locally appropriate innovations and ownership of the concepts and process. In addition, undertaking research at different institutional levels will both shorten the time between developing new knowledge and scaling it up; and, ensure that local solutions are nationally relevant and applicable. Also, by sharing the learning process among cities, replication of innovations will be accelerated.

SWITCH combines the above into a *Strategic Planning Process* that encourages and enables all stakeholders (through the LAs), to view the urban water cycle in an integrated way (holistic framework embracing all aspects of the water systems in a city), and allow the development of new strategic directions for urban water management. This strategic planning process involves a decision-making tool (CityWater), that enables the LAs to work together to develop a cross-sectrol strategic plan (e.g. including energy issues and urban planning). A key feature of the tool is that it enables cities and their stakeholders to recognize and take into account the uncertainties associated with future global change pressures, enabling them to develop a flexible and adaptable strategy. The strategic planning process has three subsequent phases. It starts with the visioning, i.e. the development of a shared vision of the future organization of the urban water system. After that the LA develops a number of plausible scenarios, i.e. a qualitative and quantitative description of the future external situation that the city and its inhabitants will face. Finally, strategies are developed which describe the various general strategic approaches which could be followed under certain scenarios to achieve the vision.

SWITCH – KEY CONCEPTS

As stated above, the main goal of SWITCH is the development of sustainable and effective water management systems for the 'city of the future'. To achieve this goal SWITCH aims to improve the scientific basis for the development and management of urban water systems, to ensure that they are robust, flexible and adaptable to a range of future global change pressures. SWITCH is based on several key concepts of urban water management including: resilience of urban water systems to global change pressures; interventions over the entire urban water cycle; reconsideration of the way water is used (and reused); and greater application of natural systems for water and wastewater treatment as discussed below:

Resilience of urban water systems to global change pressures

Projections of future global change pressures are plagued with uncertainties which cause difficulties when developing urban water management strategies that are insensitive to these global change uncertainties. SWITCH aims to develop processes and a framework that can generate optimal urban water management systems that are robust, adaptable and sustainable under future global change pressures. These flexible systems will be characterized by their ability to cope with uncertainties and will hence have the capability to adapt to new, different, or changing requirements. The framework will have embedded in it new techniques such exploratory modelling (combines the best features of traditional quantitative decision analysis with those of narrative scenario-based planning), risk assessment and real-options analysis. Risk assessment is the process of identifying, evaluating, selecting, and implementing actions to reduce risks and manage uncertainties. Real-options-based decision making recognizes the value of flexibility, or flexible alternatives. It develops decision alternatives that may provide flexibility for future decision making or develop decision alternatives that may be exercised flexibly in time to cope with uncertainty.

SWITCH is also developing new process technologies that promote a more modular approach to urban water management. Modular diversity exponentially increases the amount of possible configurations that can be achieved for urban water systems from a given set of inputs (complex adaptive systems). SWITCH will develop a diverse repertoire of alternative options for urban water systems which have internal degrees of freedom that allow their flexibility and sustainability to be optimized over time. For example, in relation to storm water management, SWITCH is focusing on small-scale decentralised measures, such as infiltration devices that have the ability to respond more flexible to changes in boundary conditions. In relation to sanitation SWITCH is shifting the paradigms in wastewater treatment from an approach with centralized mixed systems to decentralised systems based on source control and separate treatment of concentrated and diluted household wastewater flows. For water and wastewater treatment systems, SWITCH is developing methods that use the natural capacities of soil and vegetation to absorb and retain water and to take-up, transform, or otherwise treat pollutants. One of the main features of these natural systems is their adaptability to almost all conceivable applications, and improved renewal and readjustments opportunities (essential for flexibility).

Interventions over the entire urban water cycle

An important aspect of urban water systems is the interactions that take place between different components of the system (e.g. foul water from leaky sewers entering into a drinking water distribution network (Vairavamoorthy *et al.*, 2007b). It is widely recognised that it is important to consider these interactions in order to maintain an effective, efficient and safe service of water and sanitation (Vairavamoorthy *et al.*, 2007c).

An integrated urban water management (IUWM) approach involves managing freshwater, wastewater, and storm water as links within the resource management structure, using an urban area as the unit of management. The approach encompasses various aspects of water management, including environmental, economic, technical, political, as well as social impacts and implications.

SWITCH is developing CityWater to analyse the interactions across the urban water cycle for a range of management and technological solutions. The creation of the CityWater Tool was the natural consequence of the need for strong coordinated action to the building and testing of a new DSS that embeds the concepts of the paradigm shift proposed for IUWM; that explicitly accounts for the evolution of a city from its current situation; and, that would be an appropriate vehicle for all the learning alliances to explore alternative techniques and proposals for dealing with the different future scenarios applicable to their cities. CityWater has a number of component tools and models, such as CityWater Vision, CityWater Balance, CityWater Drain, CityWater System, CityWater Economic, CityWater Risks etc. It will enable optimal urban water systems to be developed, driven by sustainability criteria, while recognising uncertainties associated with global change pressures.

In addition to CityWater, SWITCH aims to develop a multi-objective integrated urban planning process that can reconcile the often conflicting objectives that define the urban form, significant opportunities exist to significantly enhance a city's landscape and environment. This process will ensure that their is integration of rivers, lakes, banks, and their landscape systems for sustainable ecosystems planning within the aesthetic design strategies for an urban environment to meet water control, recreation, hazard and environmental health objectives.

Reconsider water use

The challenge of servicing more people with water requires us to critically look into water use practices and to develop strategies that maximize the benefits of water services while minimizing the usage. Balancing the demands for water between the various sectors will need to be accompanied by the use of new and alternative resources.

SWITCH is developing innovations that promote increased recycling of wastewater that will ensure better access to safe water, reduced vulnerability to extremes and increased adaptive capacity. These technologies will ensure that water can be used multiple times, by cascading it from higher to lower-quality needs (e.g. using household grey water for irrigation), and by reclamation treatment for return to the supply side of the infrastructure.

SWITCH is developing a combination of end-use efficiency, system efficiency, storage innovations (using different managed aquifer recharge options), and reuse strategies that would reduce water demand. More specifically the research in SWITCH is focusing on: water saving concepts and technologies; increased knowledge of water quality requirements for all applications and purposes; aquifer recharge and recovery technologies; decision support systems and demand management systems to efficiently allocate and use water resources.

Natural treatment systems

SWITCH is researching new solutions to water quality problems that are typically costly to resolve and difficult to avoid in all modern cities. While artificial treatment systems are applicable for certain forms of pollution, notably where the polluted waters can be transported to a treatment plant, they are far less effective for more distributed pollution and for pollutants trapped in soils and sediments. For these more recalcitrant forms'of pollution, treatment systems that use the natural capacities of soil and vegetation to take-up, transform, or otherwise treat pollutants are preferable. This is particularly relevant where there are opportunities for opening up the natural water bodies in a city landscape to provide not only pollution control, but also improved public amenity and health.

The main features of natural treatment systems are: *simplicity* – plants design, construction and operation are very simple; *cost-effectiveness* – plants require low building, labour, operation and maintenance costs (the only limiting factor is the availability and the cost of land to place the treatment plants); *efficiency* – natural systems plants are generally efficient at removing multiple contaminants in a single system; *reliability*: natural systems are very reliable even in extreme operating conditions and are better able to absorb a variety of both hydraulic and contaminant shocks.

SWITCH aims at maximizing the exploitation of natural systems and processes for the effective management of municipal water resources, of water supply and sanitation services, and of the municipal water cycle as a whole. These natural treatment systems (NTS) are characterized by the use of naturally occurring energies as opposed to conventional treatment technologies which are highly dependent on non-renewable fossil fuel energies, and on added chemicals. On the basis of a detailed inventory to NTS, SWITCH is developing promising NTS for: drinking water treatment (e.g. Engineered Artificial Reservoir); wastewater and stormwater treatment and reuse (soil aquifer treatment and aquifer storage recovery; constructed wetland-electro-flocculation process (CW-EF); and for the stimulation of water retention and self purification capacity of water resources (in-stream remediation using 'eco-hydrology principles' and hyporheic zone assessment).

Decentralised wastewater systems

SWITCH is exploring the benefits to be gained from decentralising wastewater treatment and the development of technologies to make a decentralised approach effective. Local water reuse approaches that permit safe and productive re-use of urban water within domestic, industrial and (urban) agriculture systems has the potential both to increase the available water but also to reduce the operational costs for wastewater treatment. Assessing the validity of this potential benefit and the conditions needed to make it successful are crucial to the development of a city water system based on a decentralised water distribution and treatment model.

SWITCH is developing new solutions to urban sanitation: shifting the paradigms in wastewater treatment from an approach with centralised mixed systems to decentralised systems based on source control. In addition SWITCH is promoting the safe use of rain and wastewater for urban agriculture. Through the Learning Alliances in selected cities (Accra, Beijing, and Chongqing) a better understanding of farmers' innovations and perceptions of wastewater use are being developed that will enable the development of flexible "response scenarios" coupled with appropriate risk reducing strategies that are technically, economically, socio-culturally and politically compatible.

CONCLUSIONS

There is an urgent need for planned action to manage water resources effectively. The problems in urban areas of developing countries are of particular concern as still large sections of the community are living without safe water supply and basic sanitation services. Providing adequate urban water and basic sanitation is likely to become more challenging in the near future due to several global change pressures.

To address the challenge of urban water management for the future, there is a need for a paradigm shift. The paradigm shift should include several actions such as: flexible and robust system design capable of adapting to new, different, or changing requirements and capable of coping with uncertainties; interventions over the entire urban water cycle; reconsideration of the way water is used (and reused); greater application of natural systems for water and wastewater treatment, and innovative sanitation systems.

It is anticipated that during the next few years, SWITCH will produce knowledge, technologies, models, techniques, institutional frameworks and improved management tools for sustainable urban water management for the city of the future. An important component of SWITCH is that it aims to bring together all stakeholders involved with, or who have interest in urban water management. These multi-stakeholder learning alliances will guide and support SWITCH on the implementation of research and demonstration activities, by taking account of local problems and needs. Clearly, this will substantially contribute to a reduction in the vulnerability of cities and their capacity and preparedness to cope with global changes.

REFERENCES

Brockerhoff, M.P. (2000). Population Bulletin: 'An Urbanizing World'. *A Publication of Population Reference Bureau*, **55**(3), 1–45.

Elimelech, M. (2006). The global challenge for adequate and safe water. *Journal of Water Supply: Research and Technology-AQUA*, **55**(1), 3–10.

Hiessl, H., Walz, R. and Toussaint, D. (2001). Design and sustainability assessment of scenarios of urban water infrastructure systems. Paper presented at the *5th International Conference on Technology, Policy and Innovation*, Delft, The Netherlands, June 26–29, 2001.

Khatri, K. and Vairavamoorthy, K. (2007). Challenges for urban water supply and sanitation in the developing countries. Water for changing world-Enhancing local knowledge and capacity, June 13–15, UNESCO-IHE, Delft.

UNICEF/WHO (2004). Meeting the MDG drinking water and sanitation target – A mid term assessment of progress: United Nations Children's Fund and World Health Organisation, p. 36.

Vahala, R. (2004). European Vision for Water Supply and Sanitation in 2030. Water Supply and Sanitation Technology Platform. http://www.wsstp.org/Shared%20-Documents/

Vairavamoorthy, K., Gorantiwar, S.D. and Mohan, S. (2007a). Intermittent water supply under water scarcity situations. *Water International*, **32**(1).

Vairavamoorthy, K., Yan, J. and Gorantiwar, G.D. (2007b). Modelling the risk of contaminant intrusion in water mains. *Proceedings of the Institution of Civil Engineers, Water Management*, **160**(2), 123–132.

Vairavamoorthy, K., Yan, J., Galgale, H.M. and Gorantiwar, S.D. (2007c). IRA-WDS: A GIS- based risk analysis tool for water distribution systems. *Environmental Modelling and Software*, **22**(7), 951–965.

WHO/UNICEF (2006). Meeting the MDG drinking water and sanitation target – The urban and rural challenge of the decade. World Health Organisation and United Nations Children's Fund, p. 41.

RECLAIM WATER – managed aquifer recharge for safe indirect potable reuse

T. Wintgens[a], C. Kazner[b] and T. Melin[b]

[a]University of Applied Sciences of Northwestern Switzerland, Gründenstrasse 40, 4132 Muttenz, Switzerland (E-mail: thomas.wintgens@fhnw.ch)
[b]RWTH Aachen University, Dept of Chemical Engineering, Turmstr. 46, 52056 Aachen, Germany (E-mail: christian.kazner@avt.rwth-aachen.de; thomas.melin@avt.rwth-aachen.de)

Abstract Solutions to global water stress problems are urgently needed yet must be sustainable, economical and safe. The utilisation of alternative water sources like reclaimed municipal wastewater is one of the most obvious and promising options in integrated water resources management. Among the various beneficial uses of reclaimed wastewater Managed Aquifer Recharge (MAR) receives growing attention because it features advantages such as additional natural treatment, storage capacity to buffer seasonal variations of supply and demand as well as mixing with natural water bodies which promotes the acceptance of further uses, particularly indirect potable use. Major concerns about the safety of this exploitation route of an alternative water source are connected to microbial and chemical contaminants occurring in wastewater, among which are emerging trace organics like endocrine disrupters and pharmaceuticals. This paper presents the European RECLAIM WATER research project on safe water reclamation technologies for aquifer recharge.

Keywords indirect potable reuse, managed groundwater recharge, water reclamation

INTRODUCTION

Many countries and regions of the world are facing water scarcity and deterioration of groundwater quality caused by climate change and a continuous population growth especially in coastal areas. The acute or chronic water stress triggers the need for integrated water cycle management and the supplementation of the available freshwater resources. The utilization of alternative water sources such as seawater and brackish water desalination as well as water reclamation and reuse are mitigation options applied up to now (Bixio *et al.*, 2006; Fritzmann *et al.*, 2007). It is to note that this practice is encouraged for many Mediterranean countries as outlined by the European Commission in their publications on the management of water scarcity and drought (European Commission, 2007).

The water stress conditions in the Mediterranean region are expected to aggravate as a probable consequence of climate change impacts (EEA, 2007). Many regions are prone to experience an increase in water stress, which can be expressed with the water exploitation index. The water exploitation index is a simple parameter expressing the ratio of water abstraction to renewable freshwater resources. Besides the quantitative aspects, water stress has also a water quality dimension. In course of the implementation of the Water Framework Directive (2000/60/EC) the status of water resources has been analysed and evaluated with regard to achieving a "good ecological status" by 2015. European Union Member states had to assess the quality of the water bodies according to Art. 5 of the Water Framework Directive (WFD). The investigation in the Member States has revealed that quite a number of water bodies in the European Union are at risk of failing to meet the quality objectives by 2015, as set in the WFD. Between 30% and 75% of the groundwater bodies in UK, France, Spain, Malta and Cyprus are categorised "at risk", and an additional 15% to 50% were evaluated as being probably at risk (Figure 1).

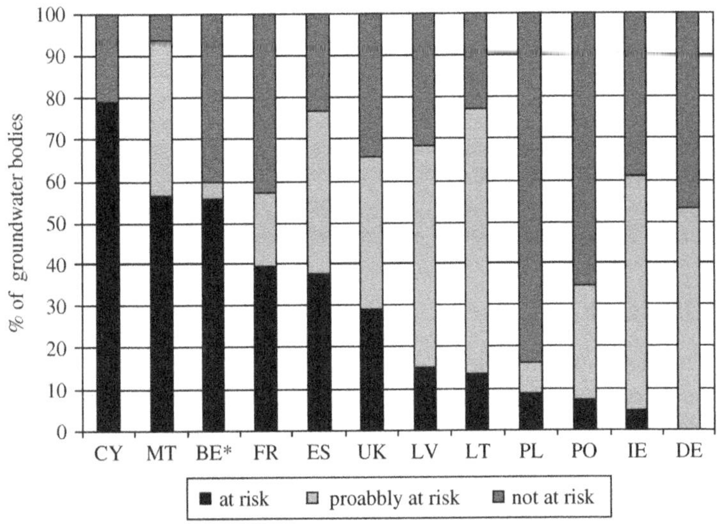

Figure 1. Status of groundwater in selected European countries (National reports on Art. 5 WFD accessed via CIRCA, 2006).

The types of pressures exerting a significant impact on water resources are either qualitative (e.g. diffuse or point pollution with nitrates and pesticides or seawater intrusion into coastal aquifers) or quantitative (e.g. over-abstraction of groundwater). In the scope of River Basin Management plans the water authorities will have to address the identified pressures and set up a programme of measures. The utilisation of alternative water resources or options such as aquifer recharge can be part of these programs. In the European policy water saving, e.g. supported and enforced by economical measures, has a priority. While water reuse is regarded as a possible water conservation measure, seawater desalination is regarded as a last resort option (European Commission, 2007).

Among the various beneficial uses of reclaimed wastewater Managed Aquifer Recharge (MAR) receives growing attention because it features advantages such as additional natural treatment, storage capacity to buffer seasonal variations of supply and demand as well as mixing of reclaimed water with natural water bodies which promotes the acceptance of further uses, particularly indirect potable use. Major concerns about the safety of this exploitation route of an alternative water source are connected to microbial and chemical contaminants occurring in wastewater, among which are emerging trace organics like endocrine disrupters and pharmaceuticals.

The concept of Managed Aquifer Recharge offers potential for various subsequent uses like irrigation, industrial process water and augmentation of public water supplies. The latter, indirect potable use, is certainly one of the most challenging water reclamation and reuse applications with a high demand in terms of safety because of the potential use as drinking water (whose quality criteria a fixed in directive 98/83/EC) and the general level of protection required for groundwater resources as laid down in the directive on the protection of groundwater against pollution and deterioration (2006/118/EC). Even if drinking water supplementation is not explicitly foreseen in an aquifer recharge project, the provision of drinking water quality in the recovered product is a common bottom line in many applications, e.g. in Israel where the Dan Region Reclamation and Soil Aquifer Treatment scheme operated by the national water company Mekorot provides "accidental drinking water quality" on a large scale.

The RECLAIM WATER project and its objectives

RECLAIM WATER is a Specific Targeted Research Project supported by the European Commission under the Thematic Priority 'Global Change and Ecosystems' of the Sixth Framework Programme (Contract-No. 018309). The project started in October 2005 and ran till December 2008. The project consortium consists of universities, research and technology institutes, as well as technology providers and small and medium size enterprises, with a total number of 20 partners from 16 countries (European Union Member States + Australia, China, Israel, Mexico, Singapore, South Africa, Switzerland).

The strategic objective of RECLAIM WATER is to develop hazard mitigation technologies for water reclamation providing safe and cost effective routes for managed aquifer recharge. It assessed different treatment applications in terms of behaviour of key microbial and chemical contaminants such as organic micropollutans. The knowledge generated in the project and the technologies developed are also suited to the needs of developing countries, which have a growing need of supplementation of freshwater resources. The participation of partners from China, Australia, Singapore, South Africa and Mexico demonstrates the anticipation of the global dimension of the water reclamation and aquifer recharge issue.

This paper outlines the methodological approach and key results of the RECLAIM WATER project.

APPROACH

RECLAIM WATER depicted water cycle schemes utilizing different alternative water sources (see Figure 2) and a range of pre-treatment options prior to infiltration or injection for aquifer recharge. Various recharge systems including direct infiltration through dug wells, infiltration basins and different geological settings (sandy to carstic aquifers) were investigated in the project.

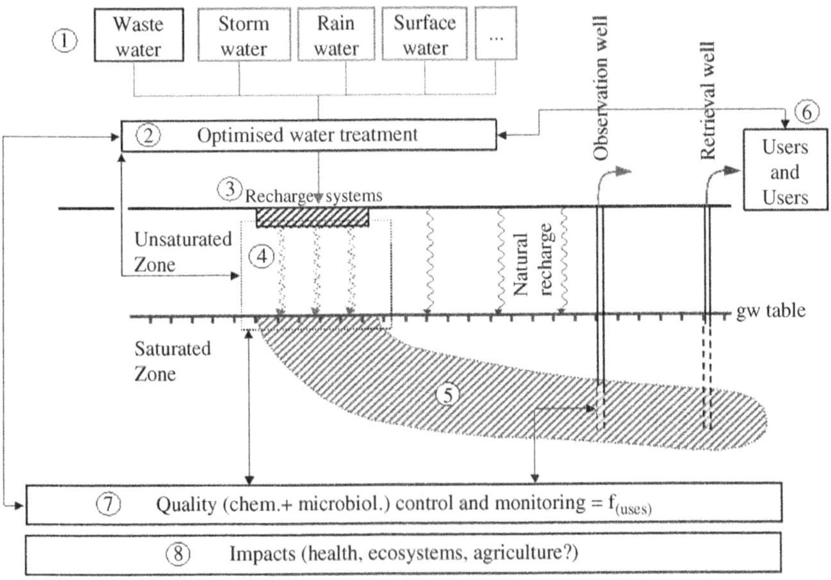

Figure 2. Concept of managed aquifer recharge and points of consideration.

The majority of the data was generated in a set of globally distributed case studies. These also served as a basis to validate technical and managerial concepts, prove their practicability and effectiveness, take into account regional and use type variety as well as to assess a number of selected treatment processes (Figure 3).

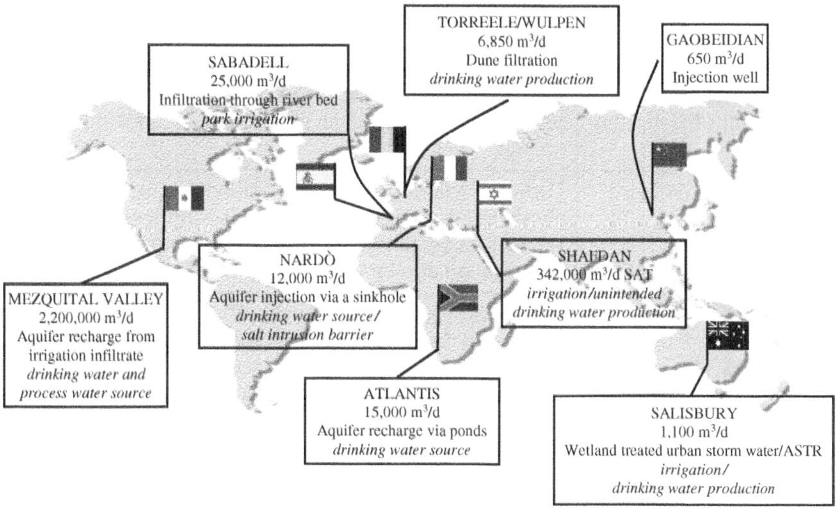

Figure 3. RECLAIM WATER case studies of MAR.

Table 1 gives an overview on the treatment technologies, recharge systems and end-uses relevant to the different case studies.

KEY RESULTS

The investigations in the RECLAIM WATER project delivered a great number of research results in a number of areas.

Water reclamation technologies

The effectiveness of both natural and intensive water reclamation technology options which can be applied prior to aquifer recharge options has been assessed. The investigations included constructed wetlands, slow sand filters, advanced primary treatment for a developing country context as well as advanced options such as membrane bioreactors (Abegglen *et al.*, 2009), dense membrane processes, ozonation (Benner *et al.*, 2008) and process combinations also for

Table 1. Water reclamation schemes investigated in RECLAIM WATER.

	Site location	Scheme description
Europe	Sabadell (near Barcelona), Spain	Secondary treated wastewater effluent discharged into a river bed where it infiltrates and is recovered. The water is then disinfected (UV) and distributed for parks irrigation.
	Nardò Salento Region, South of Bari, Italy	Secondary treated municipal effluent is transported to aquifer injection. Recharge acts as a salt intrusion barrier and resource is also used as drinking water source.
	Torrele (Wulpen) Belgium	Tertiary treated municipal effluent is upgraded by ultrafiltration and reverse osmosis, and then infiltrated via infiltration ponds to prevent salt intrusion and to recharge an aquifer used for drinking water production.
Asia	Shafdan Tel Aviv Region, Israel	Secondary wastewater from the Tel-Aviv area is recharged to an aquifer via a soil aquifer treatment (SAT) system. Recovered water is primarily used for irrigation but has accidental drinking water quality.
	Gaobeidian Beijing, China	Tertiary effluent is used for aquifer recharge. Treatment is provided by coagulation, filtration and ozonation (in test) prior to infiltration and recharge. water is used for irrigation
	NEWater Singapore	NEWater is treated 'used water' further purified using dual-membrane (microfiltration and reverse osmosis) and UV treatment. Four NEWATER factories are in operation supplementing Singapore's water supplies, part of the product water is use to augment drinking water reservoirs.
Australia	Salisbury Adelaide Region, South Australia	Wetland treated urban stormwater injected into a brackish aquifer. Water recovered via separate recovery wells. Recovered water is intended for drinking supplies and until proven will be used for irrigation.
Africa	Atlantis (near Cape Town) South Africa	Urban stormwater run-off is collected via a series of detention basins, blended with secondary treated domestic wastewater and recharged up-gradient of a production well field for augmenting the water supply. The blend of natural groundwater and recharged water abstracted from the well field is used as potable water supply for the city of Atlantis.

Table 1. Continued

	Site location	Scheme description
America	Mezquital Valley State of Mexico, Mexico	The raw wastewater from Mexico City mixed with stormwater and surface water is discharged to an irrigated area of more than 76,000 ha. About 40% of the irrigation water infiltrates into the aquifer. The water is recovered via separate wells and springs. 206 well systems, 31 springs, and 63 waterwheels are in operation. Recovered water is chlorinated and locally distributed for drinking water supply, industrial use, irrigation and other purposes (bathing, swimming, washing).

brine treatment, e.g. utilizing activated carbon adsorption plus membrane filtration as well as electro-deionisation (Ng *et al.*, 2008). A particular hybrid process investigated in the project included powdered activated carbon adsorption and Nanofiltration (Kazner *et al.*, 2008). The removal performance of this process is shown for some selected organic micropollutants in Table 2.

Table 2. Removal of selected micropollutants by powdered activated carbon (PAC) adsorption and Nanofiltration (NF, MWCO ~ 300 Da) (Kazner *et al.*, 2007).

Parameter	Raw water	Feed after PAC	PAC removal	NF permeate	Total removal
Direct NF					
Benzotriazole	10.70 μg/L	–	–	5.10 μg/L	52.3%
Carbamazepine	0.92 μg/L	–	–	0.47 μg/L	48.9%
Primidone	0.14 μg/L	–	–	0.11 μg/L	21.4%
Sulfamethoxazole	0.60 μg/L	–	–	0.23 μg/L	61.3%
NF w 50 mg/L PAC					
Benzotriazole	9.55 μg/L	0.48 μg/L	95.0%	0.20 μg/L	97.9%
Carbamazepine	1.58 μg/L	0.03 μg/L	98.0%	<0.002 μg/L	>99.9%
Primidone	0.38 μg/L	0.06 μg/L	83.7%	0.02 μg/L	93.7%
Sulfamethoxazole	0.70 μg/L	0.21 μg/L	69.6%	0.08 μg/L	87.9%

Joss *et al.* (2008) and Wintgens *et al.* (2008) have reviewed water reclamation technology options particularly with regard to micropollutants removal and suitability for indirect potable use.

Microbial contaminants in aquifer recharge schemes

Besides the standard indicator parameters for microbiological water quality such as Total bacteria counts, Total coliforms, *E. Coli*, Enteroccoci and Clostridium spores, an enhanced spectrum of parameters was considered including parasites, protozoa, pathogenic bacteria and viruses as well as antibiotic resistance genes. It could be confirmed that advanced treatment schemes, e.g. those utilising membrane filtration, could readily remove those compounds and that subsurface processes can significantly attenuate those (Böckelmann *et al.*, 2009; La Mantia *et al.*, 2008).

Chemical contaminants in aquifer recharge schemes

Particular attention has been paid to organic chemicals in the different stages of the investigated aquifer recharge schemes. Both bulk organic characterisation as well as analysis of trace organics has been carried out. The parameters included pharmaceutically active compounds and disinfection by-products as well as estrogenic activity. Analytical methods have been developed and adopted to the different matrices encountered in the sites (Fink and Ternes, 2006; Krauss and Hollender, 2008). Those methods have been used to elucidate the fate of the target compounds in different treatment processes (Schulz *et al.*, 2008; Yu *et al.*, 2009). It has been shown that the double membrane process (UF/RO) with subsequent soil aquifer treatment as used in advanced indirect potable reuse schemes is able to remove a wide range of considered compounds below the limit of quantification.

Hydro-geochemical characterisation of aquifer recharge

Sophisticated characterisation tools such as boron and lithium isotope monitoring has been utilised to characterise the hydrology and hydro-geochemistry of full scale aquifer recharge sites. Native and influenced groundwater could be clearly distinguished and input to flow modelling could be provided (Kloppmann *et al.*, 2008; Kloppmann *et al.*, 2009). Conceptual and numerical models have been set up to better describe and characterise hydrology and hydro-geochemistry in the investigated aquifer recharge sites.

Integration of results in risk assessment and management

State-of-the-art concepts for risk assessment such as QMRA and risk management (e.g. HACCP and Water Safety Plan) have been used to integrate the water quality and treatment efficiency data, putting the results into a perspective which is relevant for end-users and public perception. Comprehensive risk assessment approaches have been developed for selected full scale schemes involved in the project and the contribution of subsurface processes to risk control has been assessed (Dillon *et al.*, 2008).

DISSEMINATION AND FURTHER INFORMATION

The project outcomes are summarised in a policy brief document directed to the European Commission and other decision makers in the water sector. Further information on the project and the project results such as reports can be found on the project website: www.reclaim-water.org

CONCLUSIONS

RECLAIM WATER has provided a comprehensive assessment of different water reclamation and managed aquifer recharge methods and technologies covering a broad range of water quality parameters (including pathogens, antibiotic resistance genes and emerging contaminants), different pre-treatment and infiltration options, different end-uses (potable and non-potable), raw water sources for replenishment (from untreated wastewater to highly treated tertiary effluent and purified stormwater). The investigations covered schemes on all continents and used a coordinated methodology for assessment. Finally, RECLAIM WATER results underline that managed aquifer recharge can be a safe and reliable climate change adaptation method (Wintgens *et al.*, 2005). Technologies and methods can be tailored to the different socio-economic circumstances. In developing countries' context de-intensified natural systems, in particular MAR, can also provide a decent water quality at very low costs.

ACKNOWLEDGEMENTS

The European Commission is acknowledged for co-funding the RECLAIM WATER Project under contract number 018309 in the Global Change and Ecosystem sub-priority of the 6th Framework Programme for Research and Technological Development.

REFERENCES

Abegglen, C., Joss, A., McArdell, C.S., Fink, G., Schlüsener, M., Ternes, T.A. and Siegrist, H. (2009). The fate of selected micropollutants in a single-house MBR, *Water Research* (accepted).

Benner, J., Salhi, E., Ternes, T.A. and von Gunten U. (2008). Ozonation of reverse osmosis concentrate: Kinetics and efficiency of beta blocker oxidation. *Water Research*, **42**, 3003–3012.

Bixio, D., Thoeye, C., De Koning, J., Joksimovic, D., Savic, D., Wintgens, T. and Melin, T. (2006). Wastewater reuse in Europe. *Desalination*, **187**, 89–101.

Böckelmann, U., Dörries, H.-H., Ayuso-Gabella, N., Salgot de Marçay, M., Tandoi, V., Levantesi, C., Masciopinto, C., Van Houtte, E., Szewzyk, U., Wintgens, T. and Grohmann, E. (2009). Quantitative real-time PCR monitoring of bacterial pathogens and antibiotic resistance genes in three European artificial groundwater recharge systems. *Applied and Environmental Microbiology*, **75**(1), 154–163.

CIRCA (2006). *CIRCA – Communication & Information Resource Centre Administrator* – http://forum.europa.eu.int/Public/irc/env/wfd/library

Dillon, P., Page, D., Vanderzalm, J., Pavelic, P., Toze, S., Bekele, E., Prommer, H., Higginson, S., Regel, R., Rinck-Pfeiffer, S., Purdie, M., Pitman, C. and Wintgens, T. (2008). A critical evaluation of combined engineered and aquifer treatment systems in water recycling. *Water Science and Technology*, **57**(5), 753–762.

European Commission (2007). *Addressing the challenge of water scarcity and droughts in the European Union*. Communication from the Commission to the European Parliament and the Council. COM(2007) 414 final.

European Environment Agency (2007). *Climate change and water adaptation issues.* Technical Report EEA 2/2007.

Fink, G. and Ternes, T.A. (2006). Simultaneous determination of psychoactive drugs and their metabolites in aqueous matrices by liquid chromatography mass spectrometry. *Environmental Science and Technology*, **40**(23), 7321–7328.

Fritzmann, C., Löwenberg, J., Wintgens, T. and Melin, T. (2007). State-of-the-Art of Reverse Osmosis desalination. *Desalination*, **216**, 1–76.

Joss, A., Siegrist, H. and Ternes, T.A. (2008). Are we about to upgrade wastewater treatment for removing organic micropollutants? *Water Science and Technology*, **57**(2), 251–255.

Kazner, C., Fink, G., Ternes, T., Wintgens, T. and Melin, T. (2007). Removal of organic micropollutants by nanofiltration in combination with adsorption on powdered activated carbon for artificial groundwater recharge with reclaimed wastewater, Proceedings of the 5th IWA Micropol & Ecohazard 2007 Conference, 17–20 June 2007, DECHEMA e.V., Frankfurt/Main, Germany, pp. 259–265.

Kazner, C., Lehnberg, K., Kovalova, L., Wintgens, T., Melin, T., Hollender, J. and Dott, W. (2008). Removal of endocrine disruptors and cytostatics from effluent by nanofiltration in combination with adsorption on powdered activated carbon. *Water Science and Technology*, **58**(8), 1699–1706.

Kloppmann, W., Chikurel, H., Gaus, I., Picot, G., Guttman, Y., Aharoni, A., Guerrot, C., Millot, R. and Wintgens, T. (2009). B and Li isotopes as intrinsic tracers for injection tests in aquifer storage and recovery systems. *Applied Geochemistry* (accepted).

Kloppmann, W., Van Houtte, E., Gaus, I., Picot, G., Vandenbohede, A., Lebbe, L., Guerrot, C., Millot, R. and Wintgens, T. (2008). Monitoring reverse osmosis treated wastewater recharge into a coastal aquifer by environmental isotopes (B, Li, O, H). *Environmental Science and Technology*, **42**(23), 8759–8765.

Krauss, M. and Hollender, J. (2008) Analysis of Nitrosamines in Wastewater: Exploring the Trace Level Quantification Capabilities of a Hybrid Linear Ion Trap/Orbitrap Mass Spectrometer. *Analytical Chemistry*, **80**, 834–842.

La Mantia, R., Masciopinto, C., Levantesi, C. and Tandoi, V. (2008). Fate and transport of faecal contamination microbial indicators, pathogenic protozoa and Campylobacter in the artificially recharged fractured aquifer of Salento, Italy. *Water Science and Technology*, **57**(6), 849–856.

Ng, H.Y., Lee, L.Y., Ong, S.L., Tao, G., Viswanath, B., Kekre, K., Lay, W. and Seah, H. (2008). Treatment of RO brine-towards sustainable water reclamation practice. *Water Science and Technology*, **58**(4), 931–936.

Wintgens, T., Hochstrat, R., Melin, T., Jeffrey, P. and Salgot, M. (2005). Political and legislative framework conditions for wastewater reclamation and reuse in Europe. Proceedings 20th Annual WateReuse Symposium, September 18–21, 2005, Denver, Colorado, USA.

Wintgens, T., Salehi, F., Hochstrat, R. and Melin, T. (2008). Emerging contaminants and treatment options in water recycling for indirect potable use. *Water Science and Technology*, **57**(1), 99–107.

Yu, L., Fink, G., Wintgens, T., Melin, T. and Ternes, T.A. (2009). Sorption behavior of potential organic wastewater indicators with soils. *Water Research*, **43**(4), 951–960.

ACQWA – water resources in mountain regions sensitive to climatic change

M. Beniston, M. Stoffel and D. Cripe

Climatic Change and Climate Impacts C3i, University of Geneva, Switzerland (E-mail: Martin.Beniston@unige.ch; Douglas.Cripe@unige.ch)

Abstract The ACQWA project will assess the impacts of a changing climate on the quantity and quality of water in vulnerable mountain regions. Environmental conditions of mountain regions will be particularly affected by rapidly rising temperatures, prolonged droughts and extreme precipitation. Modeling techniques will be used to project the influence of climatic change on the major determinants of river discharge at various time and space scales. Regional climate models will provide the essential information on shifting precipitation and temperature patterns, and snow, ice, and biosphere models will feed into hydrological models in order to assess the changes in seasonality, amount, and incidence of extreme events in various catchment areas. Environmental and socio-economic responses to changes in hydrological regimes will be analyzed in terms of hazards, aquatic ecosystems, hydropower, tourism, agriculture, and the health implications of changing water quality. Adaptation and policy options will be elaborated on the basis of the model results. The methodological developments gained from a European mountain focus will be used to address water issues in regions whose economic conditions and political structures may compromise capacities to respond and adapt, such as the Andes and Central Asia where complex problems resulting from asymmetric power relations and less robust institutions arise.

Keywords climate change, hydrology, modelling, mountains, policy, water resources

INTRODUCTION

As the evidence for anthropogenic contributions to climate change becomes clearer, so too does the realization that its effects will have impacts on environmental and socio-economic systems (Beniston, 2002; Orlove *et al.*, 2008). Some regions are more vulnerable than others, both to expected physical changes and to the consequences they will have for ways of life. In particular, mountainous regions are recognized as particularly sensitive physical environments with populations whose histories and current social positions often strain local capacity to accommodate intense and rapid changes to their resource base.

In this context, the European Commission has recently funded a research project entitled ACQWA (Assessing Climate change and impacts on the Quantity and quality of WAter).

Objectives

The objectives of the ACQWA project are to assess the impacts of a changing climate on the hydrologic cyle in mountain regions, focusing on the quantity and quality of water. Special attention is paid to regions where snow- and ice melt represent a large contribution to streamflow (Barnett *et al.*, 2005). Indeed, with increasing evidence of glacier retreat, permafrost reduction and snowfall decrease in many mountainous regions, thus suggesting that climatic change may seriously affect streamflow regimes (Burlando *et al.*, 2002; Etchevers *et al.*, 2002), in turn threatening the discharge of water resources, increasing the downstream risks related to mass movements and floods, impacting hydropower generation, agriculture, forestry, tourism and aquatic ecosystems (Beniston, 2004; Brown *et al.*, 2007; Stoffel *et al.*, 2008). As a consequence, socio-economic structures of populations living downstream will be also affected, requiring better water resource management, and the development of strategies to avoid exacerbating situations of potential conflicts of interest. Already in today's climate problems of water resource appropriation are already seen to lead to tensions between different economic actors, and these tensions are likely to increase as the basic resource becomes rarer.

METHODOLOGY AND APPROACH

The ACQWA project aims to use advanced modelling techniques to quantify the influence of climatic change on the major determinants of river discharge at various temporal and spatial scales, while analyzing their impact on society and economy. A novel component of the ACQWA project is that both climatological and socio-economic feedback mechanisms will be identified and evaluated. The initial focus will be on the selection of appropriate continuous transient scenarios from the 1960s up to 2050. In comparison to many existing studies, the limitation of the modelling horizon to the middle of the 21st century allows for more realistic assessments of the progressive impact on the social, economical and political systems. These systems are expected to evolve typically in an adaptive mode on shorter time scales than the centennial ones, eventually shifting to new equilibria when forced abruptly. Using these continuous transient scenarios to provide initial and boundary conditions, regional climate models will provide the essential information on shifting precipitation and temperature patterns. In addition, snow, ice, and biosphere models will be coupled with hydrological models in order to assess the changes in basin hydrology and seasonality, including the amount, and incidence of extreme events in catchment areas. The type of extremes that will be analyzed here take into account particular weather patterns that can lead to exceptional flooding (e.g. storms of Mediterranean origin), and geomorphic hazards. These hazards include various forms of erosion and slope instabilities that often contribute to increased sediment loading and thus to problems for infrastructure close to the rivers or hydropower installations.

Environmental and socio-economic responses to changes in hydrological regimes will be analyzed in terms of hazards, aquatic ecosystems, hydropower, tourism, agriculture, and the health implications of changing water quality. Attention will also be devoted to the interactions between land use/land cover changes, and changing or conflicting water resource demands. Integration of the information from all these sectors and the impacts on economies will feed into a quantitative model of water use incorporating supply and demand. The supply side requires inputs from both the regional climate models and from socioeconomic sectors based on property, costs, and regulatory factors. Demand reflects population evolution, costs, and economic activity.

The resulting integrated model will permit the construction of plausible scenarios and allow us to evaluate various policy options for adaptation and mitigation. A significant novelty provided by the model constructs that will be implemented by the project is the ability to not only focus on the changes of the mean values, but also to assess the variability at different time and space scales. It will thus be possible to identify changes of the hydrological response to climatic change, while assessing its impact on a number of socio-economic sectors resulting from modifications of the temporal and spatial structure of the key variables describing water resources. Using this approach, inter-annual variability, distribution properties and scale issues will naturally be encompassed. Variability estimates will represent an important input for the analysis of the costs and benefits of the changes as well as the associated new risks these represent for societies. Figure 1 illustrates the various Work Packages of the ACQWA project and the segments of the study each will manage, as well as the flow of information between them.

The research on policy for adaptation/mitigation will involve not only European but also non-European partners, and will provide one of the major social-science outputs of the ACQWA project.

EXPECTED RESULTS

Model areas

The Rhone and Po river basins in the European Alps will be a common "test ground" for model investigations, where the different methodological

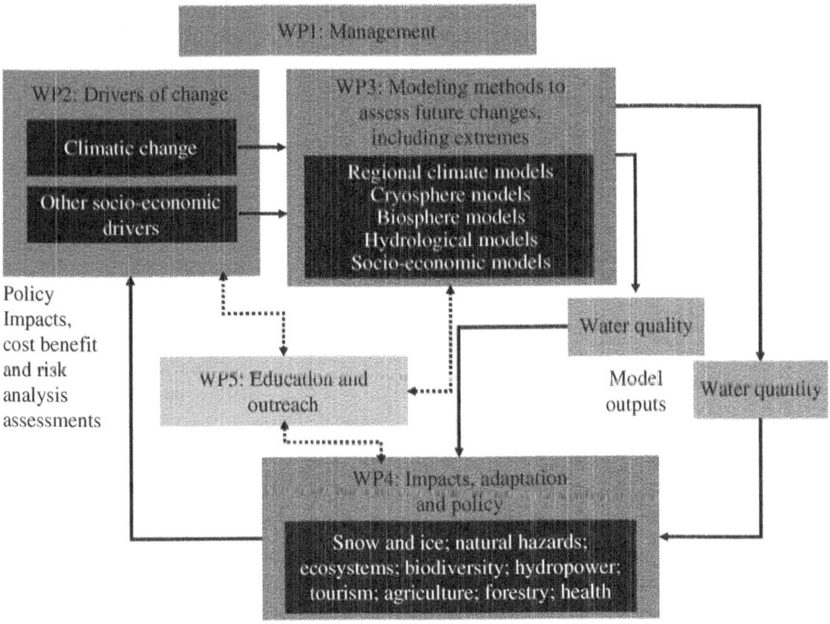

Figure 1. Flowchart illustrating the various work packages of the project.

approaches will converge to the basin scale through appropriate up- or down-scaling techniques. Both basins constitute ideal case-study areas, as they comprise all the elements of the natural environment that will be modeled (snow, ice, vegetation, hydrology). At the same time, these are highly regulated watersheds, subject to hydropower, irrigated agriculture, and tourism activities in the context of a climate that is at the borderline between Mediterranean and continental, and are therefore particularly vulnerable to climatic change. The methodologies developed over the first three years for the Alpine region will be subsequently applied to non-European study areas, in particular the Aconcagua Basin in Chile where receding glaciers pose a genuine threat to water availability, and in Central Asia (Kyrgyzstan) where the same process involving much larger glaciers could represent a significant economic opportunity (the development, for example, of hydro-power as a source of foreign income).

Water of mountain systems – an interdisciplinary investigation

The focus on water is the key element of the ACQWA proposal because it is essential for human populations as well as animal and plant communities. Water is relevant in every aspect of mountain systems, in the physical, biological and socio-economic systems. It directly influences the energy supply (hydropower), tourism (snow, water usage, glaciers), forestry and agriculture (productivity changes with changes in water supply, need for irrigation), and services from natural and semi-natural ecosystems. On the other hand, changes in any of these compartments will produce a feedback on water availability. Afforestation and deforestation processes triggered by climate change will likely affect the water balance, the generation of flood runoff and sediment production due to enhanced or damped erosion. The project aims at going beyond the current state of modeling by accounting for such feedback mechanisms, particularly focusing on the impact of changing forests on the water regimes. Increased competition for water among these sectors may arise, requiring regulatory or market mechanisms to attenuate tension and assure efficient and equitable use. Achieving sustainable water use poses particular challenges for policy making because of its nature as a public good, and also because it often has both upstream/downstream and transboundary/transnational characteristics. Any changes in climate affecting precipitation and the behavior of snow and glaciers (where relevant), will have a major influence on the seasonality, amount, and quality of surface runoff. The main changes are thus expected in the surface water systems. These systems will be comprehensively analyzed in order to quantify the changes affecting the streamflow regime, having the potential to cause uneven temporal distribution of the resource throughout the year. Thus, the project will mainly concentrate on surface waters, both because mountain regions generally depend on them rather than on groundwater, and the configuration of topography and landscape do not allow the presence of sizeable alluvial deposits necessary for groundwater systems to develop. However, an assessment of the impact on downstream river-fed groundwater systems will be carried out, though without detailed modeling of the aquifer systems, by considering the effects on the recharge of groundwater systems due to predicted altered streamflow regimes and to modified space-time precipitation patterns.

The diverse and complex issues discussed succinctly above require an interdisciplinary and holistic approach to water management as climatic and socio-economic pressures change in the future. Although the interdependencies between the components of the project are complex and involve a consortium of over 30 partners (Figure 2), the project will promote a set of methodological approaches aimed at addressing these key issues and will prepare a decision-making framework for mountain water management. Short-term solutions to current and future problems of water resource management include, at least from a technological point of view, the construction of reservoirs to buffer the effects of increasing drought. The caveat of this solution is that there is a reduction in

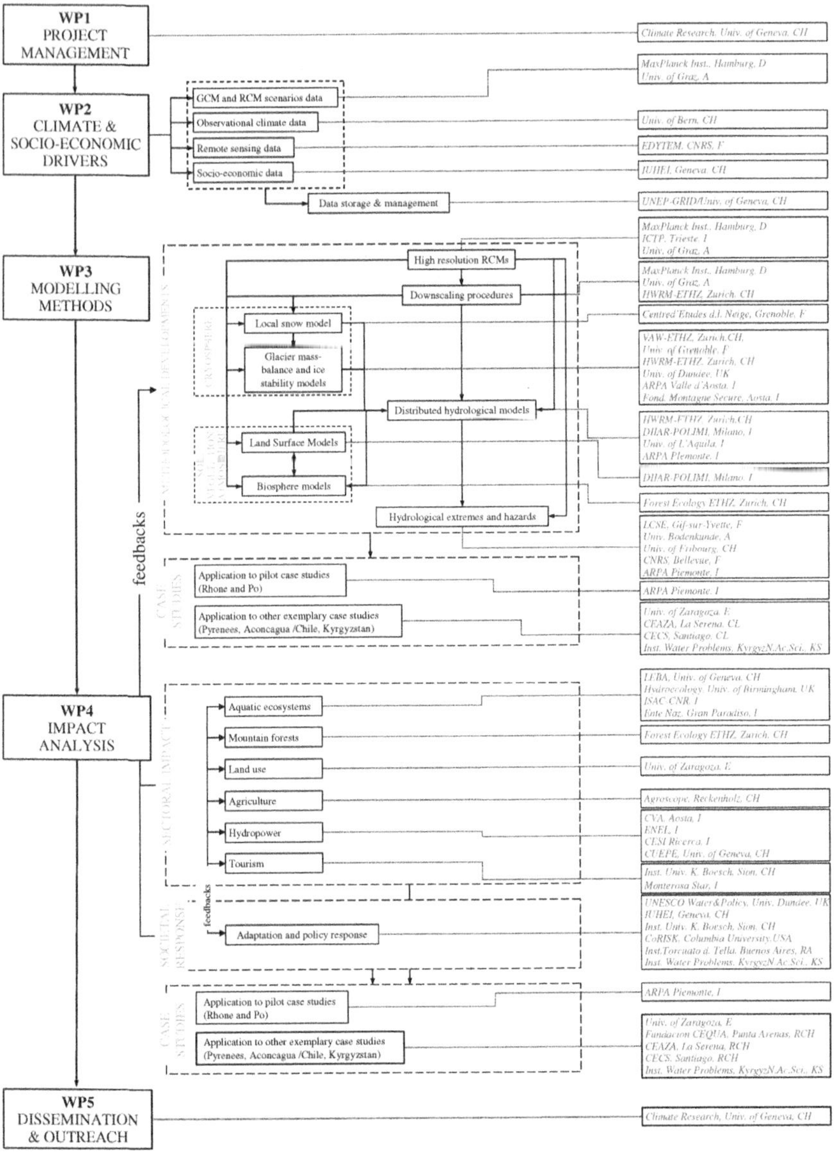

Figure 2. Interdependencies between project components and consortium partners.

the social acceptability related to the construction of new dams in many areas since artificial lakes can result in negative environmental impacts. Longer-term solutions require addressing the issues of greenhouse-gas emissions, beyond the 2012 targets imposed by the Kyoto Protocol. Changing land use in catchment areas may also be a means of adapting to the negative consequences of water availability and quality in many mountain regions in a changing climate. While the main focus is on the European case, its strengths and weaknesses are best assessed in a comparative framework. Response strategies are thus studied by contrasting the Europe Alps, with their relatively high level of economic development and robust institutional structure, with Central Asia and Chile, where standards of living and latent conflict compromise effective water management.

CONCLUSIONS

The ACQWA project is thus a truly interdisciplinary initiative that brings over 30 European, South American and Central Asian teams together from the physical, biological, social, economic and political sciences. The 5-year project was allocated 6.5 million Euros (roughly $8.75 million) and will be completed by the end of 2013. Further information can be found at: www.acqwa.ch

ACKNOWLEDGEMENTS

The European Commission is acknowledged for co-funding the ACQWA Project under the 7th Framework Programme.

REFERENCES

Barnett, T.P., Adam, J.C. and Lettenmaier, D.P. (2005). Potential impacts of a warming climate on water availability in snow-dominated regions. *Nature*, **438**, 303–309.

Beniston, M. (ed.) (2002). *Climatic Change. Implications for the Hydrological Cycle and for Water Management.* Kluwer Academic Publishers (now Springer Publishers), Dordrecht and Boston, 528 pp.

Beniston, M. (2004). *Climatic Change and Its Impacts: A Review Focusing on Switzerland.* Kluwer Academic Publishers (now Springer Publishers), Dordrecht/Heidelberg/Boston, 296 pp.

Brown, L.E., Hannah, D.M. and Milner, A.M. (2007). Vulnerability of alpine stream biodiversity to shrinking glaciers and snowpacks. *Global Change Biology*, **13**, 958–966.

Burlando, P., Strasser, U. and Pellicciotti, F. (2002). Modelling mountainous water systems between learning and speculating, looking for challenges. *Nordic Hydrology*, **33**, 47–74.

Etchevers, P., Golaz, C., Habets, F. and Noilhan, J. (2002). Impact of a climate change on the Rhone river catchment hydrology. *Journal of Geophysical Research*, 107, 10.1029/2001JD000490.

Orlove, B., Wiegandt, E. and Luckman, B.H. (2008). *Darkening Peaks. Glacier Retreat, Science, and Society*. University of California Press, Berkeley, 296 pp.

Stoffel, M., Conus, D., Grichting, M.A., Lièvre, I. and Maître, G. (2008). Unraveling the patterns of late Holocene debris-flow activity on a cone in the central Swiss Alps: Chronology, environment and implications for the future. *Global and Planetary Change*, **60**, 222–234.

Water Treatment

Development of advanced treatment trains and decision support tools to meet future drinking water challenges

J.-C. Schrotter, C. Breda, F. Vince, M.-R. de Roubin, P. Roche and P. Bréant

Veolia Environnement, Anjou recherche, Chemin de la Digue, BP 76, 78603 Maisons Laffitte Cedex, France (E-mail: jean-christophe.schrotter@veolia.com)

Abstract As a result of the anticipated stringent water quality needs, there has been increasing emphasis by the water industry on the removal of natural organic matter (NOM) and emerging pollutants from raw-water supplies. NOM and emerging pollutant removal options are Granular or Powered activated carbon (PAC or GAC) adsorption and membrane filtration with coagulation being widely used as a pre-treatment step for NOM removal. The performance of each of these technologies is discussed, with an emphasis on Nanofiltration (NF) and Reverse Osmosis (RO) systems which proved to reach high water quality even with polluted supplies. Current and advanced NF and RO treatment trains are presented. The environmental impact of technologies is also a key factor that needs to be taken into account when selecting a treatment train. A decision support method is presented to deal with this aspect.

Keywords emerging pollutants, membrane, nanofiltration, natural organic matter, reverse osmosis

INTRODUCTION

Drinking water is one of the scarcest and most valuable resources in the world. According to a study conducted by the WHO and UNICEF, around 900 million people currently have no access to clean drinking water. By the year 2025, it is estimated that half of the world‘s population will lack access to clean drinking water. Ensuring the availability of adequate amounts of clean water is set to be global issue number one in the future, and the market for the technologies required to process drinking water is one of the fastest-growing markets worldwide.

Water for public supply can be obtained from different sources: underground by wells sunk into aquifers, or from surface such as purpose-built reservoirs or lakes (collecting rainwater run-off or water from streams), rivers and sea. The safety of the water is of utmost concern – several million people die each year after consuming contaminated water. The primary aim in water treatment is the elimination of any pathogenic micro-organisms present. All the above-mentioned sources can be subject to pollution. In the case of underground water,

 TECHNEAU: Safe Drinking Water from Source to Tap. Edited by Theo van den Hoven and Christian Kazner. ISBN: 9781843392750. Published by IWA Publishing, London, UK.

polluted surface water can enter the saturation zone of an aquifer and so lead to its contamination. Pollution can come from waste tip leachate containing heavy metals and organic compounds, farm run-off containing nitrates and pesticides, and industrial wastes which may have been deliberately dumped down old coal mine shafts. River water can be affected by farm drainage, sewage works and industrial effluents, and also the run-off water from roads. Thus there is a need to ensure that the water quality is suitable for public supply, and at the same time that the cost of treatment is kept as low as possible.

Because of water stress and human concentration, industrials of the water sector will have to treat water from increasingly polluted sources without compromising the final quality. They will therefore need to face the multiple challenge of high performance in terms of quality, reasonable cost, and an environmental impact that will have to remain as low as possible. It is the goal of this paper to describe some treatment solutions for high quality water production, and decision support tools for the assessment of potable water supply systems, which take into account both economic and environmental objectives.

WATER QUALITY CHALLENGES

The goal of water treatment is to provide the consumer with water that is safe and free of unpleasant aspects, such as appearance, or taste and odour. The Drinking Water Directive (Drinking Water Directive, 1998) sets quality standards for drinking water quality at the tap (microbiological, chemical and organoleptic parameters) and the general obligation that drinking water must be wholesome and clean.

In order for these requirements to be fulfilled, the water at the outlet of the treatment plant must, not only be free of contaminants, but also stable, thereby ensuring that undesirable compounds will not develop during water distribution.

Contaminants to be removed include microbiological, mineral and organic pollutants. In order to ensure water quality at the consumer's tap, water treatment must also eliminate precursors of disinfection by-products, remove organic matter (including its biodegradable fraction) and turbidity.

Traditional water treatment train have been designed to remove the major known pollutants: microbiological pollutants, mineral and organic including micropollutants such as pesticides. However, new trends presently challenge these treatment files. First, the stringent regulation concerning bromates (10 μg/l) limits the use of ozone. Furthermore, attention has recently been raised on some emerging micropollutants. Beside the traditional organic micropollutants, the chemical as well as pharmaceutical industries produce a number of compounds that may be released in the environment and therefore be found at the inlet of drinking water plants. These compounds include: Pharmaceutical and Personal Care products (PPCPs), surfactants, industrial additives, and numerous chemicals suspected of being endocrine disruptors (Endocrine Disrupting Chemicals or EDCs). These compounds display a wide number of chemical structures and properties (type and number of chemical functions, pKa, polarity etc.) and therefore may challenge the capacity of the traditional water treatment processes. Furthermore, these molecules are designed to be biologically stable and therefore may remain for long periods of time in the environment.

Commonly found EDCs include the natural hormones 17β and 17α oestradiol and oestrone, the synthetic hormone ethyniloestradiol, as well as surfactant degradation products Nonylphenols (NP) , Octylphenol (OP) and their mono and di-ethoxylates (NPE1 and NPE2) (Ying *et al.*, 2004). Some of these compounds may not be fully eliminated by traditional wastewater treatment plants (Herry 2004a and b). Indeed, EDCs have been detected in treated wastewaters at concentration levels ranging from ng/l to µg/l (Ying *et al.*, 2004). In surface waters, estrone is the most commonly found hormone, whereas estradiol can occasionally be detected (Herry *et al.*, 2004a and 2004b; Bolong *et al.*, 2009), and in some cases nonylphenols at a concentration of few hundreds ng/l (Bolong *et al.*, 2009).

Pharmaceuticals can be classified as analgesics/anti-inflammatories, lipid regulators, beta-blokers, anti-epileptics and antibiotics. The four groups of antibiotics of interest are: sulfonamides, diamidopyridine, tetracyclines and macrolides. Pharmaceuticals have been detected in river water at concentrations up to a few hundred nanograms per liter (Stan and Heberer, 1997; Daughton and Ternes 1999; Ollers *et al.*, 2001; Beausse *et al.*, 2004; Garric *et al.*, 2005). Some of these compounds, such as antibiotics, or the antidepressant fluoxetine, or aspirin, may occasionally be found at the outlet of treatment plants, at a concentration below 50 ng/L (Beausse *et al.*, 2003; Herry *et al.*, 2004; Garric and Ferrari, 2005).

Physico-chemical treatments such as coagulation and flocculation were generally unable to remove EDCs and PPCPs. Adsorption (particularly on powdered activated carbon) and oxidation appeared as the most promising treatments. However, the application of these treatments is hindered by the concern about by-products, the effects of which are unknown (Bolong *et al.*, 2009).

The removal of Natural Organic Matter (NOM) is of paramount importance for a stable operation of the distribution system: Its biodegradable fraction participates in bacterial regrowth in distribution systems as well as biofilm formation. If post disinfection occurs, NOM is involved in the formation of potentially toxic disinfection by-products it is associated with taste and odour formation. Furthermore, it reacts with the disinfectant and results in higher disinfectant demand.

Natural Organic Matter can be divided into particulate and Dissolved Organic Matter, the latter comprising molecules from a few hundred to 100 000 Daltons which is in the colloïdal range. Beside the allochtnous and autochtonous fractions that have been present for years, a third fraction, namely anthropogenic, has been recently added, and participates in the total OM (Tondelier, 2009).

Dissolved OM comprises two fractions: the first one, Biodegradable Organic Matter (BOM) can be used by bacteria as energy and carbon source, whereas the second one is called the refractory fraction. BOM concentration can be measured by bioassays based on the measurement of DOC concentration (BioDegradable Organic Carbon or BDOC) or bacterial counts (Assimilable Organic Carbon or AOC). BOM includes both low and high molecular weight compounds. Some small molecular weight compounds can be readily biodegraded, whereas higher molecular weight compounds, such as humic substances, can be only partially biodegraded (Volk and LeChevallier, 2000). BOM has a significant impact on water quality in distribution networks. Numerous studies have linked the BOM concentration with bacterial regrowth and coliform occurrence in distribution systems. BOM, together with other parameters (such as disinfectant residual and temperature), can be used to evaluate the likelihood of coliform regrowth (Volk and Joret, 1994; LeChevallier, 2000). A threshold of 100 μg/l AOC has been derived from this work, with 82% more coliforms positive samples in networks not fulfilling this condition. BOM also has an impact as organic carbon limits microbial growth in biofilms (Gagnon *et al.*, 2000; Lehtola *et al.*, 2004). Distribution system biofilms are deleterious, in that they can release indicator organisms, and heterotrophic bacteria, and may cause taste and odor problems (Camper, 2001). In addition, other organisms can attach biofilms, which display protective properties: Legionella and Mycobacterium avium complex (MAC) are two organisms frequently isolated from biofilms. MAC levels have been correlated with AOC and BDOC levels (Falkinham, 2001).

NOM also plays an important role in the formation of Disinfection By-Products (DBPs) in the distribution network. Chlorination by-products of regulatory interest include TriHaloMethanes (THMs) and HaloAceticAcids (HAAs), with a Maximum Admissible Concentration of 100 μg/l for total THMs (ref: DW Directive). For HAAs, a Maximal Admissible Concentration of 80 μg/l is being discussed for the revision of the DWD. These products are suspected of being linked with cancer and birth defects (Boormann *et al.*, 1999; Hwang *et al.*, 2008). Many studies have shown the potential of membranes, particularly nanofiltration, in the removal of precursors of chlorination by-products (Amy *et al.*, 1990; Taylor, 1992; Siddiqui, 2000).

Among the DBPs are compounds involved in unpleasant taste and odours, namely haloanisoles, which have been shown to derive from the microbial methylation of halophenols. Chlorination of lignin, humic and fulvic acids leads to the formation of chlorinated halophenols, 2-4-6 trichlorophenol being the major and most stable compound. Within the network, chlorophenols can be biomethylated into chloroanisoles, a major source of a musty odor of drinking water. A very low concentration at the sub nanogram per liter (ng/L = ppt) level of these solutes is sufficient to be detectable by the human nose (Nystrom *et al.*, 1992; Tondelier *et al.*, 2009).

As a result of these challenges, the use of membranes has received increasing attention for treating poor quality water as they appear as the best technologies to (1) eliminate microbial as well as organic including emerging micropollutants (2) produce stable water that will remain wholesome in the distribution network.

FROM PRACTICAL EXPERIENCE TO THE DESIGN OF OPTIMISED TREATMENT TRAINS

The treatment used depends on the quality of the abstracted water. For water that has little pollution, it may only be necessary to use preliminary settlement, rapid sand filtration and chlorination, whereas poor quality water may require a more advanced treatment. An example of a treatment train which includes current treatment processes used today is shown in Figure 1.

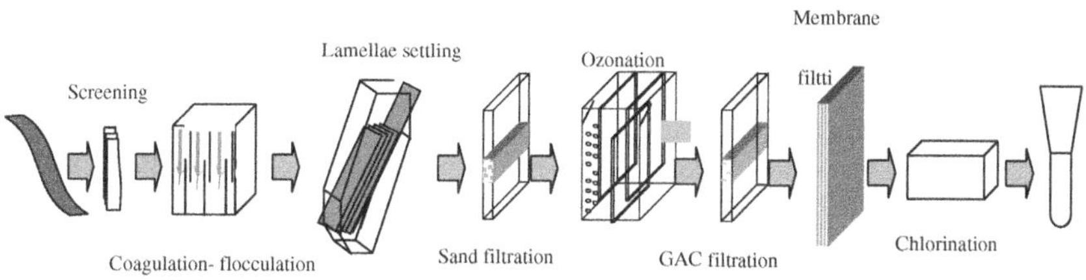

Figure 1. Extensive water treatment train for highly polluted water.

Natural organic matter removal: practical experience

TOC/DOC removal as measured with traditional analytical methods

In practice, a large proportion of natural organic matter (NOM) can be easily eliminated by coagulation-flocculation-settling operations. The preliminary jar tests make it possible to determine the dose of coagulant to use, although this is strongly dependent on the nature of the organic matter present. Nevertheless, the doses are generally close to 2 mg Fe/mg Total Organic Carbon (TOC) for Ferric salts, and 1 mg Al/mg TOC for aluminum salts. While the jar tests make it possible to achieve yields of 75% for the elimination of organic matter, in reality on site performance is closer to 50%. The fraction of NOM not eliminated by the settling step corresponds to the most hydrophilic fraction. This remains problematic and necessitates the implementation of refining or finishing processes. The refining treatment processes used so far have mainly been developed to ensure the elimination of micro pollutants and their performance in relation to NOM is still insufficient (ozone-GAC coupling, for example). In general, and although the ozone-GAC coupling increases the performance of the active carbon filters, the percentage of Dissolved Organic Carbon (DOC) eliminated is relatively low, and less than 30% after a few months of operation. Similar results can be observed with Powered Activated Carbon injection. Nanofiltration (NF) and Reverse Osmosis (RO) membranes are the best available technology that allows a more substantial reduction in organic matter than the other processes. The yields for the elimination of NOM, in the form of DOC and BDOC, are in the order of 90%. Results are summarized in the Table 1.

Table 1. Practical experience of organic matter removal with current treatment processes.

	Coagulation floculation	PAC	Ozone/GAC	NF and RO
TOC	30–70%	10–30%	<30%	>90%

Organic matter removal as analysed with advanced analytical techniques

Over the last five years LC SEC OCD analysis has been widely used to better characterise the organic matter present in the water. The system called Liquid Chromatography-Organic Carbon Detection includes a size exclusion chromatography column in order to separate hydrophilic organic molecules according to their molecular size. The separated compounds are then detected by 3 different detectors: a UV detector (absorption at 254 nm), a DOC detector (185-nm-UV-oxidation process) and a DON detector (210 nm-UV-oxidation process). Depending on the size exclusion chromatography column and the detectors response, it is possible to characterize the sample in terms of organic fractions.

For instance for DOC detector, the main groups of substances are:

- Tr = 30 min, the group of polysaccharides, proteins and colloids,
- Tr = 40–55 min, humic substances centered at 43min and "degraded" humic substances centered at 48 min, called "building blocks".
- After Tr = 55 min, the organic acids and molecules with the lowest molecular weights.

An example of organic matter removal by a 3 step treatment train: coagulation/flocculation/settling process, ozone/GAC followed by nanofiltration is presented in Figure 2. The removal rate is about 60% for the humic substances and building blocks peaks whereas the removal of polysaccharides is about 80% for coagulation/ flocculation/settling process. Humic substances are more eliminated than the building blocks as it can be seen by the relative increase of buiding block peak centered at 48 min compared to the Humic substances peak centered at 43 min. The Ozone/GAC process removes all the different substances by an additional 19%. Nanofiltration is used as a finishing step with a complete removal of polysaccharides, humic substances and building blocks fractions.

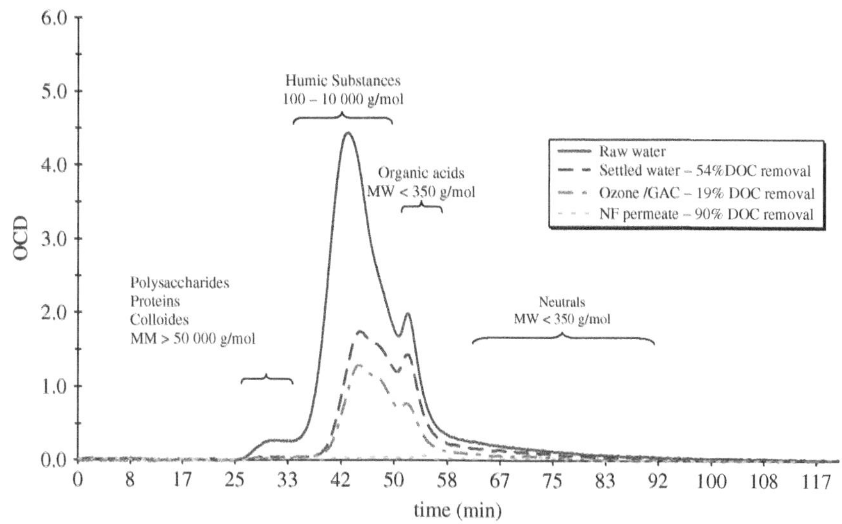

Figure 2. Example of removal rate of settling, ozone/GAC and NF processes on organic matter fingerprint.

This technique allows a better understanding of NOM removal by treatment processes and therefore helps in more accurately targeting the right treatment train according to the raw water quality and treatment objective. For instance, it is today well known that polysaccharides are one of the species responsible for membrane fouling and therefore a coagulation/flocculation/settling membrane pre-treatment is recommended.

Emerging micropollutants: practical experience

To face water contamination with emerging pollutants such as pharmaceuticals, pesticides and healthcare products, Veolia has conducted over the last five years extensive research to assess the performance of current available technologies. The main results on coagulation/flocculation/settling, PAC, ozone/GAC and membranes are presented in Tables 2–4 and 5. All results were obtained with lab-scale and pilot plant units under practical operating conditions (pH = 7): PAC dose and contact time are respectively in the range of 10–20 ppm and 60 min. Ozone concentration range from 1 to 2.5 ppm. Membrane test are performed at 10% recovery factor unless the recovery is stated. NF1 is a Dow chemical membrane NF 200 with MWCO = 200 g/mole, NF2 is a Dow chemical membrane NF90 with MWCO = 90 g/mole, RO membrane is a Brackish water membrane from Dow chemical: BW30LE.

Table 2. Antibiotic removal with current treatment processes.

Antibiotics	Coagulation flocculation	PAC	Ozone/ GAC	NF 1 at 85% recovery	NF 2	RO	Chlori- nation
Sulfamerazine	<5%	>96.1%	>97%	94.7%	>97%	>96%	>97%
Sulfametho- xazole	<5%	91.9%	>97%	>94%	>98%	>97%	>96%
Sulfachloro- pyridazine	<5%	97.8%	>98%	>95.8%	98.4%	98.4%	>98%
Spiramycine	<5%	95.6%	95.3%	>97.3%	>91%	>90%	86.6%
Roxithro- mycine	<5%	96.7%	95.2%	93.4%	95.4%	>97%	57.1%

Table 3. Hormones removal with current treatment processes.

Hormones	Coagulation flocculation	PAC	Ozone/ GAC	NF 1 at 85% recovery	NF2	RO	Chlori- nation
17 – a – Estradiol	<5%	97.8%	>94%	>94.4%	>88%	>94%	>94%
17 – b – Estradiol	<5%	94.6%	>94%	>93.5%	>89%	>95%	>94%
Ethinyl Estradiol	6.5%	87.1%	>90%	>88.6%	>79%	>95%	>94%

Table 4. Pharmaceutical removal with current treatment processes.

Pharmaceuticals	Coagulation flocculation	PAC	Ozone/GAC	NF1 at 85% recovery	NF1	NF2	RO	Chlorination
Carbamazepine	<10%	>97.8%	>90%	>95%	>90%	>96%	>96%	<10%
Metoprolol	–	>91%	>95%	94%	–	>85%	>95%	<10%
Propranolol	–	>97%	>95%	96.3%	–	>90%	>99%	–
Diclorofenac	<10%	–	>90%	–	>90%	>90%	>90%	50–90%
Bezafibrate	<10%	>87%	>88%		>90%	97%	>99%	<10%

Table 5. Pesticides removal with current treatment processes.

Pesticides	Coagulation flocculation	PAC	Ozone/GAC	NF1 at 85% recovery	NF1	NF2	RO	Chlorination
Atrazine	<10%	55%	>90%	–	>90%	>95%	>95%	<10%
DEA	<10%	50%	90%	85%	–	>96%	>96%	<10%
DIA	<10%	50%	90%	60%	–	>95%	>95%	<10%
Linuron	<10%	80%	>95%	>95%	–	–	–	–
AMPA	50%	<10%	>86%	>88%	–	>97%	>97%	>90%

According to these results, membrane technology with NF and RO, ozone/GAC and PAC are efficient processes to remove micro pollutants and emerging compounds. However activated carbon adsorption could be very expensive when large fractions of NOM are present, because adsorption sites are mainly taken by NOM, present in much higher concentrations. Organic micropollutants may be present at the ng/l or μg/l level, whereas NOM concentrations may be 10,000 times higher. The adsorption columns have to be regenerated rapidly because the column capacity is mainly used for NOM adsorption instead of micropollutants adsorption. Furthermore, new concern on by-products formation and the stringent regulation concerning bromate (10 μg/l) limit the use of ozone. Therefore for poor quality water, NF and RO system is the best available technology to ensure a high-quality drinking water; should chlorination be required to maintain a disinfectant residual in the distribution network, the produced permeate would likely be low in DBPs and would present a minimal risk of bacterial regrowth. Compared to RO, NF membranes could have some advantage for drinking water production as they can reject most organics and micro pollutants while limiting the rejection of salt and thus reducing post-treatment costs such as remineralisation. Furthermore, NF process is operated at lower applied pressure than RO process for similar working fluxes, thus limiting the energy demand. Table 6 presents raw and treated river water quality after NF treatment with classical pre-treatment (coagulation/flocculation followed by sedimentation and Sand filtration). As it can be seen, the treatment by nanofiltration provides an excellent water quality with all the studied parameters below the detection limit.

Table 6. Raw and treated river water quality after NF treatment with classical pre-treatment (coagulation/flocculation followed by sedimentation and Sand filtration).

Parameters		Raw water	NF
Turbidity	(NFU)	10 to 400	<0.01
TOC	(ppm)	2 to 3	<0.2 ppm
Pharmaceuticals:			
Diclofenac	(ng/l)	10 to 50	<10
Bezafibrate	ng/l	10 to 100	<10
Pesticides:			
Bentazone	μg/l	0.02 to 0.1	<0.02
Hydroxy atrazine	μg/l	0.02 to 0.1	<0.02
Isoproturon	μg/l	0.02 to 0.1	<0.02
Hormons:			
Estradiol-béta	ng/l	0.2 to 0.5	<0.2
Estrone	ng/l	0.2 to 0.5	<0.2
Microbiology:			
Coliforms (in 100 ml)	n/100 mL	2 to 5000	<DL
E. Coli (in 100 ml)	n/100 mL	100 to 500	<DL
Enterococus (in 100 ml)	n/100 mL	200 to 1000	<DL

Designing the most appropriate treatment train

Fouling is one of the main problems in any membrane separation and chemical pretreatment processes such as coagulation have shown to improve membrane performance by reducing the rate of membrane fouling. This emphasizes the importance of the pre-treatment that has to be designed based on raw water quality and treatment objective. For surface waters, coagulation/flocculation followed by sedimentation is usually necessary; Ultrafiltration (UF) or Microfiltration (MF) is a valuable alternative. An example of 3 treatment trains is presented in Figure 3. The main difference between the treatment trains 1, 2 and 3 is the complexity/simplicity of the NF/RO pre-treatment. Treatment train 1 is based on a classical chemically-enhanced pretreatment while treatment train 2 and 3 involve a MF or UF pre-treatment which simplify the pre-treatment. Train 2 and particularly train 3 are attractive for water treatment as they answer the demand for more efficient, compact and discreet treatment units involving a reduction of chemicals use and therefore a reduction of sludge production. They are particularly suitable for countries such as Australia where the cost or the environmental impact of the sludge disposal is high.

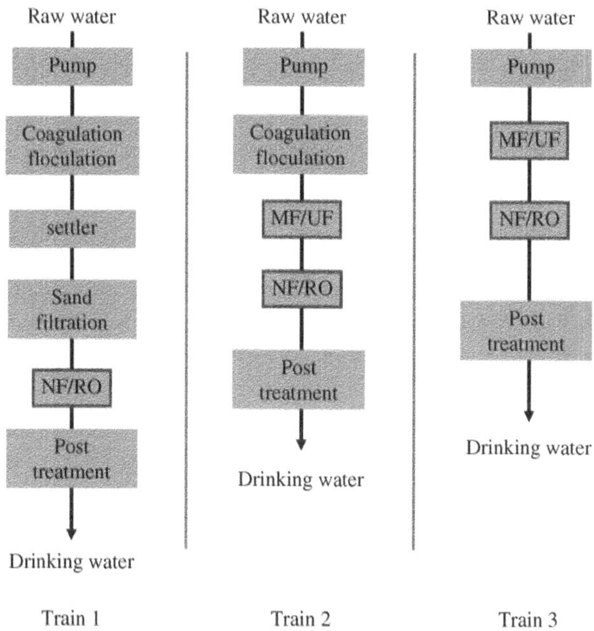

Figure 3. Treatment trains for drinking water production.

The advantages of the "Integrated Membrane Systems" (IMS) such as a MF/UF pre-treatment for NF/RO membrane systems are the followings:

- Better MF/UF filtrate quality: The colloidal fouling load to the NF/RO is reduced with significantly lower SDI (Silt Density Index) and turbidity.

- Constant and consistent MF/UF filtrate quality provided by the absolute barrier of the membrane. This is particularly useful for treating surface sources afflicted by rapidly fluctuating quality.
- Reduced sludge production due to the use of relatively less chemicals than in chemically-enhanced conventional pre-treatment processes.
- Smaller footprint due to a reduced floor space requirement for MF/UF systems
- Improved flexibility due to the modularity of membrane systems. Future expandability is easier to design in.

Nevertheless, the benefits of Integrated Membrane Systems still need to be proved as only few side-by-side pilot-plant trials have been performed so far.

Disposal and Treatment of concentrate

In general, concentrate disposal as such from drinking water production is feasible as long as a limited number of parameters such as sulphate, chloride, phosphate, iron and anti-scalant are under control. However, there is a risk that concentrate disposal will be banned in the future and therefore we should explore possibilities to treat the concentrate. The number of works published on concentrate treatment for drinking water application is limited. Some patents have demonstrated the feasibility of implementing a treatment train to selectively remove the phosphonate-based anti-scalants (Daines-Martinez *et al.*, 2008) and improve the recovery factor (Daines-Martinez *et al.*, patent pending). Recently, Bozkaya-Schrotter *et al.* (2008) studied pesticide and natural organic matter elimination on real membrane concentrates. The main results of this work entitled "Treatment of NF concentrates: organic micropollutant and NOM removal" are presented in this book.

DEVELOPING DECISION SUPPORT TOOLS

In order to foster the development of sustainable practices within the water sector, Veolia Environnement Research & Innovation has developed Eolia^{TM} Potable Water, a decision support tool dedicated to the environmental evaluation of potable water supply systems (ex: conventional treatment, desalination, wastewater reuse, water import...).

Life Cycle Assessment methodology

Eolia^{TM} Potable Water relies on the ISO 14040 standardized Life Cycle Assessment (LCA) method for the environmental performance evaluation of water treatment processes. LCA has gained a wider acceptance as the method providing the most comprehensive overview of the environmental impacts generated by an industrial process (ISO, 2000; UNEP/SETAC, 2005). LCA is a "cradle-to-grave" approach that addresses the impacts (ex: green house gases emissions) generated during all phases of the life cycle of the studied process: from the extraction of raw materials through the operation and finally the decommissioning of the process (cf. Figure 4).

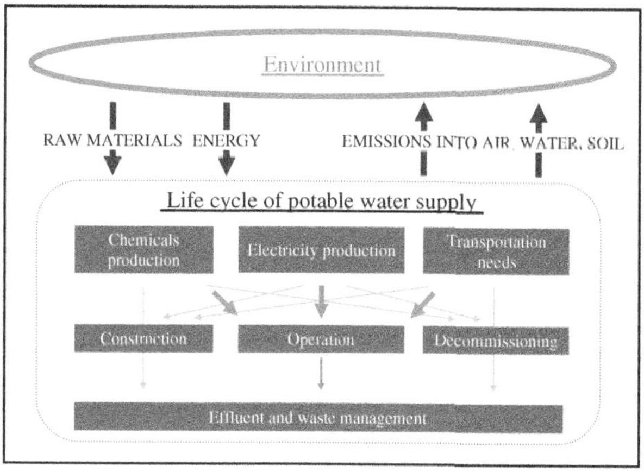

Figure 4. Life-cycle of potable water supply.

Eolia™ Potable Water facilitates the environmental evaluation of water treatment processes and potable water supply systems developed, thus allowing Veolia Environnement to systematically integrate sustainability factors during process design by:

- assessing the environmental performances of water treatment processes and identifying R&D needs;
- highlighting impact sources and improvement levers on existing potable water production plants;
- benchmarking technical proposals for future potable water supply projects (ex: groundwater transfer from a distant water basin vs. reverse osmosis desalination, comparison between membrane pretreatment options).

Presentation of the Eolia™ Potable Water tool

Eolia™ Potable Water includes 15 pre-defined potable water supply systems (e.g. conventional treatment, degraded surface water treatment using nanofiltration, wastewater effluent reclamation, seawater desalination using reverse osmosis) calibrated by water treatment experts using constructor and operating data from Veolia Water plants.

These pre-defined supply systems may be easily sized and adapted to project constraints and specific design choices. If needed, the user may build tailor-made solution using a database of process unit modules (e.g. clarification, sand or granular activated carbon filtration, ultrafiltration, cartridge filters) that may be graphically interconnected and adjusted via the graphical interface of Eolia^{TM} Potable Water.

Within Eolia^{TM} Potable Water, the environmental performances are calculated automatically on 5 criteria: greenhouse gases (GHG) emissions, energy consumption, damages to human health, damages to biodiversity (Vince, 2008). This exhaustive evaluation allows to quantify how design choices should be adapted for an improved environmental performance (ex: trade-offs between the on-site impacts of the potable water supply system and the indirect impacts of its supply chain, trade-offs between liquid discharges vs. electricity consumption).

Case studies

Eolia^{TM} Potable Water has been applied on a various range of case studies, which illustrate its flexibility and its relevance for decision support (Vince, 2007).

Identification of impact sources and penalizing processes

In Figure 5, the impacts of each process within the life-cycle of the studied supply system are quantified as a relative contribution to the overall environmental score. Sixty-seven percent of the impacts result from electricity production for intake pumping and distribution, 14% from electricity production for water treatment (including ozone production) and 10% from chemicals production (especially coagulant and activated carbon). Interestingly, the plant construction contributes to less than 4% of the impacts. Energy management on site (ex: optimized pumping systems) and a leak reduction campaign on the distribution network appear therefore to be the first measures towards impact mitigation.

Benchmark between technical alternatives

In Figure 6, the green house gases emissions of two technical alternatives for feed water pretreatment options are compared. In this example, conventional pretreatment is highlighted to generate more GHG emissions than an advanced pretreatment option using UF membrane process.

As illustrated in these examples, Eolia^{TM} Potable Water allows a quick and easy evaluation of the environmental footprint of potable water supply solutions.

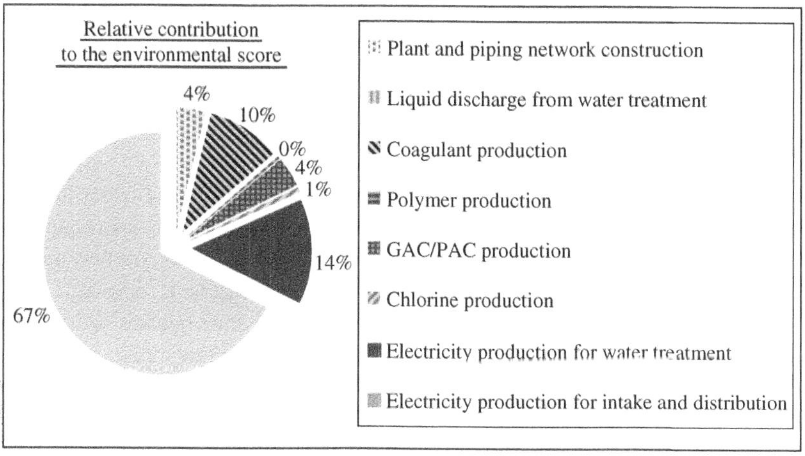

Figure 5. Relative contribution of processes to the environmental score (case study: conventional potable water supply system in UK).

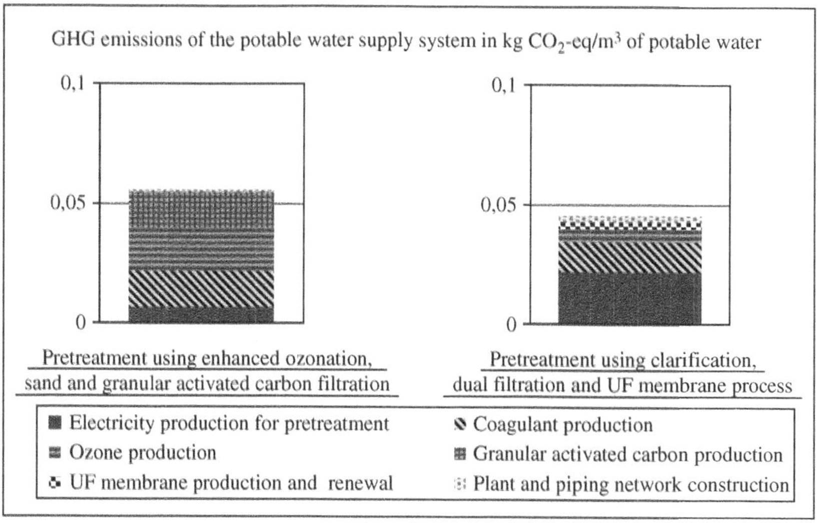

Figure 6. Comparison of the GHG emissions generated by two pretreatment options.

It will therefore be used by Veolia Environnement, in order to enhance the design practices over and above technical, economic criteria by including environmental performance indicators at the early stages of the decision-making process.

CONCLUSION

Membrane processes have gained wide acceptance in the drinking water industry because of their ability to produce a high-quality and consistent water. Of particular interest are Nanofiltration and Reverse Osmosis which are best options for Natural Organic Matter (NOM) and emerging pollutant removal. These membrane processes anticipate future challenges on water quality as attention has recently been raised on some emerging micropollutants that are poorly eliminated by conventional waste water treatment. Therefore, their concentrations in aquifers will increase in the future. However, these membrane systems require an extensive pre-treatment making the water treatment train often chemically and operationally intensive. Integrated Membrane Systems (IMS) are potentially promising examples of simplification of the process train while at the same time improving the efficiency of the treatment. Eolia™ Potable Water decision support tool can also be used to select the best treatment train with regard to environmental impact assessment. Nevertheless, the availability of more efficient, chemically free and cheaper processes will be necessary in the future. Nanotechnology offers a promising alternative to move toward that direction. The advances in new nanomaterials can help to develop cost effective, highly selective and fouling resistant membranes for water purification and desalination (Shannon *et al.*, 2008).

REFERENCES

Amy, G. *et al.* (1990). Removal of dissolved organic matter by nanofiltration. *J. Environ. Eng. – ASCE*, **116**, 1–1200.

Beausse, J., Boireau, V., Canu, M. and Herry, G. (2004). Determination of Pharmaceutical Compounds in Drinking Water Treatment Plants, Forum Labo 2004, Paris (France), 23–26 March 2004.

Beausse, J., Canu, M., Herry, G. and Enguehard, F. (2004). Pharmaceutical Compounds in Drinking Water Treatment Plants. Pittcon 2004, Chicago, USA, 7–12 March 2004.

Bolong, N, Ismail, A.F., Salim, M.R. and Matsuura, T. (2009). A review of the effects of emerging contaminants in wastewater and options for their removal. *Desalination*, **239**, 229–246.

Boorman, G.A., Dellarco, V., Dunnick, J.K., Chapin, R.E., Hunter, S., Hauchman, F., Gardner, F., Cox, M. and Sills, R.C. (1999). Drinking water disinfection byproducts: review and approach to toxicity evaluation. *Environ. Health Perspect.*, **107**(Suppl. 1), 207–217.

Bozkaya-Schrotter, B., Daines, C., Lescourret, A.S., Bignon, A., Breant, P. and Schrotter, J.-C. (2008). Treatment of trace organics in membrane concentrates I: Pesticide elimination water science and technology. *Water Supply*, **8**, 223–230.

Camper, A.K. (2000). Biofilms in drinking water treatment and distribution. In: *Biofilms: Recent Advances in their Study and Control*, L.V. Evans (ed.), Harwood Academic Publishers, pp. 311–332. ISBN 90-5823-093-7.

Daines-Martinez, C., Schrotter, J.C., Drouet, K. and Gaid, K. (2008). Membrane station waste treatment process WO2008/055827.

Daines-Martinez, C., Schrotter, J.C., Drouet, K. and Breant, P. Patent Pending.

Daughton, C.G. and Ternes, T.A. (1999). Pharmaceuticals and personal care products in the environment: Agents of subtle change. *Environ. Health Perspec.*, **107**(Suppl. 6), 907–942.

Drinking Water Directive (DWD) (1998). Council Directive 98/83/EC of 3 November 1998 on the quality of water intended for human consumption.

Falkinham, J.O., Norton, C.D. and LeChevallier, M.W. (2001). Factors influencing numbers of mycobacterium avium, mycobacterium intracellulare and other mycobacteria in drinking water distribution systems. *Appl. Environ. Microbiol.*, **67**, 1225–1231.

Gagnon, G.A., Slawson, R.M. and Huck, P.M. (2004). Effect of easily biodegradable organic compounds on bacterial growth in a bench-scale drinking water distribution system. *Can. J. Civ. Eng.*, **27**(3), 412–420.

Garric, J. and Ferrari, B. (2005). Pharmaceuticals in aquatic ecosystems; levels of exposure and biological effects: A review. *Rev. Sci. Eau.*, **18**(3), 307–330.

Herry, G., Raulet, C., Boireau, V., Canu, M. and Beausse, J. (2004a). Etude de nouveaux polluants émergents dans le milieu environnemental: Les perturbateurs endocriniens et les produits pharmaceutiques. ASEES Conference «Problématique des risques sanitaires pour la santé et l'environnement liés aux nouvelles molécules mises en cause par les directives européennes. Perturbateurs endocriniens», Strasbourg, France 25–26 Nov 2004.

Herry, G., Raulet, C. and Beausse, J. (2004b). Analysis of steroid hormones in waste-water treatment plants. 9th FECS Conference on Chemistry and the Environment: Behaviour of Chemicals in the Environment, Bordeaux (France), 29 Aug–1st Sept 2004.

Hwang, B.-F., Jouni, J.K., Jaakkola and Guo, H.R. (2008). Water disinfection by-products and the risk of specific birth defects: A population-based cross-sectional study in Taiwan. *Environ. Health*, **7**, 23.

ISO (2000). ISO 14040: Environmental Management – Life Cycle Assessment – Principles and Framework, International Standard Organization.

LeChevallier, M.W. (2000). Conditions favouring colifirm and HPC bacterial growth in drinking water and on water contact surfaces. In: WHO Heterotrophic Plate Counts and Drinking Water Safety, J. Bartram, C. Cotruvo, M. Exner, C. Fricker and A. Glasmacher (eds.), IWA Publishing, London, UK.

Lehtola, M., Juhna, T., Miettinen, I.T., Vartiainen, T. and Martipainen, P.J. (2004). J. IND. *Microbiol. Biotechnol.*, **31**, 489–494.

Lundqvist, J. and Gleick, P. (2000). Sustaining our water into the 21st Century. Stockholm: Stockholm Environment Institute.

MEA (2005). Ecosystems and Human Well-Being. Millenium Ecosystem Assessment.

Nixon, S.C., *et al.* (2000). Une utilisation durable de l'eau européenne? État, perspectives et résultats. Agence Européenne pour l'Environnement.

Nyström, A., Grimvall, A., Krantz-Rülcker, C., Sävenhed, R. and Akerstrand, K. (1992). *Wat. Sci. Tech.*, **25**, 241.

Ollers, S., Singer, H.P., Fassler, P. and Muller, S.R. (2001). Simultaneous quantification of neutral and acidic pharmaceuticals and pesticides at the low ng/l in surface and waste water. *J. Chrom. A*, **911**, 225–234.

Shannon, M.A., Paul, W., Bohn, M.E., Georgiadis, J.G., Marinas, B.J. and Mayes, A.M. (2008). Science and technology for water purification in the coming decades, *Nature*, **452**, 301–310.

Siddiqui, M., Amy, G., Ryan, J. and Odem, W. (2000). Membranes for the control of natural organic matter from surface waters. *Water Res.*, **34**(13), 3355–3370.

Siddiqui, M. (2000). Membrane technology for the control of natural organic matter from surface waters, water quality and technology for the 21st century, MI Section, AWWA.

Stan, H.J. and Heberer, H.S. (1997). Pharmaceuticals in the aquatic environment. *Analysis*, **25**(7), 20–23.

Taylor, J., *et al.* Reduction of disinfection by product Precursors by nanofiltration, USEPA, PB92-149269/AS.

Tondelier, C., Thouvenot, T., Genin, A. and Benanou, D. (2009). On-tap passive enrichment, a new way to investigate off-flavor episodes in drinking water. *J. Chrom. A* (in press).

UNEP/SETAC (2005). Life Cycle approaches: the road from analysis to practice. UNEP/SETAC Life Cycle Initiative.

UNESCO (2003). The United Nations World Water Development Report 1: Water for people, water for life. World Water Assessment Programme.

UNESCO (2006). The United Nations World Water Development Report 2: Water, a shared responsibility. World Water Assessment Programme.

United Nations Development Programme (2006). The Millennium Development Goals: Report. New York: Department of Economic and Social Affairs.

Vince, F. (2008). Design approach for the improvement of both the economic and the environmental performances of potable water supply systems. PhD Thesis in Energy (EPFL – Lausanne).

Vince, F., *et al.* (2007). LCA tool for the environmental evaluation of potable water production, *EDS Conference "Desalination and Environment"* (Halkidiki – Greece: European Desalination Society).

Volk, C. and LeChevalier, M. (2000). Assessing biodegradable organic matter. *J. AWWA*, **92**(5), 64–76.

Volk, C. and Joret, J.C. (1994). Paramètres prédictifs de l'apparition des coliformes sur les réseaux de distribution. *Rev. Sci. Eau*, **7**, 131–152.

Volk, C.J. and LeChevallier, M.W. (1999). Impacts of the reduction of nutrient levels on bacterial water quality in distribution systems. *Appl. Environ. Microbiol.*, **65**(11), 4957–4966.

Wenzel, H., *et al.* (2002). Process integration design methods for water conservation and wastewater reduction in industry. Part 3: Experience of industrial application. *Clean Technol. Environ. Policy*, **4**, 1625.

Ying, G.G., Kookana, R. and waite, T.D. (2004). Endocrine disrupting chemicals (EDCs) and pharmaceutical and personal care products (PPCPs) in reclaimed water in Australia, report of the Australian water conservation and reuse research program. http://www.clw.csiro.au/awcrrp/stage1files/AWCRRP_1H_Final_27Apr2004.pdf

Development and evaluation of a new concept for drinking water treatment: the OBM process

K. Azrague, S.W. Osterhus and T. Leiknes

Dep. of Hydraulic and Environmental Engineering, The Norwegian University of Science and Technology (NTNU), S.P. Andersensvei 5, N-7491 Trondheim, Norway
(E-mail: kamal.azrague@ntnu.no; stein.w.osterhus@sintef.no; torove.leiknes@ntnu.no)

Abstract A new concept of drinking water treatment combining ozonation with biofiltration and membrane filtration (OBM) has been investigated for NOM containing water. In-line mixer was found to be superior to diffusers allowing a faster transfer of ozone from the gas to the liquid phase. In comparison to stainless steel column packing media, ceramic media improves the decomposition of ozone into hydroxyl radical. Rapid NOM fractionation showed that the feed water NOM was dominated by the VHA, with considerable less SHA. Ozonation removed more than 50% of the VHA producing CHA and NEU. An average of approximately 75% of the BDOC produced during the ozonation was removed in the biofilter. Except for the GAC biofilter media, which was superior and removed the BDOC very efficiently, only minor differences between media of KMT, quarts sand, anthracite, and calcium carbonate were observed. Two types of membranes have been used for the final filtration step; a polymeric ultrafiltration membrane; and a ceramic microfiltration membrane. Severe irreversible fouling occurred with the ceramic microfiltration membrane. The ultrafiltration membrane worked well with approximately 600 h of operation before chemical cleaning was needed for a flux of 60 LMH and 400 h for a flux of 80 LMH. The OBM performed well with an average removal of 63% for UV-absorbance (UV254), 79% for color, and 28% for DOC when treating NOM containing surface water. The measured parameters in treated water comply with the Norwegian drinking water standard.

Keywords biofiltration, drinking water treatment, membrane filtration, ozonation

INTRODUCTION

In areas with water shortage they are often forced to utilize water sources having severe quality problems. The sources may contain humic substances which impart a brown/yellow color to the water (Back and Christman, 1963), can complex with metals (Reuter and Perdue, 1977) and organic pollutants such as pesticides (McCarthy, 1989), can serve as electron donor for the growth of heterotrophic bacteria in distribution systems (LeChevallier *et al.*, 1996), and most significantly, are precursors for mutagenic halogenated compounds formed after chlorination (disinfection by-products, DBPs) (Rook, 1977). In addition to humics, the water may contain micro-pollutants coming from the anthropogenic

 TECHNEAU: Safe Drinking Water from Source to Tap. Edited by Theo van den Hoven and Christian Kazner. ISBN: 9781843392750. Published by IWA Publishing, London, UK.

activities. Organic micro-pollutants are a mixture of dissolved and non-dissolved compounds which are potentially persistent in the environment and have adverse effects on human health even at low concentrations.

To improve the water quality, surface water may be treated in different ways. Several processes, such as sand filtration, coagulation followed by separation, activated carbon filtration, etc are commonly used technologies to produce drinking water. However, most of these conventional drinking water treatments processes produce excessive sludge waste, and they are not sufficient to remove synthetic organic compounds from the water.

Ozone is recognized as an alternative to commonly used disinfectants in water treatment such as chlorine, chloramines, and chlorine dioxide. Therefore, it reduces or eliminates the need for chlorination and decreases formation of chlorinated disinfection by-products (Speitel *et al.*, 1993). Since the ozone is a strong oxidant it can also be used to remove metallic ions, micro-pollutants and compounds causing odour and bad taste (Camel and Bermond, 1998). In addition, by breaking down the structure of natural organic matter (NOM) enhancing the transformation of higher molecular weight compounds into lower molecular weight compounds it converts non-biodegradable organic matter to biodegradable organic carbon (BDOC) (Yavich *et al.*, 2004), and thereby also removing color and UV absorbance. For some recalcitrant micro-pollutant, ozone can be combined with other oxidant (H_2O_2) (Legrini *et al.*, 1993) or catalyst (Legrini *et al.*, 1993; Cooper and Burch, 1999) leading to the formation of hydroxyl radicals which are more powerful oxidants and able to break down most of the organic compounds.

In order to remove the biodegradable fraction coming from the ozonation and to avoid the growth in the network, ozonation is often followed by biofiltration (Melin and Odegaard, 2000). The biofiltration can also be used to remove certain micro-pollutant which are not removed in the ozonation step but are bio-degradable, such as MTBE (Lee and Cho, in press). The biofiltration step will produce/release sludge, bacteria and particles which should be removed from the water before distribution.

It is well known that membrane filtration is an effective method to remove particles, micro-organisms and organic matter from drinking waters. Compared with conventional treatment methods, membrane processes (i) can provide better quality water, (ii) minimize disinfectant demand, (iii) are more compact, (iv) provide easier operational control and less maintenance, and (v) generate less sludge.

Consequently, as a part of the EU-project TECHNEAU, and on the basis of the MBR technology, a novel process combination consisting of the three steps oxidation (ozonation), biodegradation and membrane filtration (the OBM-process) has been investigated for the drinking water treatment.

METHODS

Experimental set-up

A schematic sketch of the process configuration and the pilot plant used for the investigations is shown in Figure 1. The pilot plant consisted of the following steps; two ozone columns in series, two biofilters in parallel followed by an immersed Zenon ZW10 ultrafiltration membrane reactor module. The ozone was generated by Ozone Solutions CDO-8000E ozone generator, and added to the feed water in counter-current packed bubble column (∅15 cm, h = 400 cm). The ozonated water was then distributed to a second similar column (not packed) to finalize the chemical oxidation, to provide a desired CT-value (Concentration × Time) for pre-disinfection, and to remove residual ozone prior to the biofilters. The two parallel biofilters (each with ∅12 cm, h = 400 cm) were filled with KaldnesTM K1 carrier (KMT) (220 cm was filled), offering 500 m^2/m^3 effective area in filter configuration for the growth of biofilm. The pilot plant was operated with a flow of 90–100 $L.h^{-1}$ resulting in empty bed contact times (EBCT) of ca. 35 minutes for each of the ozone columns and a total of ca. 30 minutes for the biofilters with a filter velocity of ca 4.5 $m.h^{-1}$. The membrane system used was a Zenon ZW10 membrane pilot module. The hollow fiber membrane module was made with a PVDF ultra-filtration (UF) membrane that has a nominal pore size of 0.04μm and a surface area of 0.93 m^2.

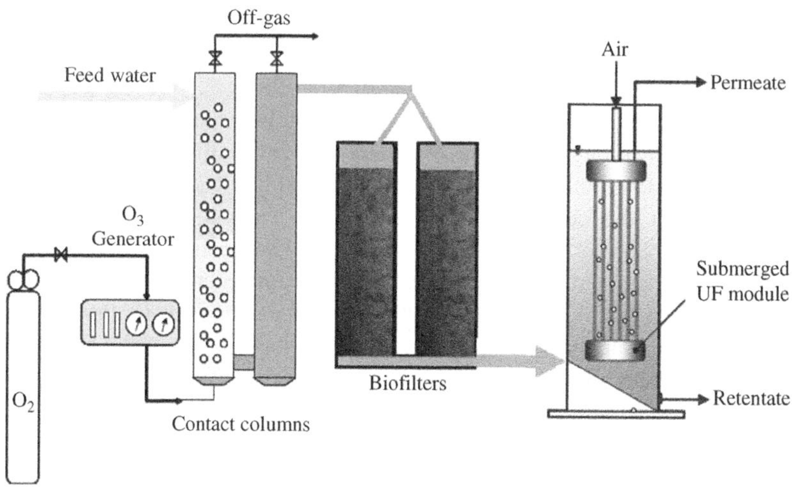

Figure 1. Schematic sketch of the OBM pilot plant.

The membrane system was run with an outside-in configuration, and the module was submerged in a 30 L tank. The system was run at fluxes of 60, 70 and 80 LMH, with backwash every 29.5 minutes which lasted for 30 seconds at a rate of 1.5 times the operating flux. Air scouring was applied only during backwash. The recovery was 98%.

Additional configurations were also tested; including a ceramic microfiltration (MF) membrane unit (supplied by Metawater Co.), which had a nominal pore size of 0.1 μm and a surface area of 0.43 m^2 and was operated with a flux of 140 LMH and backwash of 5 seconds every hour; a co-current packed ozonation bubble column; and biofilter columns with different filling media. For the ozonation bubble columns two types of packing media were used, one stainless steel media, resistant to oxidation, and one Raschig ring media made of ceramic (min 92% $Al_2O_3 + SiO_2$ with max 1.0% Fe_2O_3) and with a total bed contact time of 17.5 minutes each. The biofilters were filled with four different media: anthracite, calcium carbonate, activated carbon (GAC) and KaldnesTM K1 carrier (KMT) and with an empty bed contact time of 20 minutes each.

Three types of feed water were used: 1) tap water with a color of 15 mg Pt/L, 2) a mixture of tap water and NOM concentrate with a final color of 50 mg Pt/L and 3) distilled water (NOM free water). The NOM-concentrate used was a regeneration solution from an anion exchanger for NOM-removal at a drinking water treatment plant in Kongsvinger, Norway. Prior to the ozonation, pH of the feed water was adjusted to the desired pH by adding HCl 0.6% solution and to desired total inorganic carbon (TIC) by adding $NaHCO_3$. The applied ozone dose was 0.18 ± 0.01 mg O_3/mg Pt.

Analyses

Prior to measurements of color and UV_{254}, the samples were adjusted to pH 7 and then filtered through 0.45 μm Millipore membrane filters. The color was measured at 410 nm and UV_{254} was measured using Hitachi U-3000 Spectrophotometer. For the color profile in the ozone column, the samples were quenched with Na_2SO_3 prior to the color analysis. Suspended solids (SS) were analyzed by filtering through a Whatman GF/C 1.2 μm according to the Standard Methods for water and wastewater, 1998 (Clesceri *et al.*, 1998). Particle size distribution (PSD) analysis of the water in the various stages of the membrane process was done using laser diffraction spectroscopy (Beckman Coulter LS230). The concentration of ozone in the in-gas and off-gas was measured with an online BMT 964 Ozone Analyzer. The residual ozone concentration in the water was analyzed by the indigo method (Clesceri *et al.*, 1998). Turbidity was measured using HACH 2100N Turbidimeter. Alkalinity and pH were analyzed using Metrohm Titroprocessor 726 equipped with sample changer 717 and Dosimat 685. Dissolved organic carbon (DOC) samples was filtered using 0.45 μm Millipore membrane filters prior to analysis using Tekmar Dohrmann Apollo 9000 total organic carbon analyzer. The specific UV absorbance, SUVA, was obtained by dividing the UV_{254} absorbance (m^{-1}) by the DOC. The conductivity was measured using WTW microprocessor conductivity meter LF537.

Biodegradable dissolved organic carbon (BDOC) was determined in columns as a 9 h BDOC using similar set-up as for rapid BDOC (Yavich *et al.*, 2004). The set-up used in this work consisted of 6 columns in series run in single pass mode. A fractionation unit described by Chow *et al.* (2004) was used to determine the concentrations of four NOM fractions: very hydrophobic acids (VHA), slightly hydrophobic acids (SHA), charge hydrophilics (CHA) and neutral hydrophilics (NEU) (Chow *et al.*, 2004).

The performance of the membrane module was determined by measuring the transmembrane pressure (TMP) for constant flux operation. The developments of TMP at different fluxes were measured continuously using an online pressure transducer connected to a data acquisition system. Permeate water flow rates were measured with a flowmeter and its temperature with a temperature transducer, both connected via a control panel and logged continuously.

RESULTS AND DISCUSSION

Ozonation

Effect of the type of mixer

Different mixer systems have been tested, a counter-current bubble column, and a co-current bubble column either with a diffuser or an in-line ozone mixer. The concentration profiles throughout the columns of the ozone concentration are shown in Figure 2.

In the counter current bubble column, the ozone concentration is high in the lower part of the column, slightly above where the ozone is added, and then the concentration approaches zero towards the inlet of the column. The concentration decreases towards the outlet of the first column and the inlet of the second column, and then continues to decompose throughout the second column were no gas bubbles are present and consequently no gas transfer occurs. The ozone concentration profile of the first column is a result of a mix of different processes which comprises of transfer of ozone from the gaseous phase to the liquid phase and decomposition of ozone. In a counter-current column, the transfer rate is highest in the lower part of the column (due to a larger driving

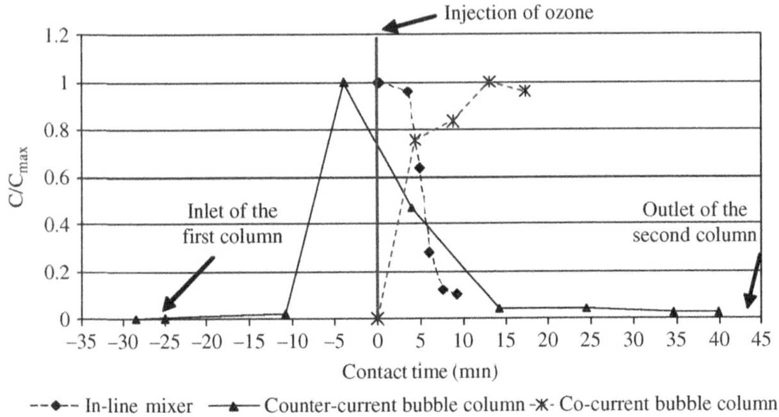

Figure 2. Ozone concentration after an in-line mixer (Cmax = 0.50 mg/L); in co-current diffuser bubble column (Cmax = 0.04mg/L); and in a counter-current diffuser bubble column (Cmax = 0.23 mg/L).

force), while the potential for ozone decomposition is highest in the upper part of the column.

For a co-current bubble column, the change in ozone concentration can be ascribed to a mix of different processes which comprises of transfer of ozone from the gaseous phase to the liquid phase and then decomposition of ozone. The ozone concentration reaches a constant level where the transfer equals the decomposition. At longer contact times the concentration will decrease and approach zero as the transfer ceases.

The in-line mixer has shown to be more efficient giving a much faster transfer of ozone from the gas phase to the liquid phase as much smaller bubbles are created than when diffusers are used. The in-line mixer does not only improve the transfer rate but also the maximum concentration of ozone reached in the water phase. Thus, the size of the ozone system can be reduced for the same efficiency. This system hasn't been used in the OBM pilot study but it should be preferred to a diffuser.

Effect of the packing media

Different test have been carried out in a co-current bubble column to compare ceramic versus stainless steel as packing media, and it has been found that the ceramic increases the rate of ozone decomposition (Figure 3a). Indeed, when using the ceramic, the ozone concentration reaches more or less the same maximum values as for the stainless steel packing media, however, instead of being constant versus time (or length of bubble column) at low pH, the concentrations decrease with time. Nevertheless, the ceramic packing media has a smaller effect at high pH because the ozone decomposition due to high pH is already more pronounced (see next paragraph). This decomposition leads to the formation of hydroxyl radical (stronger oxidant than ozone) which has been documented by a faster degradation of parachlorobenzoic acid (pCBA). Its degradation rate increased by about 10–80% when using ceramic media (Azrague *et al.*, 2008). Thus, the ceramic can be used to increase the formation of hydroxyl radicals and thereby lead to enhanced DOC removal and improved removal of micro-pollutant recalcitrant to ozone.

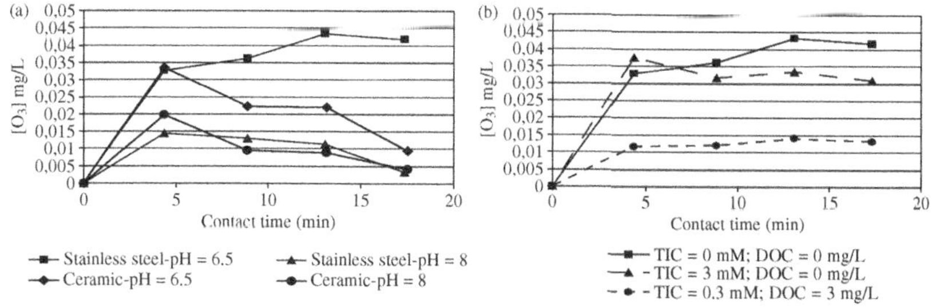

Figure 3. (a) Ozone concentration versus time in distilled water at different pH with stainless steel and ceramic packing media. (b) Ozone concentration versus time in distilled water at different TIC and DOC concentrations with stainless steel packing media.

Effect of water qualities

The ozonation is largely dependent on the packing media but also on the water quality. The effects of pH, NOM and TIC on the ozonation have been investigated.

The ozone concentration versus contact time at two different pHs for a co-current and counter-current bubble column are shown in Figure 3a and 4a. The ozone stability is shown to be pH dependent. At high pH the ozone gets decomposed into hydroxyl radicals (Azrague *et al.*, 2008) which, like mentioned before, improves the removal of DOC and some types of micro-pollutants. However the color removal is reduced (Figure 4b) since the ozone is more selective than the hydroxyl radical for removing the double bonds and therefore the color. This ozone decomposition phenomenon due to high pH is described by (Staehelin and Hoigne, 1982).

TIC acts like initiator for hydroxyl radical formation, which is according to Hoigné and co-workers (Hoigné and Bader, 1976; Staehelin and Hoigne, 1982), stating that compounds capable of inducing the formation of superoxide ion $O_2^{\cdot -}$ from

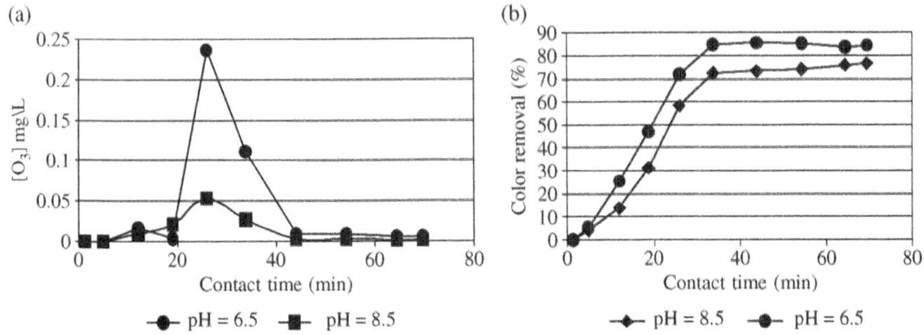

Figure 4. Ozone concentration (a) and color removal (b) in a counter-current ozone bubble column at pH 6.5 and 8.5 (DOC = 6.5 mg/L, color = 50 mg Pt/L, and dose = 0.18 mg O_3/mg Pt).

an ozone molecule will consequently lead to hydroxyl radical formation. The result in Figure 3b, shows that the ozone concentration is slightly lower in presence of TIC compared to without TIC, indicating that the TIC is inducing the formation of hydroxyl radicals. On the other hand, carbonate and bicarbonate are known to be radical scavengers, and since $CO_3^{\cdot-}$ (1,78V/ENH) is much less reactive than HO· (2,8V/ENH), carbonate will inhibit the degradation by hydroxyl radical (Legrini *et al.*, 1993). Thus, as it has been reported by (Azrague *et al.*, 2008), the degradation of pCBA is much slower in presence of 3 mmol/L of TIC than without, even though more ozone is decomposed (Figure 3b).

DOC, which in this case mainly consists of NOM, is like TIC affecting the ozone stability and the oxidation efficiency. Indeed, NOM is an organic matter which will behave as a pollutant being oxidized, and therefore consume oxidants. Nevertheless, in contrary to the presence of TIC, presence of NOM improves the pCBA's removal rates at pH 6.5 (Figure 3b). At pH 8, no significant increase was found. Indeed, NOM act similar as TIC, but is in a much larger extent, an initiator for the ozone decomposition, and less of a scavenger than TIC. Consequently, the degradation rates of pCBA are increased in the presence of NOM, while for the presence of TIC the same rates are decreased (Azrague *et al.*, 2008). According to Pi *et al.* (2005) this increase in degradation rates is due to the decomposition of ozone which is strongly promoted by the presence of double bonds because they result in the formation of the intermediate product hydrogen peroxide and leads to formation of hydroxyl radicals.

NOM ozonation

In addition to the effect on disinfection and on micro-pollutants, one of the main effects of ozonation concerns the NOM oxidation. Ozonation causes structural changes to the NOM and in particular to the humic fractions. The changes include a strong and rapid decrease in color (≈80%) and UV_{254} (≈60%) due to loss of aromaticity and depolymerisation, a small reduction in DOC (≈10%) , a slight decrease in the high apparent molecular weight fractions, a slight increase in the smaller fractions, a significant increase of the carboxylic functions, and the formation of ozonation by-products. These by-products include small molecular weight compounds like aldehydes, ketoacids, carboxylic acids, etc, which may be easy biodegradable (Carlson and Amy, 1997). However, most of the ozonation by-products are unknown. Several of the ozonation by-products formed are easily biodegradable, resulting in an increase in for example BDOC after ozonation, which can be seen in Table 1.

Table 1. Examples of DOC and BDOC after ozonation and biofiltration using different biofilter medias (DOC = 2.8 mg/L, color = 15 mg Pt/L, pH = 8.0, and dose = 0.18 mg O_3/mg Pt).

	Feed	Ozonated	Biofiltered water (EBCT = 20 min)				
Parameter	water	water	KMT	Sand	Anthracite	CaCO3	GAC
DOC, mg/L	2.8	2.5	2.1	2.1	2.0	2.2	1.4
BDOC, mg/L	0.1	0.8	0.4	–	–	–	0.1

Figure 5 shows an example of a comparison of rapid NOM fractionation of the feed water, and the water after ozonation, and biofiltration using KMT or GAC biofilter media. The feed water NOM is dominated by the VHA, with considerable less SHA. The VHA and SHA are predominantly composed of higher molecular weight humic and fulvic acids. Ozonation removes more than 50% of the VHA producing CHA (mainly proteins, amino acids, anionic polysaccharides, etc.), and NEU (mainly carbohydrates, aldehydes, ketones, alcohols, etc.). Shift in oxidation pathway from direct to indirect (i.e. by introducing a catalyst or increasing pH) will give a larger variety of by-products of a smaller average size (Myllykangas *et al.*, 2002) resulting in a slightly higher BDOC fraction and CHA and NEU fraction.

Biofiltration

An average of approximately 75% of the BDOC produced during the ozonation was removed in the biofilter (using KMT-media with an EBCT ~30 min. and temp ~18°C) of the pilot plant. However, the percent removal was slightly higher,

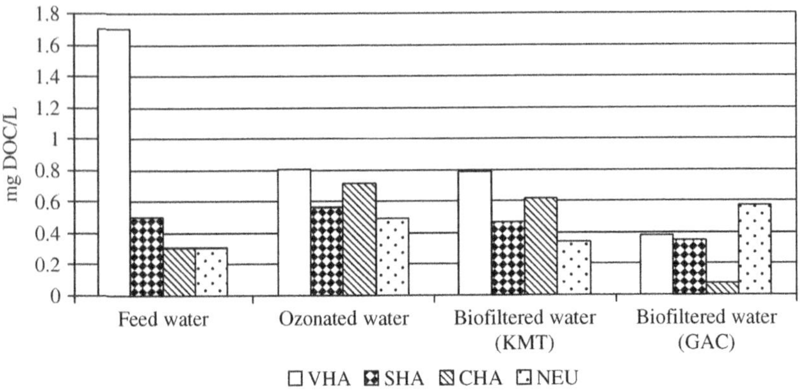

Figure 5. Rapid NOM fractionation of the feed water, and after ozonation, and biofiltration using KMT or GAC biofilter media (DOC = 2.8 mg/L, color = 15 mg Pt/L, pH = 8.0, and dose = 0.18 mg O_3/mg Pt).

and the BDOC removal rate constant was much higher (about 2 times higher) at pH 8.0 than at 6.5. This is caused by a shift towards the indirect oxidation pathway at pH 8.0, resulting in production of smaller and more easily biodegradable by-products, which also is in agreement with Myllykangas *et al.* (2002).

Different biofilter media were investigated, and only minor differences between media of KMT, quarts sand, anthracite, and calcium carbonate were observed. These tests were run at EBCT of 20 min, which was insufficient to reduce the BDOC below 0.3–0.4 mg/L, as can be seen from Table 1. However, the granular activated carbon efficiently removed the BDOC (Table 1), mainly by removing the charged hydrophilic fraction, CHA (Figure 5). It is likely that this partly was due to improved biodegradation, because GAC has been reported to both maintain much larger amounts of biomass (Niquette *et al.*, 1998) and to better tolerate adverse conditions (Liu *et al.*, 2001) compared to other media. Nevertheless, although, the GAC had been run with a throughput of approx. 10,000 bed volumes before the test, it is clear that its adsorption capacity was not exhausted, seen by the high VHA removal and DOC removal in general.

GAC would normally be the preferred biofilter media, not only because it is very efficient in supporting biomass and biodegradation, but also due to its adsorption properties which are very useful in a start-up phase or in other situation where the biodegradation is less efficient, and since it decomposes possible residual ozone preventing it from entering the biofilter and destroying biofilm.

Membrane filtration

The membrane is mainly used here to assure particles and pathogens removal and is largely responsible for the turbidity removal. Two types of membrane have been investigated: a microfiltration ceramic membrane and an ultrafiltration submerged polymeric membrane. Tests using microfiltration ceramic membrane showed severe fouling, and consequently most of the membrane work concentrated on ultrafiltration which worked very well. It is, however, possible that ceramic microfiltration may be improved by using for example coagulants.

The ultrafiltration membrane has shown to perform very well with long runs (600 hours with flux 60 LMH and 400 hours with flux 80 LMH) and the mechanical cleaning seems to remove the reversible fouling accumulated on the membrane. When considering the effect of feed water quality in Table 2, it is clear that the effect of increased NOM content (measured as DOC) on TMP and ΔTMP (irreversible and reversible fouling) is relatively small despite of the fact that the particle load (measured as SS) is largely affected.

Table 2. The development in TMP and ΔTMP at different feed water DOC.

@ steady state	@ ~ 3.5 mg DOC/L	@ ~ 6.5 mg DOC/L
SS after backwash	1 mg/L	10 mg/L
Turbidity after backwash	2 NTU	3 NTU
ΔTMP	−26 mbar	−32 mbar
d(TMP)/dt	−0.51 mbar h^{-1}	−0.54 mbar h^{-1}

The pH of the feed water is also affecting the performances of the membrane (Table 3). Indeed, the experiments at pH 6.5 gives a higher irreversible fouling than at pH 8.5, which is caused by the different oxidation pathway and consequently by the change in size and hydrophobicity of the by-products. On the other hand, the reversible fouling is less important when the pH is low which is probably due to a different biomass production due to the changes in ozonation by-products and consequently due to their biodegradability.

Table 3. Development in TMP and ΔTMP for Flu 80 LMH at different feed water phs.

pH	pH 6.5	pH 8.5
d(TMPinitial)/dt	−1.37	−1.02
d(ΔTMP)/dt during dead end	−1.12	−1.67
d(ΔTMP)/dt at steady state	0	0
ΔTMP value at steady state	−77	−92

CONCLUSION

The OBM process has shown to achieve a good color removal (>80%) during ozonation even in high NOM water. The biofiltration stage produces bio-stable water, with a very low sludge production. In contrary to ceramic microfiltration, ultrafiltration using polymeric membranes has worked very well as it could be operated at high fluxes (80 LMH) in an immersed configuration, with a high recovery (98%), still permitting operational time of approx 400 hrs before chemical cleaning was needed (approx. 600 hrs of operation at flux of 60 LMH). By modifying the operation mode (air scouring) and reactor design, the process can probably be further optimized.

The OBM process is suited for medium to low NOM containing water, and acts as an efficient hygienic barrier producing high quality water. It has shown to be very robust and flexible, and it may be designed for removal of several micro pollutants.

ACKNOWLEDGEMENTS

The authors wish to thank the European Union project TECHNEAU for the financial support, Zenon Environmental Inc. and Metawater Co. for supplying the membrane modules.

REFERENCES

Azrague, K., Osterhus, S.W. and Biomorgi, J.G. (2008). Degradation of pCBA by catalytic ozonation in natural water. In: *Proceeding of the IWA-Chemical Industries Conference (CD)* Beijing, China.

Back, A.P. and Christman, R.F. (1963). Characteristics of colored surface waters. *Journal of American Water Works Association*, **55**(1963), 753–770.

Camel, V. and Bermond, A. (1998) The use of ozone and associated oxidation processes in drinking water treatment. *Water Research*, **32**(11), 3208–3222.

Carlson, K. and Amy, G. (1997). The formation of filter-removable biodegradable organic matter during ozonation. *Ozone Science and Engineering*, **19**, 179–199.

Chow, C.W.K., Fabris, R. and Drikas, M. (2004). A rapid fractionation technique to characterise natural organic matter for the optimisation of water treatment processes. *J. Water Supply Res. T.*, **53**(2004), 85–92.

Clesceri, L.S., Eaton, A.D. and Greenberg, A.E. (1998). *Standard Methods for the Examination of Water and Wastewater (20th)*. American Public Health Association/ American Water Works Association/Water Environment Federation, Washington DC, USA.

Cooper, C. and Burch, R. (1999). An investigation of catalytic ozonation for the oxidation of halocarbons in drinking water preparation. *Water Research*, **33**(18), 3695–3700.

Hoigné, J. and Bader, H. (1976). The role of hydroxyl radical reactions in ozonation processes in aqueous solutions. *Water research*, **10**, 377–386.

LeChevallier, M.W., Welch, N.J. and Smith, D.B. (1996) Full-scale studies of factors related to coliform regrowth in drinking water. *Applied and Environmental Microbiology*, **62**(7), 2201–11.

Lee, E.-H. and Cho, K.-S., Effect of substrate interaction on the degradation of methyl tert-butyl ether, benzene, toluene, ethylbenzene, and xylene by Rhodococcus sp. *Journal of Hazardous Materials*, In Press, Corrected Proof.

Legrini, O., Oliveros, E. and Braun, A.M. (1993). Photochemical processes for water treatment. *Chemical Reviews*, **93**(2), 671–698.

Liu, X., Huck, P.M. and Slawson, R.M. (2001). Factors Affecting Drinking Water Biofiltration. *Journal of American Water Works Association*, **93**(12), 90–101.

McCarthy, J.F. (1989). Bioavailability and toxicity of metals and hydrophobic organic contaminants. In: *Aquatic Humic Substances: Influence on Fate and Treatment of Pollutants*, Suffet, I.H. and MacCarthy, P. (eds), American Chemical Society, Washington, DC, pp. 263–279.

Melin, E.S. and Odegaard, H. (2000). The effect of biofilter loading rate on the removal of organic ozonation by-products. *Water Research*, **34**(18), 4464–4476.

Myllykangas, T., Nissinen, T.K., Rantakokko, P., Martikainen, P.J. and Vartiainen, T. (2002). Molecular size fractions of treated aquatic humus. *Water Research*, **36**(12), 3045–3053.

Niquette, P., Prevost, M., Servais, P., Beaudet, J.F., Coallier, J. and Lafrance, P. (1998). Shutdown of BAC filters: effects on water quality. *Journal of American Water Works Association*, **90**(12), 53–61.

Pi, Y., Schumacher, J. and Jekel, M. (2005). Decomposition of aqueous ozone in the presence of aromatic organic solutes. *Water Research*, **39**(1), 83–88.

Reuter, J.H. and Perdue, E.M. (1977). Importance of heavy metal-organic matter interactions in natural waters. *Geochimica et Cosmochimica Acta*, **41**(2), 325–334.

Rook, J.J. (1977). Chlorination reactions of fulvic acids in natural waters. *Environmental Science & Technology*, **11**(5), 478–482.

Speitel, G.E., Symons, J.M., Diehl, A.C., Sorenson, H.W. and Cipparone, L.A. (1993). Effect of ozone dosage and subsequent biodegradation on removal of DBP precursors. *Journal of American Water Works Association*, **85**(5), 86–95.

Staehelin, J. and Hoigne, J. (1982). Decomposition of ozone in water: rate of initiation by hydroxide ions and hydrogen peroxide. *Environmental Science and Technology*, **16**(10), 676–681.

Yavich, A.A., Lee, K.-H., Chen, K.-C., Pape, L. and Masten, S.J. (2004). Evaluation of biodegradability of NOM after ozonation. *Water Research*, **38**(12), 2839–2846.

Direct and hybrid ceramic membrane filtration in water treatment

E.R. Cornelissen[a], B. Hofs[a], U. Muller[b], E.F. Beerendonk[a] and S.G.J. Heijman[a,c]

[a]KWR Watercycle Research Institute, Groningenhaven 7, 3430 BB Nieuwegein, The Netherlands (E-mail: emile.cornelissen@kwrwater.nl)
[b]DVGW Water Technology Center (TZW), Karlsruher Str. 84, D-76139 Karlsruhe, Germany
[c]Delft University of Technology, Faculty of Civil Engineering and Geosciences, Department of Sanitary Engineering, P.O. Box 5048, 2600 GA Delft, The Netherlands

Abstract Ceramic MF/UF membranes are attracting increasing interest in water treatment. Membrane fouling and removal of particles, micro-organisms (both bacteria and viruses) and organic matter (precursors for disinfection by-products) are investigated for direct ceramic MF/UF and hybrid ceramic MF/UF (coagulation followed by ceramic MF/UF and activated carbon treatment followed by ceramic MF/UF) in water treatment on the basis of literature and experimental research. Direct ceramic MF/UF was carried out using dam water after rapid sand filtration leading to severe flux decline due to membrane fouling of hydrophobic organic matter and polysaccharides. In-line coagulation prior to ceramic MF/UF using 4 mg-Fe/L could control membrane fouling using Twente canal water (turbidity of 14 NTU and TOC of 11 mg/L) as a result of cake-layer formation on the membrane surface preventing irreversible fouling. In-line activated carbon followed by ceramic MF/UF using Delft canal water (turbidity of 11 NTU and TOC of 15 mg/L) was used to evaluate organic matter removal (TOC and UVA254). Particle and bacterial removal using direct and hybrid ceramic MF/UF on the three water types was very effective ($>95\%$) as a result of size exclusion. Removal of MS2-phages was lower for direct ceramic MF/UF (0–3 log) compared to ceramic MF/UF preceded by 4 mg-Fe/L coagulation (>4 log) because of aggregation of viruses and/or virucidal activity of the coagulant. Based on literature data, removal of organic matter for coagulation followed by ceramic MF/UF is relatively poor ($<50\%$), varies strongly for direct ceramic MF/UF (18–83%) because of membrane fouling, and is good for activated carbon filtration followed by ceramic MF/UF (60–99%).

Keywords ceramic MF/UF, membrane fouling, organic matter removal, particle removal, virus removal

INTRODUCTION

Drinking water demands an increased reduction of turbidity and a reduction of chemicals for water disinfection, which boosted the development of membranes in developed countries (EPA, 2001). Partly as a result of this, the application of membrane filtration in water treatment processes is steadily increasing worldwide (IDA, 2006). Microfiltration (MF) and ultrafiltration (UF) processes offer the possibility of the removal of particles, colloidal material, micro-organisms (bacteria and viruses) and precursors for disinfection by-products (e.g. organic matter) in drinking water production and wastewater treatment. Compared to alternative separation processes such as coagulation/flocculation and sand filtration, MF/UF processes are compact, modular, require only moderate amounts of chemicals and can be easily automated. Commercial MF and UF membranes are either polymeric or ceramic, however, the application in water treatment is dominated by polymeric membranes. Ceramic MF/UF membranes are applied in industrial processes, such as clarification of natural fruit juices (Fokumoto *et al.*, 1998), wastewater filtration from pulp and paper production (Luonsi *et al.*, 2002; Park *et al.*, 2006) wastewater filtration in the textile industry (Voigt *et al.*, 2001; Weber and Seidel 2003) and dairy processing (Grangeon and Lescoche 2000; Zapata-Montoya *et al.*, 2006) Ceramic MF/UF membranes offer a better mechanical, thermal and chemical stability compared to polymeric MF/UF membranes resulting in longer lifetimes (Tsuru, 2001) Furthermore, compared to polymeric MF/UF membranes, ceramic MF/UF membranes can be operated at higher fluxes and higher recoveries (Kanaya *et al.*, 2007) and can be cleaned with more extreme cleaning agents for a better fouling control. The application of ceramic MF/UF in water treatment was limited by high membrane costs for ceramic membranes (approximately 750 €/m^2) (Ciora and Liu, 2003). Recent advances in the production of ceramic monolithic MF membrane modules resulted in a decrease in membrane cost (Ciora and Liu, 2003) [http://www.metawater.co.jp/] leading to an attractive alternative for polymeric membranes (Lee and Cho, 2004). Controlling membrane fouling and achieving high removal capacities for particles, micro-organisms, organic matter and precursors for disinfection by-products remains a challenge for (ceramic) MF/UF processes. In this paper, these topics will be investigated for direct ceramic MF/UF and hybrid ceramic MF/UF processes (i.e. coagulation followed by ceramic MF/UF, activated carbon treatment followed by ceramic MF/UF) in water treatment on the basis of literature and experimental research.

BACKGROUND

Ceramic MF/UF membranes used in water treatment are aluminum oxide (Bottino *et al.*, 2001; Lerch. *et al.*, 2005; Matsui *et al.*, 2005; Matsushita *et al.*, 2005; Shirasaki *et al.*, 2009) titanium dioxide (Lee and Cho, 2004) zirconium dioxide (Ericsson and Tragardh, 1997) and a titanium dioxide/zirconium dioxide combination (Bodzek and Konieczny 1998; Klomfas and Konieczny, 2004; Konieczny *et al.*, 2006). The most commonly applied ceramic membrane material is aluminum oxide (α-Al_2O_3) because of the stability in alkaline and acidic environments. Aluminum hydroxide is the ground material and is transferred into α-Al_2O_3 at elevated temperatures (>1000°C) in the production process. Sintering of α-Al_2O_3 particles leads to an increase in the average pore diameter of the membrane, resulting in a minimum pore diameter of approximately 80 nm, because of instability of the separating layer at low pore diameters. When smaller diameters are required other ceramic materials can be applied, such as titanium dioxide (TiO_2), zirconium dioxide (ZrO_2) or silicon dioxide (SiO_2) [10]. The pore size of silicon dioxide can relatively easily be adjusted during production. In contrast to titanium dioxide and zirconium dioxide, however, silicon dioxide is only stable in aqueous environments at low to medium pH. The stability of silicon dioxide can be improved by adding other metal oxides. Alternative ceramic membrane materials are silicon carbide (SiC), silicon nitride (Si_3N_4) and aluminum nitride (AlN), although the application of these materials in ceramic MF/UF membranes is limited.

The charge of ceramic MF/UF membranes plays an important role in fouling by particles, micro-organisms, organic matter and precursors for disinfection by-products. Membrane fouling and deposition of silica particles of uncharged ceramic MF/UF membranes was found to be higher than for charged ceramic MF/UF membranes (Nazzal and Wiesner, 1994). The charge of membranes is pH dependent and can be determined by zeta-potential measurements (de la Rubia *et al.*, 2006) and membrane surface titration. At a pH of 7, aluminum oxide is positively charged (+10mV at pH of 7 for α-Al_2O_3 (Nazzal and Wiesner, 1994)), titanium dioxide and zirconium dioxide are moderately negatively charged (−30mV at pH of 7 for TiO_2 (Lee and Cho, 2004) and (−5mV at pH of 7 for ZrO_2/TiO_2) and silicon dioxide and silicon carbide are strongly negatively charged. The water chemistry, the presence of particles, micro-organisms and organic matter, and membrane fouling will ultimately alter the membrane charge during ceramic membrane filtration.

Commercial ceramic MF/UF membrane modules are available as monoliths, plate membranes and hollow fiber membranes (Chan and Brownstein, 1991; Strathmann, 2000). Ceramic monolith membranes are produced in one part by extrusion. Ceramic elements have a low surface area. For full-scale applications multiple ceramic elements are inserted within stainless steel housings. Ceramic monolith membranes consist of considerable higher packing densities (e.g. up to 25 m^2) compared to the alternative ceramic membrane modules (Chan and Brownstein, 1991). Ceramic plate membranes are produced as discs and are housed in a plate-and-frame configuration. The membrane discs can be individually inspected, cleaned or replaced which facilitates a better fouling control. Flow distribution in plate membranes is, however, not optimal and the packing density is relatively limited. Ceramic hollow fiber membrane modules have the highest packing density (Peters *et al.*, 2005). Hollow fiber membranes need to be potted into a membrane housing, which is a weak point in the construction with the risk of leakage. Ceramic MF membranes used in water treatment are predominantly multi channel monoliths (Bodzek and Konieczny, 1998; Yusaka, 1998; Bottino *et al.*, 2001; Matsuyi *et al.*, 2003; Klomfas and Konieczny, 2004; Lerch. *et al.*, 2005; Matsui *et al.*, 2005; Matsushita *et al.*, 2005; Konieczny *et al.*, 2006; Shirasaki *et al.*, 2009).

Ceramic MF/UF membranes are operated either in cross-flow or in (semi) dead-end mode. In cross-flow operation the feed flow is tangential to the permeate flow, while in (semi) dead-end operation the feed flow is perpendicular to the permeate flow. In dead-end mode retained solutes accumulate on the membrane surface to form a cake-layer with an additional resistance to feed flow. A periodical backwash will temporarily revert the negative effects of the build-up of the cake-layer. In ceramic MF/UF both operational modes are applied, cross-flow operation with different cross-flow velocities ranging from 0.1 to 4 m/s (Bodzek and Konieczny, 1998; Lee and Cho, 2004) and dead-end operation with different backwash frequencies ranging from 20 minutes to 17 hours (Ericsson and Tragardh, 1997; Klomfas and Konieczny, 2004; Lerch *et al.*, 2005; Matsushita *et al.*, 2005; Konieczny *et al.*, 2006; Shirasaki *et al.*, 2009) were found.

DIRECT CERAMIC MF/UF

Materials and methods

Dam water after rapid sand filtration was taken as raw water for direct ceramic MF/UF. The average feed water quality was characterized during the research period by a turbidity of 0.7 NTU, a UVA254 of 6.6 m^{-1}, a UVA436 of 0.36 m^{-1} and a pH of 6.6. The organic matter predominantly consisted of humic substances and hydrolysates of humic substances (32–47%), followed by low molecular weight acids (17–25%), neutral substances (17–19%), hydrophobic organic matter (4–13%) and polysaccharides (2–6%) as was determined by liquid chromatography followed by organic carbon detection (LC-OCD).

Different ceramic MF/UF membranes were used from four suppliers, using different ceramic materials, pore sizes and channel diameters (Table 1). The ceramic MF/UF membranes were operated in semi dead-end mode at a constant trans membrane pressure (TMP) of 2 bar, while the flux during filtration was monitored on-line. The filtration time was 5 hours for membrane A–C and 2.5 hours for membrane D at a backwash flux of approximately 5,000 L/m^2h.

The backwash duration was 60 seconds and 30 seconds for respectively membrane A-C and membrane D, using alternating water and pulsed air. No chemically enhanced backwash was performed.

Table 1. Ceramic membrane properties for direct ceramic MF/UF experiments.

Supplier type	Membrane material	Pore size [nm]	Channel diameter [mm]	Number of channels [#]	Membrane area [m^2]	Pure normalized water flux [L/m^{2}h.bar]
A	α-Al_2O_3/TiO_2	50	6	7	0.13	1440
B	α-Al_2O_3	50	3.3	19	0.20	250
C-1	α-Al_2O_3/TiO_2	50	3.3	19	0.20	–
C-2	α-Al_2O_3/TiO_2	10	3.3	19	0.20	515
D	SiC/SiO_2	200	3.4	37	0.43	1674

Flux decline behavior was investigated for membrane C-1 and virus removal was investigated for membranes A, B, C-2 and D using dam water after rapid sand filtration, ozonation and flocculation. F-specific bacteriophages (MS2) were spiked into this feed water at a concentration of approximately 10^7 FPU/mL. The average size of the MS2 bacteriophage is 24–26 nm (Shirasaki *et al.*, 2009).

Results and discussion

Normalized pure water flux values indicated low membrane resistances for membranes A and D, while the highest membrane resistances were found for membranes B and C-2 (Table 1). The normalized pure water flux values did not correspond to the pore size values, possibly as a result of differences in porosity and membrane structure between the different membranes.

Severe flux decline was found for membrane C-1 on direct ceramic MF/UF with dam water after rapid sand filtration (Figure 1). Similar severe flux decline on direct ceramic MF/UF was reported on both surface water (Bottino *et al.*, 2001; Klomfas and Konieczny, 2004) and well water (Bodzek and Konieczny, 1998). LC-OCD analysis of waste streams from membrane cleaning (1 h soaking in NaOH and 3 mg/L NaOCl at pH = 11) indicated that both the hydrophobic organic matter and polysaccharides contents increased 16–17% indicating a contribution of these fractions to irreversible fouling, e.g. pore blocking and adsorption. This was also found for irreversible fouling of polymeric MF/UF during surface water treatment (Kimura *et al.*, 2004). Higher concentration levels of polysaccharides in the CIP-waste stream of a polymeric

UF-membrane were found (17–33%), however, compared to the ceramic membrane. Virgin pure water fluxes of the fouled membrane C-1 were restored after 1 hour soaking in a 6.7 mg/L ozone solution (results not presented here).

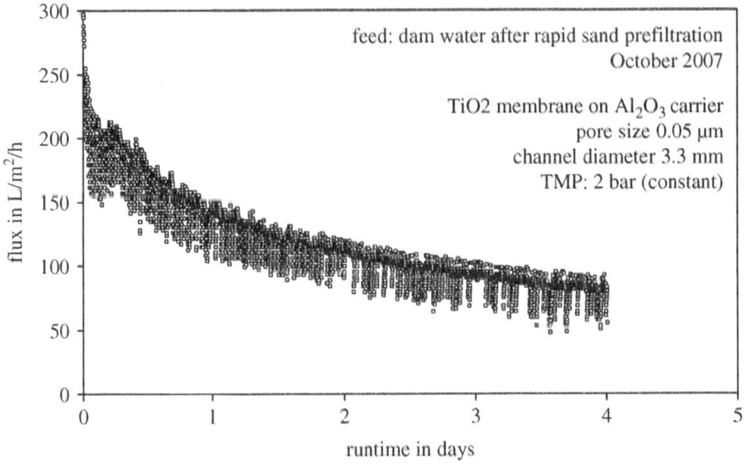

Figure 1. Flux decline in time of membrane C-1 on direct ceramic MF/UF with dam water after rapid sand filtration.

Removal of MS2 phages was determined for the different membranes A, B, C-2 and D (Table 2). A log removal of MS2-phages between 0.3–3.1 was obtained depending on the ceramic membrane material and pore size. The highest MS2-phage log removal was obtained for the membrane C-2 with the smallest pore size, while the lowest MS2-phage log removal was found for the membrane D with the largest pore size. Charge effects did not play a prominent role, since negatively charged MS2-phages (Shirasaki *et al.*, 2009) were not electro-statically repulsed by negatively charged SiC/SiO_2 membranes (see §2 background). Furthermore, membrane A and B had similar pore sizes, however the largest log removal was found for the more positively charged α-Al_2O_3 membrane (see §2 background).

Removal of organic matter in direct ceramic MF/UF was found to range between 18% and 83% based on DOC depending on water type, membrane characteristic and process conditions (Bodzek and Konieczny, 1998; Bottino *et al.*, 2001; Lee and Cho, 2004; Klomfas and Konieczny, 2004). Particulate removal in direct ceramic MF/UF was found to be higher than 99% based on turbidity as

Table 2. MS2 bacteriophage removal efficiency of different ceramic MF/UF membranes during ceramic MF/UF with dam water after rapid sand filtration, ozonation and flocculation.

Supplier type	Membrane material	Pore size [nm]	MS2 Feed Conc. [FPU/mL]	MS2 Filtr. Conc. [FPU/mL]	MS2 Removal [Log]
A (*)	α-Al_2O_3 / TiO_2	50	2.4×10^7	7.2×10^5	1.5
B	α-Al_2O_3	50	2.4×10^7	3.3×10^5	1.9
C-2	α-Al_2O_3 / TiO_2	10	2.8×10^7	2.2×10^4	3.1
D (*)	SiC / SiO_2	200	1.5×10^5	7.8×10^4	0.3

* After CIP (alkaline cleaning at pH 12–13 with 3 mg/L Chlorine followed by acidic cleaning at pH 2 with a commercial membrane cleaner containing nitric and phosphoric acid).

expected (Bodzek and Konieczny, 1998; Bottino *et al.*, 2001; Endter *et al.*, 2004; Lee and Cho, 2004; Klomfas and Konieczny, 2004). Bacterial (*E.coli*) removal in direct ceramic MF/UF was generally found to be higher than 96% (Bodzek and Konieczny, 1998; Bottino *et al.*, 2001; Endter *et al.*, 2004; Lee and Cho, 2004), although a lower removal was reported of 77% which was ascribed to non sterile testing conditions (Bodzek and Konieczny, 1998).

COAGULATION FOLLOWED BY CERAMIC MF/UF

Materials and methods

Twente Canal water was used as feed water for ceramic MF/UF preceded by in-line coagulation. Physical (temperature and turbidity), chemical (pH, organic and inorganic matter concentrations) and biological variations in the feed quality were influenced by waste water discharge and seasonal fluctuations. During the research period (July–November 2005) variations were measured in pH (7–7.7), turbidity (2.5–14 NTU), total iron (0.2–0.8 mg/L), TOC/DOC (8.5–11 mg/L), UVA254 (23–32 1/m) and bacterial counts, e.g. *E.coli* (<1–26 CFU/100mL). Coagulation was carried out in a static mixer with 2 and 4 mg-Fe/L ($FeCl_3$) supplied by a feed pump followed by in-line flocculation with a velocity gradient (G) of 300–500 1/s and a mixing time (t) of 80–150 seconds.

Three multichannel (55 × 2.5mm) monolith α-Al_2O_3 type ceramic MF membranes modules (Metawater Co., Ltd.) were used with a total membrane area of $1.2m^2$. The mean membrane pore size was 100nm. The ceramic MF membrane modules were operated in semi dead-end mode at a constant flux of 120 L/m^2h after a flux optimization study (Schippers *et al.*, 2007), while the TMP was monitored on-line (recovery 96.7%). The filtration time was 1 hour. A backwash was performed using a backwash flux of approximately 5,000 L/m^{2}h at 5 bar (2–20 seconds), followed by an air flush at 2 bar (2–3 seconds). After 12 filtration cycles a chemically enhanced backwash (CEB) procedure was carried out using HCl and 100 mg/L H_2O_2 at pH = 2 after a CEB optimization study (Schippers *et al.*, 2007).

Flux decline behavior and virus removal was investigated for the ceramic MF membranes pretreated by $FeCl_3$ coagulation and flocculation. F-specific bacteriophages (MS2) were spiked into this feed water at a concentration of approximately 10^9 FPU/mL using 0, 2 and 4 mg-Fe/L. Samples were taken from the feed and permeate to evaluate the MS2-phage removal. CIP waste was collected after 16 hours of soaking at 30°C in 1wt% citric acid followed by 3,000 mg/L NaOCl after CIP at the end of the pilot run. The CIP waste stream was analyzed on organic and inorganic compounds.

Results and discussion

Stable fluxes were obtained with ceramic MF preceded with 2 and 4 mg-Fe/L coagulation and flocculation for approximately 3 days at a membrane flux of 120 L/m^{2}h and a filtration cycle time of 1 hour using CEB (Figure 2). Reversible fouling caused by reversible deposition of particles, precipitates and/or organic matter, expressed by the TMP difference before and after a backwash, is higher for 2 mg-Fe/L (17 kPa) compared to 4 mg-Fe/L (15 kPa). Higher concentrations of coagulant lead to thicker cake-layers which are easier to remove by backwashing. Irreversible fouling caused by a permanent deposition of particles, precipitates and/or organic matter, expressed by the difference between the current TMP_t and the initial TMP_0, did not occur for both experiments (approximately 0 kPa/day).

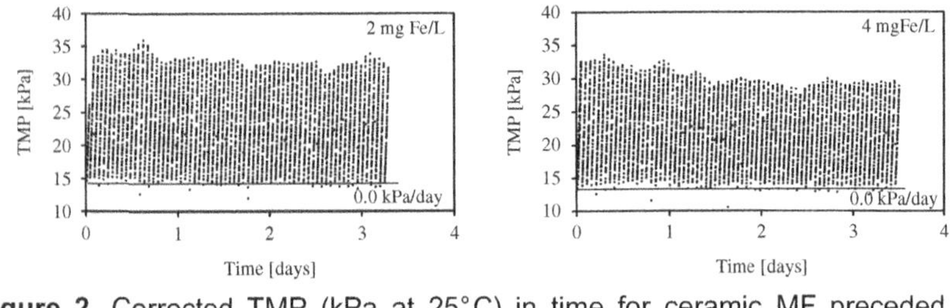

Figure 2. Corrected TMP (kPa at 25°C) in time for ceramic MF preceded by 2 mg-Fe/L and 4 mg-Fe/L coagulation at a flux of 120 L/m^{2}h and a filtration time of 1 hour (CEB : HCl and 100 mg/L H_2O_2 at pH = 2).

At the end of the filtration experiments, the ceramic MF membrane module was soaked in citric acid and sodium hypochlorite to recover the membrane performance and to determine the membrane foulants. The membrane performance was fully recovered (results not presented here) and the highest concentrations of organic matter (expressed as TOC), Si and Fe were found in the CIP-waste (45–55 mg/L), followed by Ca and Al (20–25 mg/L). This indicated the occurrence of particulate (Fe, Si) and organic matter membrane fouling in the ceramic membrane.

Log removal of MS2 phages was determined as a function of time within one filtration cycle for ceramic MF preceded by 0, 2, and 4 mg-Fe/L and varied from 1.1 to 6.2 (Figure 3). Log removal increased in time, as a result of a cake-layer growth accumulating on the membrane surface and accumulation of foulants within the internal membrane structure accounting for an increase in virus removal in time (Shirasaki *et al.*, 2009). The coagulation dose strongly affected the virus removal for ceramic MF preceded with $FeCl_3$, as a result of improved aggregation of viruses exceeding the membrane pore size (Ericsson and Tragardh, 1997) or an improved virucidal activity of $FeCl_3$ at higher concentrations (Shirasaki *et al.*, 2009). Similar log-reductions for both Qβ and MS2 bacteriophages were obtained for coagulation with poly aluminum chloride (PACl) as a function of filtration time, coagulation dose and time and membrane pore size (Ericsson and Tragardh, 1997; Matsui *et al.*, 2003; Shirasaki *et al.*, 2009). Without coagulation a log reduction of viruses was measured from 1.1 to 2.8, possibly as a result of cake-layer formation of particles from the feed water, which was in contrast with results from literature (Ericsson and Tragardh, 1997; Matsui *et al.*, 2003; Shirasaki *et al.*, 2009).

The turbidity and total Fe concentration in the filtrate was below the detection limits after $FeCl_3$ coagulation followed by ceramic MF (Table 3). Particle removal was found to be very high (>97%) in accordance with literature using PACl coagulation (Ericsson and Tragardh, 1997; Yuasa, 1998; Lerch *et al.*, 2005). Lerch *et al.* (2005), Ericsson and Tragardh (1997) and Yuasa (1998) indicating accumulation of these compounds with flocs larger than the membrane pore size. Similar results were obtained for bacteria, which were not observed in the filtrate and removed to a high degree (>97%) because of a similar removal mechanism. Removal of organic matter, expressed as TOC, DOC and UVA254, was low (<17%) since these compounds were dissolved and did not adsorb onto the cake-layer and/or membrane material. Similar results (50% DOC and 20% UVA254 removal by coagulation ceramic MF/UF) were obtained using coagulation with PAC (Yuasa, 1998).

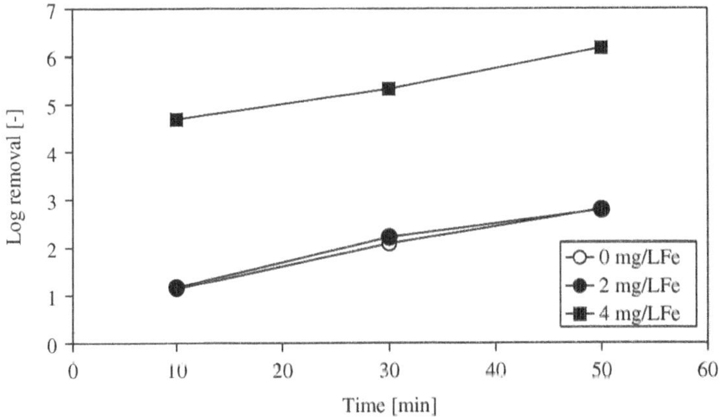

Figure 3. Log removal of MS2 phages in time within one filtration cycle of ceramic MF preceded with $FeCl_3$ coagulation (flux = 120 L/m^2h; filtration time = 1 hour and CEB = HCl and 100 mg/L H_2O_2 at pH = 2).

Table 3. Removal of particles, organic matter and bacteria with ceramic MF preceded with $FeCl_3$ coagulation (flux = 120 L/m^2h; filtration time = 1 hour and CEB = HCl and 100 mg/L H_2O_2 at pH = 2).

Parameter	Feed	Filtrate	Removal
Turbidity [NTU]	5.7	<0.1	>98%
Total Fe [mg/L]	0.4	<0.01	>97%
TOC [mg/L]	10.0	8.7	14%
DOC [mg/L]	9.7	8.7	11%
UVA254 [1/m]	27.1	22.6	17%
Coliforms@37°C [CFU/100mL]	<1840	<1	>99%
E-coli [CFU/100mL]	<26	<1	>96%

ACTIVATED CARBON TREATMENT FOLLOWED BY CERAMIC MF/UF

Materials and methods

Delft canal water was used as feed water for ceramic MF/UF preceded by in-line powdered activated carbon (PAC) treatment, with a TSS of 25 mg/L, turbidity of 11 NTU, TOC of 15 mg/L and UVA254 of 50 1/m. PAC (Norit SX ultra) was pulverized to obtain sub micron activated carbon (S-PAC) with an average particle size of 0.8 μm (Heijman *et al.*, 2008). One multichannel ceramic MF membranes module (Metawater Co., Ltd.) was used with a total membrane area of 0.4 m^2 (see §4 materials and methods). Automated dead-end filtration experiments were carried out at a constant flux of 150 L/m^2h and a filtration time of 1 hour. The S-PAC was dosed into the feed water of the ceramic MF membrane during the first 4 minutes of the filtration cycle (at a flow of 8 L/h of 7.5 g/L S-PAC stock solution). A backwash was performed using a backwash flux of approximately 1,700 L/m^2h at 5 bar (16 seconds), followed by an air flush at 2 bar (30 seconds) and a forward flush at 300 L/m^2h (60 seconds). CEB and CIP procedures were respectively carried out daily and weekly. Organic matter removal (DOC and UVA254) was investigated for the ceramic MF membranes pretreated by S-PAC treatment.

Results and discussion

Flux decline behavior or TMP build-up could not be evaluated because of technical problems with the application of S-PAC. The S-PAC dosing did not lead to an even distribution of S-PAC onto the ceramic membrane surface and/or the S-PAC was already flocculated before dosing (Heijman *et al.*, 2008). From literature, the increase in TMP when using in-line dosing S-PAC was found to be similar to the increase in TMP for in-line dosing of PAC, both combined with coagulation (Matsui *et al.*, 2005). Flux decline was observed for in-line granular activated carbon (GAC) and PAC dosing in ceramic MF/UF treatment of river water, but was less pronounced as the flux decline observed for direct ceramic MF/UF treatment (Klomfas and Konieczny, 2004). S-PAC coatings are expected to prevent membrane foulants, specifically hydrophobic organic carbon compounds and polysaccharides, to reach the (ceramic) membrane surface to cause pore blocking or adsorption (Heijman *et al.*, 2008). No information was available on the effects of CEB and CIP on ceramic MF/UF membranes preceded by S-PAC/PAC/GAC from experiments and literature.

Initial high organic matter removal was found using S-PAC dosing before ceramic MF (84% UVA254 and 81% DOC removal). For a S-PAC dose of 4.2 g, a 50% DOC (UVA254) breakthrough was reached within 10 (12.5) minutes (Figure 4). From literature, a high organic matter removal from surface water was found for PAC/GAC treatment followed by ceramic MF/UF (Klomfas, G. and Konieczny, K. 2004). DOC and UVA254 removal was higher for pre-treatment with PAC (94% UVA254 and 99% DOC removal) compared to GAC (87% UVA254 and 87% DOC removal). Particle removal based on turbidity was found to be very high ($>$91%) in accordance with literature using PAC/GAC treatment followed by ceramic MF/UF (Ericsson and Tragardh, 1997; Bodzek and Konieczny, 1998; Klomfas and Konieczny, 2004) indicating accumulation of particles with activated carbon particles larger than the membrane pore size. No data on virus removal using activated carbon treatment prior to ceramic MF/UF could be found.

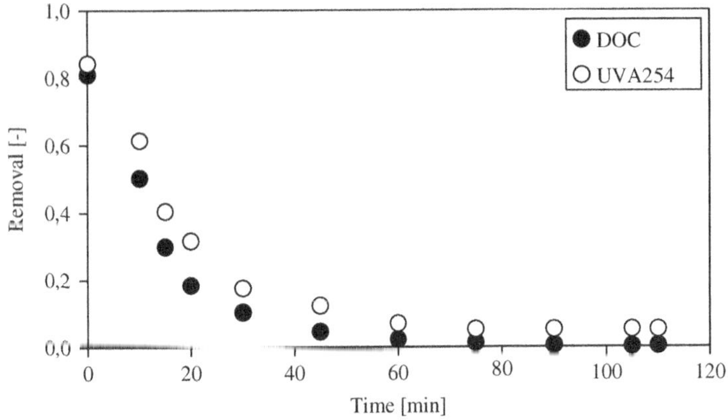

Figure 4. DOC and UVA254 removal in time with S-PAC dosing before ceramic MF (flux = 150 L/m^{2}h; filtration time = 2 hours).

CONCLUSIONS

- Fouling control was difficult in direct ceramic MF/UF and flux decline occurred as a result of irreversible fouling, e.g. pore blocking and adsorption, mainly due to hydrophobic organic matter and polysaccharides. Flux decline was observed in direct MF/UF for different types of ceramic membranes, with different surface charge and pore sizes.
- Ceramic membrane fouling was controlled by in-line $FeCl_3$ coagulation/flocculation prior to ceramic MF/UF in combination with a CEB using HCl and H_2O_2, due to cake-layer formation on top of the membrane surface preventing irreversible membrane fouling. Similar results were found in literature for in-line PACl coagulation/flocculation.
- From literature it was found that flux decline could only partly be controlled by (submicron) activated carbon treatment prior to ceramic MF/UF.
- Particle and bacterial removal using direct and hybrid ceramic MF/UF was effective (lower than detection limit); a >95% removal was observed, as a result of size exclusion.
- From literature it was found that the removal of organic matter was relatively low for coagulation followed by ceramic MF/UF, and higher for

direct ceramic MF/UF and ceramic MF/UF preceded by in-line activated carbon treatment. For direct ceramic MF/UF membrane fouling probably led to denser cake-layers able to retain a larger organic matter fraction. Activated carbon adsorbed (part of the) organic matter before ceramic MF/UF resulting in a higher removal of organic matter.

- MS2-phages removal was found approximately 0 log for ceramic MF and approximately 3 log for ceramic UF during direct filtration and depends mainly on the pore size of the ceramic membrane.
- Removal of MS2-phages for coagulation followed by ceramic MF/UF depended on the coagulant dose. For 4 mg-Fe/L $FeCl_3$ the Surface Water Treatment Rule [1] (log removal >4) was met, as a result of improved aggregation of viruses and/or a virucidal activity of the coagulant.

ACKNOWLEDGEMENTS

This work was financially supported by the Joint Research Project Techneau of the European Union (contract number 018320) and by the Dutch Joint Research Program in which all the Dutch water companies participate. The authors would like to thank Sidney Meijering (KWR) and Harry van Wegen (KWR) for constructing the ceramic MF/UF pilot. Simon Bakker (Vitens) is acknowledged for his role in the ceramic MF/UF pilot studies. Anthony Bichii (IHE) and J. Hamad (IHE) are acknowledged for carrying out the experimental work to fulfill the requirements for their thesis. Mr. Yambe Katsuhiro (Metawater Co., Ltd.) is acknowledged for his editorial work.

REFERENCES

Bodzek, M. and Konieczny, K. (1998). Comparison of various membrane types and module configurations in the treatment of natural water by means of low-pressure membrane methods. *Separation and Purification Technology*, **14**(1–3), 69–78.

Bottino, A., Capannelli, C., Del Borghi, A., Colombino, M. and Conio, O. (2001). Water treatment for drinking purpose: ceramic microfiltration application. *Desalination*, **141**(1), 75–79.

Chan, K. and Brownstein, A.M. (1991). Ceramic membranes – growth prospects and opportunities. *Am. Cer. Soc. Bul.*, **70**(4), 703–707.

Ciora, R. J. and Liu, P.K.T. (2003). Ceramic membranes for environmental related applications. *Fluid. Part. Sep. J.*, **15**(1), 51–60.

de la Rubia, A., Rodriguez, M. and Prats, D. (2006). pH, Ionic strength and flow velocity effects on the NOM filtration with TiO2/ZrO2 membranes. *Separation and Purification Technology*, **52**(2), 325–331.

IDA (2006). 19th IDA worldwide desalting plant inventory, Oxford: Media analytics Ltd.

Endter, A., Schleifenheimer, M. and Voigt, I. (2004). Nanofiltrationsanlage zur dezentralen Trinkwasseraufbereitung. *Filtrieren und Separieren*, **18**(1), 12–15.
EPA – Environmental Protection Agency (2001). *Low-pressure membrane filtration for pathogen removal: application, implementation and regulatory issues*. Environmental Protection Agency, Office of Water: Cincinnati.
Ericsson, B. and Tragardh, G. (1997). Treatment of surface water rich in humus – Membrane filtration vs. conventional treatment. *Desalination*, **108**(1–3), 117–128.
Fokumoto, L.R., Delaquis, P. and Girard, B. (1998). Microfiltration and ultrafiltration ceramic membranes for apple juice clarification. *J. Food. Sci.*, **63**(8), 845–850.
Grangeon, A. and Lescoche, P. (2000). Flat ceramic membranes for the treatment of dairy products: comparison with tubular ceramic membranes. *Lait.*, **80**, 5–14.
Heijman, S.G.J., Hamad, J.Z., Kennedy, M.D., Schippers, J. and Amy, G. (2008). Submicron powdered activated carbon used as a pre-coat in ceramic micro-filtration, in Membranes in Drinking Water Production and Waste Water Treatment, AWWA/EDS: Toulouse, p. 8.
Kanaya, S., *et al.* (2007). The world largest ceramic membrane drinking water treatment plant, in AWWA conference on membrane technology. Tamba Bay.
Kimura, K., Hane, Y., Watanabe, Y., Amy, G. and Ohkuma, N. (2004). Irreversible membrane fouling during ultrafiltration of surface water. *Water Research*, **38**(14–15), 3431–3434.
Klomfas, G. and Konieczny, K. (2004). Fouling phenomena in unit and hybrid processes for potable water treatment. *Desalination*, 163(1–3), PII S0011-9164(04)00101-8.
Konieczny, K., Bodzek, M. and Rajca, M. (2006). A coagulation – MF system for water treatment using ceramic membranes. *Desalination*, **198**(1–3), 92–101.
Lee, S. and Cho, J. (2004). Comparison of ceramic and polymeric membranes for natural organic matter (NOM) removal. *Desalination*, **160**(3), 663–665.
Luonsi, A., Laitinen, N., Beyer, K., Levänen, E., Poussade, Y. and Nyström, M. (2002). Separation of CTMP mill-activated sludge with ceramic membranes. *Desalination*, **146**(1–3), 399–404.
Lerch, A., Panglisch, S., Buchta, P., Tomita, Y., Yonekawa, H., Hattori, K. and Gimbel, R. (2005). Direct river water treatment using coagulation/ceramic membrane microfiltration. *Desalination*, **179**(1–3), 41–50.
Matsui, Y., Fukuda, Y., Inoue, T. and Matsushita, T. (2003). Effect of natural organic matter on powdered activated carbon adsorption of trace contaminants: characteristics and mechanism of competitive adsorption. *Water Research*, **37**(18), 4413–4424.
Matsui, Y., Murase, R., Sanogawa, T., Aoki, N., Mima, S., Inoue, T. and Matsushita, T. (2005). Rapid adsorption pretreatment with submicrometre powdered activated carbon particles before microfiltration. *Water Science and Technology*, **51**(6–7), 249–256.
Matsushita, I., Matsui, Y., Shirasaki, N. and Kato, Y. (2005). Effect of membrane pore size, coagulation time, and coagulant dose on virus removal by a coagulation-ceramic microfiltration hybrid system. *Desalination*, **178**(1–3), 21–26.
Nazzal, F.F. and Wiesner, M.R. (1994). PH and ionic-strength effects on the performance of ceramic membranes in water filtration. *Journal of Membrane Science*, **93**(1), 91–103.
Park, J.Y., Choi, S.J. and Park, B.R. (2006). Effect of N2-back-flushing in multichannels ceramic microfiltration system for paper wastewater treatment. *Desalination*, **202**, 207–214.

Peters, T.A., Fontalvo, J., Vorstman, M.A.G., Benes, N.E., van Dam, R.A., Vroon, Z.A.E.P., van Soest-Vercammen, E.L.J. and Keurentjes, J.T.F. (2005). Hollow fiber microporous silica membranes for gas separation and pervaporation. Synthesis, performance and stability. *Journal of Membrane Science*, **248**, 73–80.

Schippers, D., Bakker, S.H., Brummel, D. and Heijman, S.G.J. (2007). Particle removal from surface water with ceramic microfiltration, in particle separation 2007 – from particle characterisation to separation processes, IWA: Toulouse, France, p. 8.

Shirasaki, N., Matsushita, T., Matsui, Y., Kobuke, M. and Ohno, K. (2009). Comparison of removal performance of two surrogates for pathogenic waterborne viruses, bacteriophage Q beta and MS2, in a coagulation-ceramic microfiltration system. *Journal of Membrane Science*, **326**(2), 564–571.

Strathmann, H. (2000). Membrane separation processes: current relevance and future opportunities. *Aiche Journal*, **47**(5), 1077–1087.

Tsuru, T. (2001). Inorganic porous membranes for liquid phase separation. *Separation and Purification Methods*, **30**(2), 191–220.

Voigt, I., Stahn, M., Wöhner, St., Junghans, A., Rost, J. and Voigt, W. (2001). Integrated cleaning of coloured waste water by ceramic NF membranes. *Separation and Purification Technology*, 25, 509–512.

Weber, J. and Seidel, T. (2003). Characteristics and application of ceramic nanofiltration membranes. *Annals NY Acedam Sci*, **984**, 178–193.

Yuasa, A. (1998). Drinking water production by coagulation-microfiltration and adsorption-ultrafiltration. *Water Science and Technology*, **37**(10), 135–146.

Zapata-Montoya, J.E., Gaudix, E.M. and Gaudix, A. (2006). Long-term effects of chemical cleaning in the performance of ultrafiltration ceramic membranes. *Desalination*, **200**(1–3), 316–318.

Treatment of NF concentrates: organic micropollutant and NOM removal

B. Bozkaya-Schrotter, C. Daines-Martinez, J.-C. Schrotter and P. Bréant

Veolia Environnement Anjou-Recherche, chemin de la Digue 78600 Maisons-Laffitte (E-mail: bengu.bozkaya@veolia.com; catherine.daines@veolia.com; jean-christophe.schrotter@veolia.com; philippe.breant@veolia.com)

Abstract Treatment of nanofiltration membrane system concentrates was studied by eliminating mainly the pesticides and the natural organic matter. Processes tested for this purpose include adsorption, ozonation, coagulation and flocculation, GAC filtration, biological treatment, advanced ozonation and ion exchange. Some of these processes were coupled either to improve or to achieve complete pesticide and the natural organic matter elimination. Combination of ozonation and adsorption, GAC filtration and biological treatment proved to be efficient processes for this application. Concentrate treatment will continue at industrial scale using aforementioned techniques. On the other hand although efficient, H2O2/O3 and ion exchange processes were economically not viable.

Keywords concentrate, nanofiltration, organic matter, pesticide

INTRODUCTION

Drinking water demand increases everyday with the increasing population of the world. Naturally, this leads to an augmentation in the number of drinking water production sites. At the same time, water scarcity is very real and with the global warming threats the need of the protection of the existing drinking water resources is more obvious than ever. In recent years, there has been an orientation to cheaper and greener processes. Water treatment market continuously seeks for processes that can produce high quality water with lower fixed and operational costs and less waste production. Nanofiltration is one of the techniques that can meet these multiple objectives. It is becoming a common technique for surface water treatment for drinking water production purposes. These systems however, produce a waste, so called "concentrates" that is the 15 to 25% of the feed. Currently, they are discharged back into the source or when it is possible, waste integration techniques apply, e.g. used as gray or irrigation water.

Concentrate composition significantly varies with the raw/membrane feed water quality that depends on the industrial and agricultural activities of the region, weather conditions etc. It contains natural organic matter (NOM), micro-pollutants, various salts typically 5 to 6 times more concentrated than in its feed.

 TECHNEAU: Safe Drinking Water from Source to Tap. Edited by Theo van den Hoven and Christian Kazner. ISBN: 9781843392750. Published by IWA Publishing, London, UK.

Concentrate also contains acidifying, e.g. sulphates and chlorides and/or anti-scaling agents e.g. phosphonates added during the pre-treatment stages. The pH is generally higher than its feed due to high rejection of carbonate.

In order to protect existing drinking water resources, it is necessary to reduce the concentrations of these components to keep the ecological balance of the aquatic systems intact. Some patents have already demonstrated the feasibility of implementing a treatment scheme to selectively remove the phosphonate-based anti-scalants [Daines *et al.*, 2008] and improve the recovery factor [Daines *et al.*, 2008] in the concentrate.

In this study, our aim was to treat the concentrate prior to its rejection by focusing mainly on the removal of pesticides and NOM in order to obtain a composition similar to nanofiltration feed water quality. Total elimination of the pesticides and a partial elimination of the natural organic matter, was targeted.

Pesticides are widely used mainly for agricultural purposes for crop protection purposes or for the elimination of weed on motorways or railways. They are however, toxic, carcinogenic and mutagenic to living organisms and they also pollute surface and ground water resources [Ormad *et al.*, 2008]. European legislation limits the total concentration of pesticides in surface waters used for drinking water production to 5 μg/L. In this work, eight most commonly detected pesticides in French surface waters, i.e. atrazine, sulcotrione, bentazone, isoproturon, diuron, glyphosate, acetochlore, some of their main metabolites were selected as the model micropollutants.

Natural organic matter particularly creates problems for the water supply industry. It colours the water, increases the disinfection demand, introduce the problem of bio-film formation in the distribution system and potential formation of harmful chemical disinfection by-products as a result of chlorination. Concentrating NOM in the sources used for potable water production is usually negligible. However, it becomes a problematic especially when the source is relatively small and when its flowrate significantly reduces during the summer.

Conventional water treatment techniques, i.e. adsorption, oxidation by ozone and carbozone (Jans and Hoigné, 1998), biological treatment, advanced oxidation, ion exchange were tested and the impact of operational parameters, e.g. temperature, pH was investigated.

METHODS

Membrane concentrates

Real membrane concentrates were regularly collected from a river water treatment site in France. Nanofiltration pre-treatment of this site is composed of clarification, ozonation, filtration antracite + sand, anti-scaling agent addition and prefiltration (5 μm). Concentrate samples are spiked with 10 μg/L of each of the studied pesticides, i.e. atrazine, sulcotrione, bentazone, diuron, glyphosate, acetochlore (Cluzeau Info Labo, France) and isoproturon (Sigma Aldrich, France), to enable the quantitative analysis and obtain various pesticide concentration levels detected in the surface waters at different times of the year.

Concentrate has a complex composition and it is very difficult to precisely identify each of its components. Therefore, in order to have an idea on NOM and mineral concentrations some of its properties were analysed quantitavely and tabulated in Table 1.

Table 1. Properties of concentrates tested.

pH (at 22°C)	Conductivity (μS/cm)	THCa (mg/L)	THMg (mg/L)	TOC (mg/L)	COD (mg/L)
7.5–8.08	1590–2000	100–400	30–70	4–17.74	10.2–40.6

Adsorption

Adsorption of the NOM onto powdered activated carbons was tested. Previously Picasorb 16 (PICA, France) was used for pesticide elimination [Bozkaya-Schrotter *et al.*, 2008]. It was also used for the removal of the natural organic matter. In addition, an other PAC from a different producer (CECA Carbons, France) with similar properties was also tested (see Table 2). Powdered activated carbons were kindly supplied by PICA and CECA Carbons, France. They are wood based carbons activated either thermally or chemically or both methods.

Table 2. Properties of Picasorb-16 and BB105.

Carbons	Picasorb16	BB105
Precursor	Wood	Wood
Activation	Thermal	Physical
Apparent Density(kg/m^3)	0.41	0.45
Median particle size (μm)	15–35	10–40
BET	ca.1000 ^{2}	ca. 1000 m^2
Micropore Area	757 m^2/g	–
Micropore Volume	0.364 cm^3/g	–
Median Pore Diameter	8.56 Å	–
Iode indice	980 mg/g	1000 mg/g

Some of their physical properties are as follows: density: 0.22–0.45 kg/m^3, Iode indice: 700–1500 mg/g. All experiments were carried out under atmospheric conditions at ca. 20°C. A mechanical agitator (AGITELEC, France) was employed for adsorption experiments. In order to separate the solution and the carbon, vacuum filter with a 0.45 μm filter paper (WATMAN, UK) is utilised. Selected amount of concentrate sample and the carbon are placed in a funnel and agitated for contact times between 2 and 30 minutes. Once the selected contact time is elapsed, sample is filtered and analysed as described in the following section.

Oxidation by ozone and combination of PAC and O_3

Oxidation by ozone has been carried out to observe the participation of oxidation in NOM and pesticide degradation during the combined PAC + O_3 tests. OZOTEST [Roche and Prados, 1995] method was employed for the ozonation tests in this study. Concentrate sample is placed into a 1L glass funnel and ozone is introduced in the system via a 100 ml gas tide syringe (Thermofina, France). BMT 802X ozone generator (Messtechnik, Berlin) was used to produce ozone (0–120 g/Nm^3) from pure oxygen. Ozone dosage was varied between 3 and 30 mg/L. Funnel is then agitated by a mechanical agitator (AGITELEC, France) and once the selected contact time is elapsed, concentrate is collected and analyzed. Ozone residual in liquid and gas phases was measured via indigo colorimetric [Bader and Hoigné, 1981] and iodometry methods, repectively. Oxidation by ozone is combined with adsorption process by adding PAC in the system prior to ozone injection. The amount of PAC varied between 30 and 3000 mg/L. PAC was separated by vacuum filtration.

Quantitative analysis of pesticides and NOM

Quantitative analysis of pesticides was carried out by a laboratory specialised in environmental analyses based in Paris, "Centre d'Analyses Environnementales" (CAE). NOM concentration was determined by TOC (BIORITECH O.I. Analytical 1010, France), COD (LCK414, HACH-Dr Lange, Germany) and UV absorbance at 254 nm (DR 4000, HACH-Dr Lange, Germany) analyses.

Biological treatment

A laboratory scale experimental rig was constructed for biological degradation of the NOM in the concentrate. A photo of the experimental system can be seen in Figure 1. Two cylindrical PVC columns were filled with 180 ml used GAC particles. GAC sample was collected from a drinking water production site, where it was used in the biological filter since 2004 and saturated with NOM. The column diameter and height and the height of carbon bed were 3.80 cm, 25 cm and 16 cm, respectively. Polyethylene raschig rings were used as a support medium for GAC particles. A peristaltic pump (Masterflex, Cole-Palmer, US) was used for liquid transfer and silicon tubing was used in the system. To avoid formation of preferential routes in the carbon bed, two cylindrical columns (d = 1 cm, open to atmophere) were installed after each column. A polyethylene filter was placed at the bottom of the column to block carbon particles in the column.

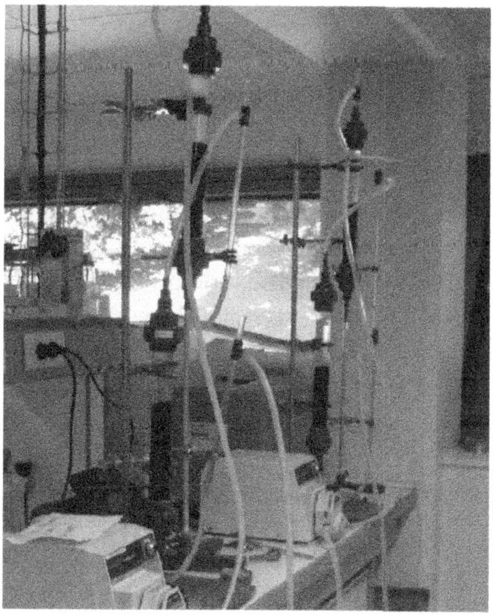

Figure 1. Photo of the GAC filtration and biological treatment experimental systems.

The flowrate of the pump was adjusted to ca. 4 ml and the concentrate was continuously pumped through both columns connected in series. Under these conditions the contact time was extremely high, i.e. 50 min. In order to reduce the contact time, the column size was modified and smaller columns are installed in the system. These columns had 1.6 cm diameter, 37 cm height and 16.25 cm carbon bed height. With the use of these columns contact time was reduced to ca. 16 min in the system. Everyday, at the exit of the second column 200 ml of sample was collected and NOM elimination was determined by chemical oxygen demand (COD), total organic carbon (TOC) and UV absorbance analyses.

GAC physical adsorption

The experimental system used for GAC filtration was identical to the one used for biological treatment. The carbon used was fresh and it was the same type of carbon used in the biological application. Experimental rig and the amount of carbon and support material used were also identical to biological treatment system. Similarly, the system was modified to decrease the contact time. COD, TOC and UV absorbance of these samples were also measured. It is unfortunately not possible to give the commercial name of the GAC, as it is a taylor made carbon. Some of the properties can be listed as follows. It contains, micro-, meso and macropores, iode indice is 1000 mg/g, apparent density is 0.18–0.26 g/cm^3, and the BET surface area is ca. 1500 m^2/g.

RESULTS AND DISCUSSION

Adsorption, oxidation and carbozone

Elimination of the pesticides in the concentrate was studied in detail. It was concluded that in order to completely eliminate selected pesticides it was necessary to couple adsorption and ozonation processes. Operational parameters were further optimised. Figure 2 shows the pesticide elimination yields of adsorption and ozonation and carbozone with optimised carbon and ozone dosages, i.e. 30 and 3 mg/L. High elimination levels, i.e. $>80\%$ were obtained in concentrate with the combined system with 10 minute-contact-time.

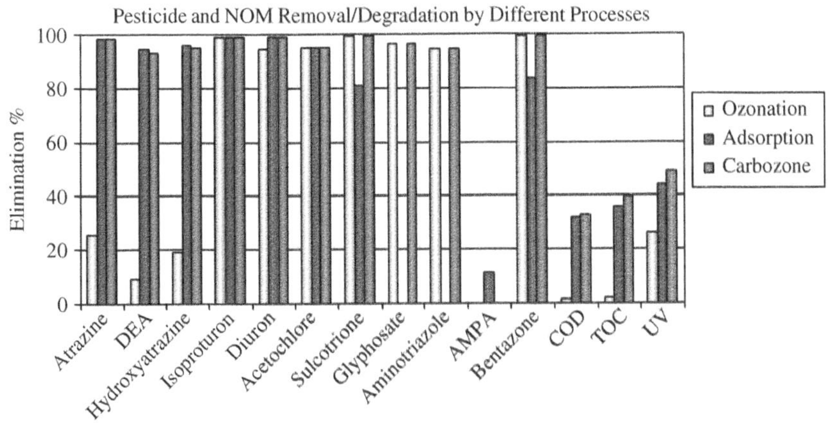

Figure 2. Pesticide elimination by ozonation, adsorption and carbozone. 30 mg/L PAC, 3 mg/L ozone, tc = 10 min. Concentrate: COD = 10.2 mg/L, TOC = 4 mg/L, UV = 0.35.

NOM removal/degradation was also achieved with the combined system with low PAC and ozone dosages, i.e. 30 and 3 mg/L, as seen in Figure 2, however elimination yield was quite low. The major fraction of the NOM is removed by adsorption; hence the PAC dosage had the greatest impact on the process efficiency. Therefore, in order to improve NOM removal, PAC dosage was increased to 100 mg/L.

NOM elimination kinetics by adsorption and combined systems is shown in Figure 3. Pesticide elimination kinetics was also determined for the combined system of 100 mg/L of Picasorb 16 and 6 mg/L ozone with 2 and 5 minutes of contact times. It was observed that 2 minutes achieved sufficient removal of the studied pesticides as seen in Figure 3. Influence of the concentrate temperature was also studied to observe any possible impact of the seasonal temperature changes on the combined system's performance. Similar NOM removal yields obtained from experiments carried out at 5 and 25°C.

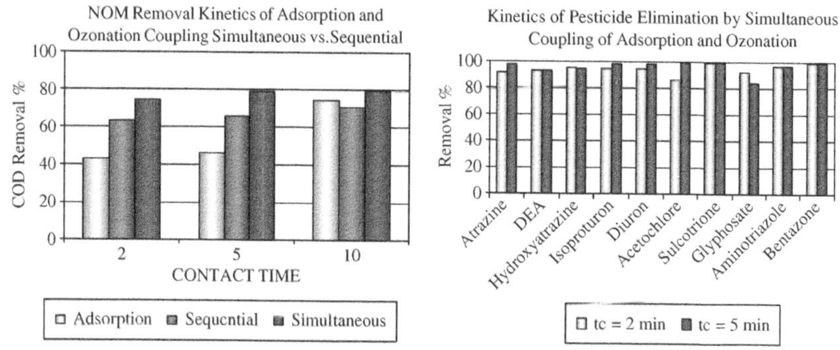

Figure 3. Kinetics of NOM and pesticide elimination, 100 mg/L PAC and 6 mg/L ozone. Concentrate: COD = 18.4 mg/L, TOC = 9.5, UV = 0.863.

Impact of ozone on adsorption capacity, basic surface properties and the possibility of re-use of carbon were also studied in adsorption and carbozone systems. In the case of using high ozone concentrations, i.e. 30 mg/L, significant decrease in the adsorption capacity of the PAC used in the simultaneous combination system was observed compared to the PAC used in adsorption system. However, at real experimental conditions, i.e. 3 mg/L ozone, the impact of ozone on carbon was negligible. This decrease observed with high ozone concentration can be due to modifications in the carbon morphology by gasification at high ozone concentrations and/or in the oxygen sites of the PAC. The results of low temperature nitrogen adsorption tests confirm significant modification on the carbon morphology due to the presence of ozone, e.g. BET surface decreased from 1359 m^2/g to 720 m^2/g.

Improved OH radical activity in NOM degradation was observed at basic conditions by several authors [Faria *et al.*, 2006; Sanchez-Polo *et al.*, 2006]. In order to reduce scavenging effect of the HCO_3^- ions, concentrate solutions are basified with NaOH (30%) solution. Oxidation experiments with 6 mg/L of ozone were carried out at various pH levels, i.e. 8.35, 10 and 12. No significant change in the ozonation efficiency was observed as ozone was readily consumed in the concentrate and it was necessary to increase ozone dosage to create OH radicals. Highest NOM degradation yields were observed following ozonation at pH 12. However, it was observed that a part of NOM, i.e. 20% COD, was precipitated during basification step without ozonation, most probably due to encapsulation by $CaCO_3$ formed.

Biological treatment

The carbon in the filter had been in use since 2004. Therefore it was assumed that the carbon was already saturated with the natural organic matter. It was also assumed that only NOM removal activity was biological degradation. It is known that to observe stable biological activity, the living organisms should adapt to new environmental conditions, solution's composition, temperature etc. This may usually take several weeks. In our case, however, this adaptation time was only a week, as the concentrate solutions were collected from the same plant as the GAC. Following one week of continuous operation, stable NOM degradation was observed as seen in Figure 4I. These results were obtained with the first columns and high contact times. Previously, biodegradable fraction in this solution (BDNOM) was determined as 20% as seen in Figure 4I. The COD, TOC and UV results showed improved results compared to biodegradable fraction. This is most probably due to adsorption of NOM onto carbon caused by the increased contact time, i.e. 50 min.

Biological treatment was also combined with ozonation to increase the biodegradability of the NOM in the concentrate. NOM in the concentrate was degraded by ozonation prior to biological treatment. Pre-ozonation was carried out for two weeks however, no visible increase in the degradation of NOM was observed with 2 mg/L ozone dosage. It was clearly seen that 2 mg/L ozone dosage was not sufficient to observe any augmentation in the NOM elimination.

We carried out a study to determine maximum ozone dosage applicable without leaving any residual in the concentrate prior to biological treatment. According to this study 6 mg/L ozone was readily consumed by the concentrate with 10 minute-contact-time leaving no ozone residual in the liquid phase. Figure 4II shows NOM elimination determined by COD analysis. The values obtained between 14/11/2008 and 18/11/2008 show the early results obtained

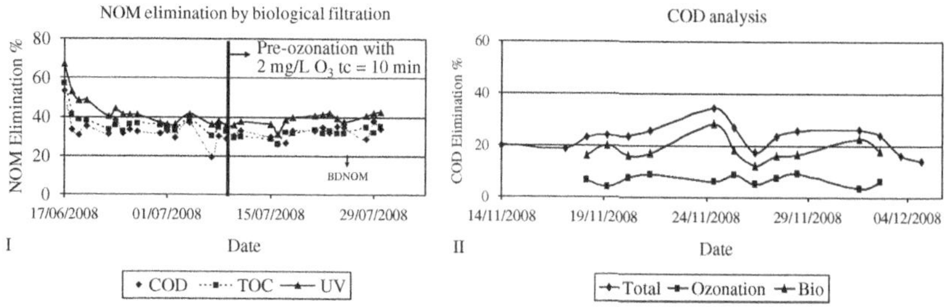

Figure 4. I. Biological degradation of NOM. Concentrate, tc = 50 min, COD = 22.9–23.5 mg/L, TOC = 10.3–10.6 mg/L, UV = 0.71–0.72 mg/L. II. Determination of NOM elimination by combined ozonation and biological treatment, tc = 16 min. (pre-ozonation), ozonation and biological treatment with the help of COD analysis. Concentrate COD = 29.7–32.9 mg/L, TOC = 13.9–14.6 mg/L, UV = 0.89–0.91.

without pre-ozonation. COD reduction under these conditions (tc = 16 min) was ca. 20%. With the application of ozone, a slight increase in the COD and TOC analysis was observed. By looking at these figures, we can conclude that the slight increase in the NOM climination is not due to improved degradation of biodegradable fraction of the NOM but by the ozone degradation itself.

Following this study, ozonation step was carried out between the two biologically activated GAC columns, with 16 minute-contact-time. It was stated in literature that ozonation between the two columns may increase the overall process efficiency and reduces the ozonation costs [Fahmi *et al.*, 2003]. An additional oxidation assessment was carried out to observe ozone consumption by the concentrate previously treated by the first column. To avoid any presence of ozone residual, 4 mg/L ozone dosage was applied for this study. Although one would expect to see an improved elimination of the NOM, no augmentation in the system performance was observed. Overall the process efficiency was surprisingly lower than the efficiency of the combined system in which ozonation was carried out before the biological treatment. Oxidation products may have less affinity to biofilm surface. In addition, if there is still adsorption taking place in the columns, formation of polar substances with ozonation may also reduce adsorption efficiency. These components have less affinity to carbon hence they are weakly adsorbed on the carbon surface.

Following two months of operation, the performance of the system in pesticide elimination was also tested. Concentrate was spiked with 10 μg/L of each pesticide and one week later, pesticide analysis was carried out in the effluent of the filter. It was previously observed that pesticides were poorly degraded on GAC-bio filters (Foster *et al.*, 1991).

In this study the pesticides identified previously as polar were poorly degraded. These results indicate that pesticides were mostly removed by adsorption onto carbon. However, glyphosate removal was significantly high. This may be due to its adsorption on the biofilm rather than onto GAC. Bentazone degradation was also surprisingly low. Bentazone may be transferred from the carbon surface into the solution if the carbon filter is saturated.

Pesticide elimination in the combined system was also observed. Pesticide analysis was carried out following the ozonation step, ozonation + biological treatment, and biological treatment. The results are shown in Figure 5. It was observed that when biological treatment is combined with ozonation complete pesticide elimination was achieved.

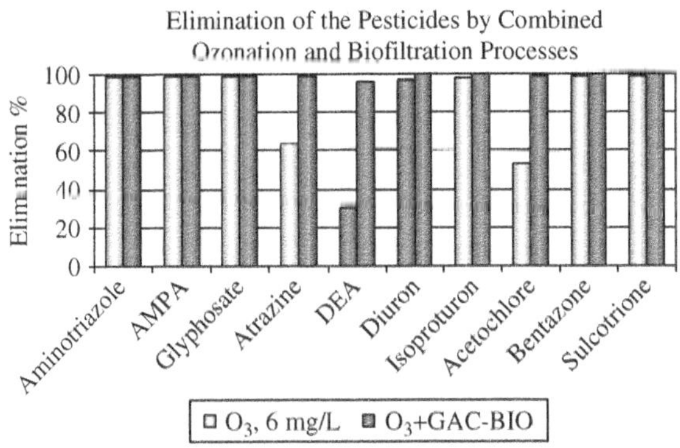

Figure 5. Pesticide elimination by biological treatment, oxidation by ozone and combination of these two processes, tc = 16 min. Concentrate COD = 29.7–32.9 mg/L, TOC = 13.9–14.6 mg/L, UV = 0.89–0.91.

GAC physical filtration

NOM removal was studied by testing GAC filtration method. Initially, with 50 minute-contact-time NOM removal was 80%. Following two months of operation diameter NOM removal was reduced to ca. 60%. COD, TOC and UV analyses were regularly carried out to observe NOM removal. Results obtained are shown in Figure 6. When the column diameter was changed contact time in the system was re-adjusted to 16 minutes, a sharp decrease was observed in the NOM removal. Under these conditions COD, TOC and UV removal was ca. 20% as seen in Figure 6.

GAC filtration process was then combined with ozonation. The aim was mainly to eliminate polar pesticides present in the concentrate that cannot be removed by GAC filtration. Any possible increase in the NOM removal was also observed. Pre-ozonation was carried out both with and without presence of ozone residual. Firstly ozonation was carried out with 4 mg/L of ozone, then

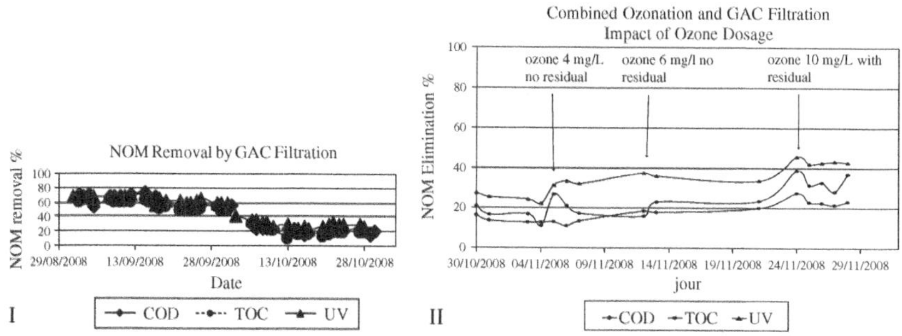

Figure 6. NOM removal by GAC filtration with 50 and 16 minutes-contact-times. Concentrate COD = 20.3–22.7 mg/L, TOC = 9.2–9.9 mg/L, UV = 0.56–0.6. II. NOM Removal by combined GAC filtration and ozonation tc = 16 minutes. Concentrate: COD = 18.8–19.3 mg/L, TOC = 9–9.2 mg/L, UV = 0.58–0.59.

since we run out of concentrate and had to collect a new concentrate batch, oxidation assessment was carried out again and ozone dosage was modified. The study was carried out with 6 mg/L of ozone. When we look at the Figure 6 showing the results of COD, TOC and UV analyses and detailed analysis of NOM removal and degradation, wee see that ozonation without O_3 residual does not significantly improve NOM degradation/removal in the system. With residual however, we see that the COD elimination was almost double. TOC elimination also slightly increases however the difference is still not significant.

Pesticide analysis was also carried out following combined filtration and ozonation tests. Results are shown in Figure 7. It was observed that apart from bentazone, all other pesticides were removed/degraded. GAC column maybe saturated with bentazone and it tends to desorb in the concentrate solution.

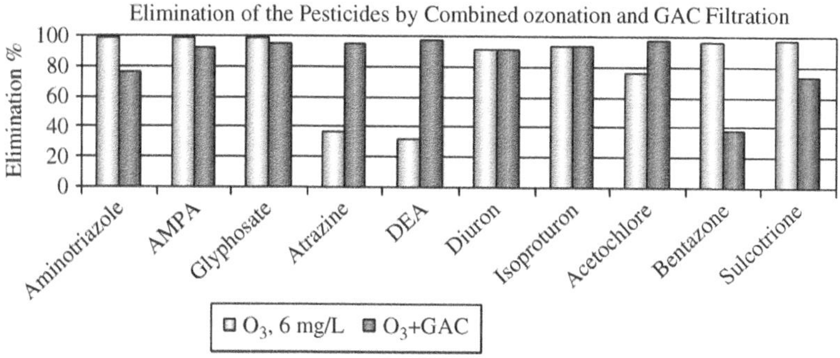

Figure 7. Pesticide elimination by the combined ozonation and GAC filtration, tc = 16 min. Concentrate: COD = 31.1 mg/L, TOC = 14.63mg/L, UV = 0.893.

CONCLUSIONS

Treatment of nanofiltration membrane system concentrates was studied by eliminating mainly the pesticides and the natural organic matter. Processes tested for this purpose were adsorption, ozonation, coagulation and flocculation, GAC filtration, biological treatment, advanced ozonation and ion exchange. Some of these processes were coupled either to improve or to achieve complete pesticide and the natural organic matter elimination.

Oxidation of all of the pesticides by ozone can be achieved with relatively high ozone dosages, i.e. 30 mg/L even with short contact times, i.e. 5 minutes. However as high dosages are economically not viable and NOM can not be degraded, high ozone dosages can not be applied at industrial scale. Studied ozone dosages were mostly between 3 and 10 mg/L. With the application of 3 mg/L with between 5 and 10 minute contact times, most of the pesticides were degraded, with the exception of atrazine and its metabolites. Natural organic matter elimination under these conditions was almost negligible.

It was observed that adsorption onto PAC was an efficient method to remove pesticides with low water solubility. However, in order to eliminate polar pesticides it was necessary to combine adsorption with ozonation process. Combining adsorption with ozonation resulted into accelerated NOM removal and complete pesticide elimination.

Coagulation and flocculation achieves removal of pesticides that are polar and phosphonate based, i.e. glyphosate and AMPA. With the application of 150 mg/L of coagulant and with the regulation of pH at 5.5, all these pesticides are successfully removed.

GAC filtration proved to be a suitable method for both pesticide and organic matter removal. Naturally, after a certain time of operation GAC activity was reduced.

Biological treatment is also a promising option for this study due to its reduced costs and simple operation. Small NOM fraction that is biodegradable can easily be degraded by this method. It was observed that in addition to biological activity, when the contact time was very high, a part of the NOM was adsorbed on the carbon pores that were still accessible. In addition to degradation by biological activity and adsorption onto carbon, it is quite likely that some of the pesticides are bonded on the biofilm surface.

A preliminary study was carried out to observe the performances of H_2O_2/O_3 and ion exchange processes in NOM removal. Although efficient, compared to processes aforementioned above, these processes proved to be very expensive.

This project continues with industrial scale applications of PAC and GAC filtration, biological treatment and their combination with ozonation.

Cost analysis of these processes is currently being performed and the results will soon be published in the next paper.

REFERENCES

Bader H. and Hoigné, J. (1981). Determination of ozone in water by indigo method. *Water Res.*, **15**, 449.

Bozkaya-Schrotter, B., Daines, C., Lescourret, A.S., Bignon, A., Breant, P. and Schrotter, J.-C. (2008). Treatment of trace organics in membrane concentrates I: pesticide elimination *Water Sci. & Tech.: Water Supply*, **8**, 223–230.

Daines, C., Schrotter, J.C., Drouet, K. and Gaid, K. (2008). Membrane Station Waste Treatment Process WO2008/055827.

Fahmi, Nishijima, W. and Okada M. (2003). Improvement of DOC removal by multi-stage AOP-biological treatment. *Chemosphere*, **50**, 1043–1048.

Faria, C.C., Órfao, J.J.M. and Pereira, M.F.R. (2006). Ozone decomposition in water catalyzed by activated carbon: influence of chemical and textural properties *Ind. & Eng. Chem. Res.*, **45**, 2715–2721.

Forster, D.M., Rachwal, A.J. and White, S.L. (1991). New treatment processes for pesticides and chlorinated organics control in drinking water. *J. Inst. Water Env. Management*, **5**(4), 466–477.

Jans, U. and Hoigné. (1998). Activated carbon and carbon black catalyzed transformation of aqueous ozone into OH-radicals *J. Ozone: Science and Engineering*, **20**, 67–90.

Ormad M.P., Miguel N., Claver A., Matesanz J.M. and Ovelleiro, J.L. (2008). Pesticides removal in the process of drinking water production. *Chemosphere*, **71**, 97–106.

Roche, P. and Prados, M. (1995). Removal of pesticides by use of ozone or hydrogen peroxide/ozone. *Ozone Science & Engineering*, **17**, 657–672.

Sanchez-Polo, M., Salhi, E., Rivera-Utrilla, J. and von Gunten, U. (2006). Combination of ozone with activated carbon as an alternative to conventional advanced oxidation processes. *Ozone: Science and Engineering*, **28**, 237–245.

Applications of quantitative structure-activity relationships for rejection modelling of organic solutes by nanofiltration membranes

V.A. Yangali-Quintanilla[a], M.D. Kennedy[b] and G.L. Amy[b]

[a]Delft University of Technology, Stevinweg 1, 2628CN Delft, The Netherlands
(E-mail: v.a.yangaliquintanilla@tudelft.nl)
[b]UNESCO-IHE, Institute for Water Education, Westvest 7, 2611AX Delft, The Netherlands

Abstract To achieve practical rejection prediction of organic solutes by nanofiltration membranes, a quantitative structure activity relationship (QSAR) approach was used to generate a model. The QSAR model was generated after implementing principal component analysis (PCA) and multiple linear regressions on an experimental database of 106 rejection cases generated after 8 nanofiltration experiments with clean and fouled nanofiltration membranes using water containing a cocktail of 14 organic solutes. PCA was performed with a database of variables related to membrane characteristics, filtration operating conditions and physical-chemical solute properties. Membrane characteristics were represented by variables of hydrophobicity (contact angle), salt rejection, and surface charge (zeta potential); variables for solute properties were hydrophobicity (log K_{ow}, log D), polarity (dipole moment), and size (molar volume, molecular length, molecular depth, equivalent width, molecular weight); and operating conditions variables were flux, pressure, cross flow velocity, back diffusion mass transfer coefficient, hydrodynamic ratio (J_0/k), and recovery. To validate the QSAR model, an external experimental database was used to quantify rejection prediction of organic solutes by other nanofiltration membranes. Results of the model application on an external database showed that R^2 correlation coefficient of 0.88 was obtained for measured rejections vs. predicted rejections.

Keywords modelling, nanofiltration, organic solutes, QSAR

INTRODUCTION

It has been recognized that nanofiltration is an effective barrier against emerging contaminant such as pharmaceuticals, endocrine disruptors, pesticides and personal care products (Kiso *et al.*, 2001; Kimura *et al.*, 2003; Yoon *et al.*, 2006). Emerging contaminants are present in surface water bodies, sewage treatment plant effluents, and are even present in finished drinking water (Boyd *et al.*, 2003). Obviously, the situation is of public concern, mainly because the possible health effects are yet not totally understood (Jones *et al.*, 2005; Pomati *et al.*, 2006). In practice, some water utilities were aware of the situation and have implemented treatment stages that will

 TECHNEAU: Safe Drinking Water from Source to Tap. Edited by Theo van den Hoven and Christian Kazner. ISBN: 9781843392750. Published by IWA Publishing, London, UK.

improve the water quality; applications include use of granular activated carbon or membrane treatment (Drewes *et al.*, 2006; Verliefde *et al.*, 2007). However, the diverse physicochemical nature of organic solutes and its degree of affinity to the materials or processes used for treatment not always guarantee acceptable removals. In membrane treatment applications, different transport/rejection mechanisms for emerging contaminants have been identified, explicitly size/steric exclusion, hydrophobic adsorption, partitioning, and electrostatic repulsion are result of interactions between solutes and membranes (Kiso *et al.*, 2001; Kimura *et al.*, 2004; Bellona and Drewes, 2005). A quantitative structure activity relationship (QSAR) is a method that relates an activity of a set of compounds quantitatively to chemicals descriptors (structure or property) of those compounds (Sawyer *et al.*, 2003). In the field of removal of compounds by membrane filtration, the principle of QSAR is to relate removal (rejection) to a function of solute structure (hydrophobicity, size), operating conditions and membrane properties (hydrophobicity, exclusion). QSAR has the objective of prediction but maintaining a relationship to mechanistic interpretation. To explain, QSAR can best be attained by comparative studies, that is, studies among QSAR to obtain lateral support for a new study and comparison with work from mechanistic interpretation to facilitate elucidation of reaction mechanisms; which makes it possible to compare models from a database and to draw analogies with models from other databases (Hansch *et al.*, 2001). This research uses the concept of QSAR analysis to quantify an activity, compound rejection by a membrane, in terms of solute physicochemical properties, membrane characteristics (salt rejection, pure water permeability, molecular weight cut-off, charge, hydrophobicity) and operating conditions (pressure, flux, cross flow velocity, back diffusion mass transfer coefficient, recovery).

METHODS

Acetaminophen, ibuprofen, phenacetin, phenazone, carbamazepine, naproxen, caffeine, sulfamethoxazole, metronidazole (pharmaceuticals) were analyzed according to the protocols described by Sacher *et al.* (2001, 2008). The information for analyses of estrone, 17β estradiol, bisphenol A, nonylphenol, atrazine (endocrine disruptors) was provided by Yangali-Quintanilla *et al.* (2009). The pharmaceuticals and endocrine disruptors were analyzed by Technologiezentrum Wasser (TZW, Germany). More information about the experimental setup, experimental conditions, compounds and membranes can be found elsewhere (Yangali-Quintanilla *et al.*, 2009).

A flow chart of the methodology used to build the QSAR model is shown in Figure 1. The methodology to develop a general QSAR equation to describe rejection was performed in three phases. The first phase was of data analysis.

The data comprised rejections of a cocktail of 14 compounds present in a water matrix. Clean and fouled Filmtec NF-90 and NF-200 membranes were used for a total of 8 filtration experiments. The variables considered as compound descriptors were molecular weight (MW), solubility, log K_{ow}, log D, dipole moment, molar volume, molecular length, molecular width, molecular depth and equivalent width (eqwidth); variables describing membrane characteristics were molecular weight cut-off (MWCO), pure water permeability (PWP), salt rejection (SR), charge of the membrane as zeta potential (ZP), and hydrophobicity as contact angle (CA); variables describing operating conditions were operating pressure (P), cross flow velocity (v), back diffusion mass transfer coefficient (k), flux (J), ratio of pure water permeation flux J_0 and back diffusion mass transfer coefficient (J_0/k) and recovery. The second phase was of variables reduction using principal component analysis. The third phase corresponded to the regression analysis. The validation of the general QSAR model was implemented by predicting rejections of an external database of rejections experiments performed with different compounds, membranes and operating conditions. PCA and MLR were performed with SPSS Statistics 16.0.

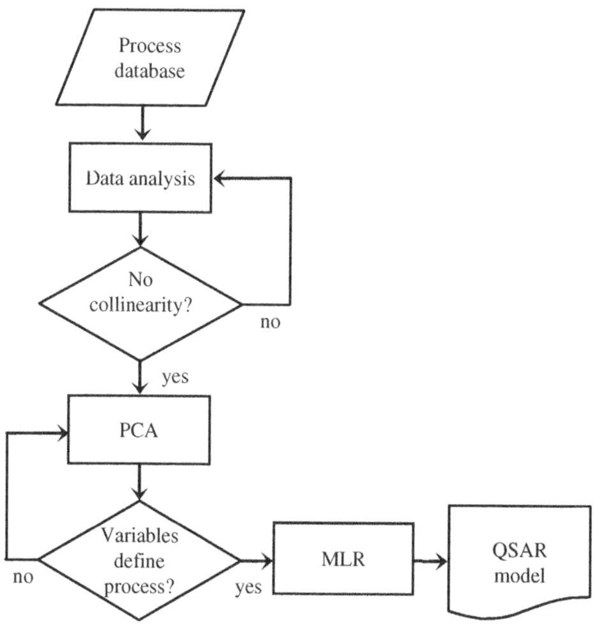

Figure 1. Flow chart for building of QSAR model.

RESULTS AND DISCUSSION

Principal component analysis allowed identification of 3 components represented by 11 variables describing three relations (components) namely membrane/operating-conditions (component 1: flux, pure water permeability, salt rejection, zeta potential, mass transfer coefficient, cross flow velocity), hydrophobicity/size (component 2: length, log K_{ow}, log D) and size (component 3: equivalent width, depth). The initial database of 21 variables was reduced to 11 variables grouped in 3 components that account for 89.3% explanation of total variance.

The next step was the implementation of multiple linear regression (MLR) using the new set of variables. The use of MLR after PCA presents the advantage of a more simplified modelling approach. Considering the previous explanation the variables log K_{ow}, log D, length, depth, eqwidth, PWP, SR, ZP, v, k, J were used to model rejection. The final regression resulted in variables SR, eqwidth, log D, length and depth. The model presented a squared correlation coefficient (R^2) value of 0.75 with a statistical F-test of 60.21, and a statistical significance of 0.00%. To limit the size of solutes referred with variables length, eqwidth (equivalent width) and depth; the rejection equation model was arranged to allow a maximum rejection of 100%, which can be explained after observing positive coefficients for equivalent width and length. This situation may also be detrimental for rejection predictions of ionic compounds of medium to large size (0.6 to 1.2 nm as equivalent width) that mostly are rejected due to electrostatic repulsion and less steric hindrance. Therefore, the defined QSAR equation was:

$$rej=\begin{cases}100\,\mathit{if}\;rej\geqslant 100\\ rej\end{cases}\quad \mathit{with}\quad \begin{aligned}rej=&265.150\,eqwidth-117.356\,depth+81.662\,length\\ &-5.229\log D+1358.090SR-1447.817\end{aligned}$$

where, eqwidth, depth and length have units in nanometres.

The QSAR model equation can be mechanistically interpreted; rejections increased with solute size; that was explained by the effect of length and equivalent width. The mechanism of steric hindrance due to size exclusion has been recognized as a main cause of rejection in many studies (Ozaki and Li, 2002; Nghiem *et al.*, 2005). It has been recognized by some studies (Kiso *et al.*, 2001; Kimura *et al.*, 2004) that hydrophobicity influenced rejection after adsorptive interactions with the membrane. In that sense, the QSAR model shows that log D decreased rejection, which shows that the effect of hydrophobicity reduces rejection due to adsorption and subsequent partitioning mechanisms. Interestingly, it was observed that log D was assuming the role of hydrophobicity for neutral and ionic compounds; compounds with high log D values adsorbed to the membrane and partitioned after saturation. On the other hand, ionic compounds presented log D values depending on the pH of the solution, thus did not adsorbed when log D was very low or negative. The role of depth in the equation compensated size exclusion contributions of length and equivalent width in a final rejection. The equation also shows that salt rejection was a parameter that incorporated steric/size hindrance and electrostatic repulsion effects related to charge of the membrane and operating conditions. Measured rejection values versus modelled rejections of the data used to generate the model are shown in Figure 2.

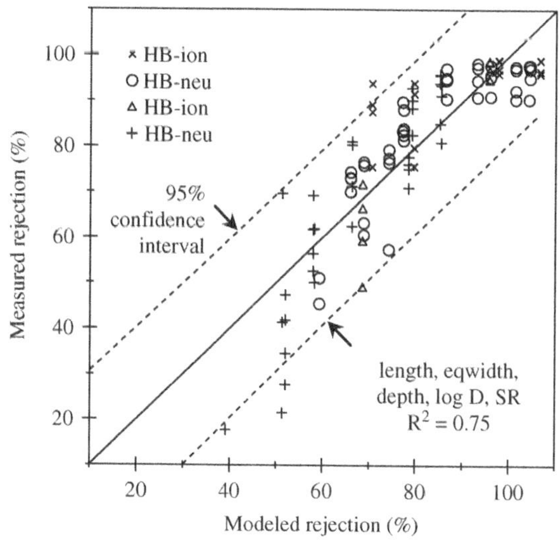

Figure 2. Measured vs. modeled rejections with QSAR model equation.

An external validation database was used to validate the QSAR model. Part of the database was compiled from information provided by Kim *et al.* (2007); for rejections of dichloroacetic acid, trichloroacetic acid, ibuprofen, diclofenac, clofibric acid, naproxen, chloroform, primidone, perchloroethene and carbon-tetrachloride by a Filmtec NF-90. Other part of the database was compiled from information provided by Verliefde *et al.* (2008) for rejections of 37 solutes (atenolol, atropine, clenbuterol, dikegulac, metoprolol, metribuzin, monomethyl-phtalate, n-acetyl-l-tyrosine, salbutamol, sotalol, terbutaline, 2-methoxyethanol, aminopyrine, antipyrine, cyclophosphamide, ethanol, glucose, metoxuron, pentox-ifylline, sucrose, clofibric acid, diclofenac, fenoprofen, gemfibrozil, ibuprofen, ketoprofen, naproxen, atrazine, bentazon, carbamazepine, chlorotoluron, diuron, estrone, isoproturon, monolinuron, simazin, metobromuron) by nanofiltration membranes Trisep TS-80 and Desal HL. Figure 3 shows plotted results of measured rejections vs. predicted rejections, an acceptable R^2 correlation coefficient was found (0.88).

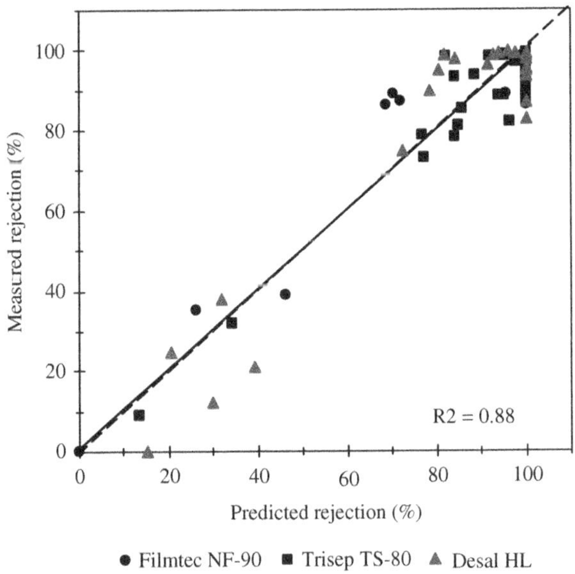

Figure 3. Measured rejections vs. predicted rejections for external validation data.

CONCLUSION

In conclusion, principal component analysis reduced the number of initial variables in 3 components describing three relations: membrane/operating-conditions (component 1: flux, pure water permeability, salt rejection, zeta potential, mass transfer coefficient, cross flow velocity), hydrophobicity/size (component 2: length, log K_{ow}, log D) and size (component 3: equivalent width, depth) explaining 89.3% of total variance. The subsequent use of multiple linear regressions after PCA was able to develop a linear equation that explains rejection. We also conclude that the QSAR model equation identified that the most important variables that influence rejection of organic solutes were log D, salt rejection, equivalent width, depth and length. Rejection increased by size/steric hindrance effects; but hydrophobicity decreased rejection due to adsorption and partitioning mechanisms. Salt rejection incorporated steric hindrance and electrostatic repulsion effects that were related to the membrane structure and operating conditions.

REFERENCES

Bellona, C. and Drewes, J.E. (2005). The role of membrane surface charge and solute physico-chemical properties in the rejection of organic acids by NF. *Journal of Membrane Science*, **249**(1–2), 227–234.

Boyd, G.R., Reemtsma, H., Grimm, D.A. and Mitra, S. (2003). Pharmaceuticals and personal care products (PPCPs) in surface and treated waters of Louisiana, USA and Ontario, Canada. *The Science of the Total Environment*, **311**(1–3), 135–149.

Drewes, J.E., Amy, G., Kim, T.-U., Xu, P., Bellona, C., Oedekoven, M. and Macalady, D. (2006). *Rejection of Wastewater-Derived Micropollutants in High-Pressure Membrane Applications Leading to Indirect Potable Reuse Effect of Membranes and Micropollutant Properties*, WateReuse, Alexandria, VA.

Hansch, C., Kurup, A., Garg, R. and Gao, H. (2001). Chem-bioinformatics and QSAR: A review of QSAR lacking positive hydrophobic terms. *Chemical Reviews*, **101**(3), 619–672.

Jones, O.A., Lester, J.N. and Voulvoulis, N. (2005). Pharmaceuticals: A threat to drinking water? *Trends in Biotechnology*, **23**(4), 163–167.

Kim, T.-U., Drewes, J.E., Summer, R.S. and Amy, G. (2007). Solute transport model for trace organic neutral and charged compounds through nanofiltration and reverse osmosis membranes. *Water Research*, **41**(17), 3977–3988.

Kimura, K., Amy, G., Drewes, J.E., Heberer, T., Kim, T.-U. and Watanabe, Y. (2003). Rejection of organic micropollutants (disinfection by-products, endocrine disrupting compounds, and pharmaceutically active compounds) by NF/RO membranes. *Journal of Membrane Science*, **227**(1–2), 113–121.

Kimura, K., Toshima, S., Amy, G. and Watanabe, Y. (2004). Rejection of neutral endocrine disrupting compounds (EDCs) and pharmaceutical active compounds (PhACs) by RO membranes. *Journal of Membrane Science*, **245**(1–2), 71–78.

Kiso, Y., Sugiura, Y., Kitao, T. and Nishimura, K. (2001). Effects of hydrophobicity and molecular size on rejection of aromatic pesticides with nanofiltration membranes. *Journal of Membrane Science*, **192**(1–2), 1–10.

Nghiem, L.D., Schäfer, A.I. and Elimelech, M. (2005). Pharmaceutical retention mechanisms by nanofiltration membranes. *Environmental Science & Technology*, **39**(19), 7698–7705.

Ozaki, H. and Li, H. (2002). Rejection of organic compounds by ultra-low pressure reverse osmosis membrane. *Water Research*, **36**(1), 123–130.

Pomati, F., Castiglioni, S., Zuccato, E., Fanelli, R., Vigetti, D., Rossetti, C. and Calamari, C. (2006). Effects of a complex mixture of therapeutic drugs at environmental levels on human embryonic cells. *Environmental Science & Technology*, **40**(7), 2442–2447.

Sacher, F., Lange, F.T., Brauch, H.-J. and Blankenhorn, I. (2001). Pharmaceuticals in groundwaters: Analytical methods and results of a monitoring program in Baden-Württemberg, Germany. *Journal of Chromatography A*, **938**(1–2), 199–210.

Sacher, F., Ehmann, M., Gabriel, S., Graf, C. and Brauch, H.-J. (2008). Pharmaceutical residues in the river Rhine—results of a one-decade monitoring programme. *Journal of Environmental Monitoring*, **10**(5), 664–670.

Sawyer, C.N., McCarty, P.L. and Parkin, G.F. (2003). *Chemistry for Environmental Engineering and Science*, McGraw-Hill, New York, pp. 479–520.

Verliefde, A., Cornelissen, E., Amy, G., Van der Bruggen, B. and Van Dijk, H. (2007). Priority organic micropollutants in water sources in Flanders and the Netherlands and assessment of removal possibilities with nanofiltration. *Environmental Pollution*, **146**(1), 281–289.

Verliefde, A.R.D., Cornelissen, E.R., Heijman, S.G.J., Verberk, J.Q.J.C., Amy, G.L., Van der Bruggen, B. and Van Dijk, J.C. (2008). The role of electrostatic interactions on the rejection of organic solutes in aqueous solutions with nanofiltration. *Journal of Membrane Science*, **322**(1), 52–66.

Yangali-Quintanilla, V., Sadmani, A., McConville, M., Kennedy, M. and Amy, G. (2009). Rejection of Pharmaceutically Active Compounds and Endocrine disrupting compounds by clean and fouled nanofiltration membranes. *Water Research*, in press, doi:10.1016/j.watres.2009.02.027.

Yoon, Y., Westerhoff, P., Snyder, S.S. and Wert, E.C. (2006). Nanofiltration and ultrafiltration of endocrine disrupting compounds, pharmaceuticals and personal care products. *Journal of Membrane Science*, **270**(1–2), 88–100.

Function and relevance of aquifer recharge techniques to enable sustainable water resources management in developing or newly-industrialized countries

G. Grützmacher[a], I. Hülshoff[a], B. Wiese[a], Y. Moreau-Le Golvan[a], C. Sprenger[b], G. Lorenzen[b] and A. Pekdeger[b]

[a]KompetenzZentrum Wasser Berlin, Cicerostr. 24, D-10709 Berlin
(E-mail: gesche.gruetzmacher@kompetenz-wasser.de)
[b]Freie Universität Berlin, Hydrogeologie, Malteserstr. 74 100, D-12249 Berlin

Abstract The use of bank filtration for drinking water treatment in Europe dates back to the days of beginning industrialization in the 19th century. With regard to improved source water quality in Europe, the millennium development goals and global climate change, aquifer recharge (AR) and bank filtration (BF) need to be reassessed in terms of sustainability and their role within an integrated water resource management. Based on the IC-NASRI study comprising 194 drinking water facilities worldwide integrating aquifer recharge techniques in their treatment system, a typical AR/BF site would be located in Central Europe alongside a river and is characterized by: a sandy gravel aquifer with a hydraulic conductivity of 2×10^{-3} m/s, a maximum aquifer thickness of 30 m, 175 m travel distance from bank to well, a travel time of 70 days and by vertical well operation with a daily capacity of 55.000 m^3. A literature survey conducted within the TECHNEAU project demonstrated that for substances highly relevant to newly-industrialized or developing countries (e.g. pathogens) the removal efficiency is good. Hydro-chemical analyses from three study sites in Delhi support these results. However, it was also shown that poor surface water quality, saline groundwater or subsurface conditions leading to mobilization of trace metals like iron, manganese or arsenic may limit the applicability of AR/BF without further post-treatment. Climate change might affect the performance of AR/BF worldwide, impairing source water quality and influencing removal efficiency. However, other factors like changes in demography or land-use can impact the systems by far more severely.

Keywords aquifer recharge, climate change, drinking water treatment, river bank filtration, vulnerability

INTRODUCTION

Bank filtration (BF) has been used for drinking water production in Europe since industrialization in the 19th century. First indications for a catchment of bank-filtered water were found in England as early as 1810 (BMI 1985). Whereas bank-filtered water was first abstracted unintentionally by pumping groundwater

 TECHNEAU: Safe Drinking Water from Source to Tap. Edited by Theo van den Hoven and Christian Kazner. ISBN: 9781843392750. Published by IWA Publishing, London, UK.

from alluvial aquifers (as believed to be still true for many sites in Europe and abroad), the systematic installation and operation of bank filtration schemes commenced in the 1870s. Since then, many modifications with respect to quality (post-treatment) and quantity (enhancement of recharged volumes by aquifer recharge via infiltration ponds) have been observed. However, bank filtration is still regarded as one of the most important steps within the multi-barrier approach recommended by the WHO and has been recognized as particularly useful for parasite removal in the USA in the past decade (Weiss *et al.*, 2005). Within the scope of improving source water quality in Central Europe, global climate change and millennium development goals, bank filtration and aquifer recharge need to be reassessed with respect to sustainability and their role within an integrated water resource management.

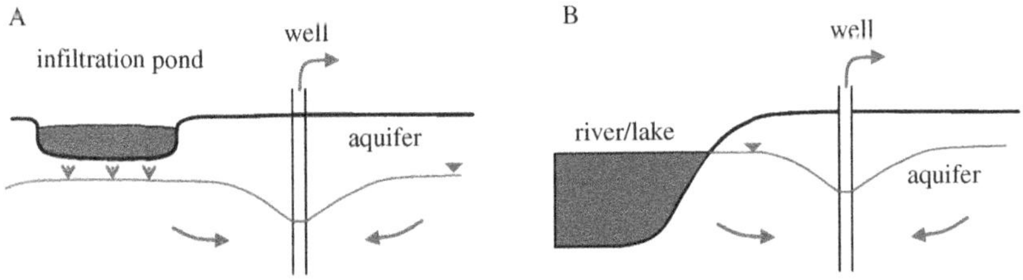

Figure 1. Generalized schemes for aquifer recharge (AR) via infiltration ponds (A) and bank filtration (B).

The EU co-funded project "TECHNEAU – Technology Enabled Universal Access to Safe Water" (2005–2010) addresses the major challenges that water supply is facing, including the function and relevance of aquifer recharge techniques with a main focus on bank filtration to enable sustainable water resources management, especially in developing or newly industrialized countries. The objective is to highlight the benefits and limitations of this technique to support water suppliers in their decisions to integrate MAR schemes into their system and to assess their vulnerability to global climate change. This is achieved by field investigations at three field sites in Delhi (India) carried out by the Free University of Berlin (FUB), the Indian Institute of Technology, Delhi (IITD) and by desk studies conducted by the Berlin Center for Competence (KWB). This paper presents intermediate results of the field investigations and is supplemented by results from the KWB project IC-NASRI with an emphasis on the BF field sites in Berlin (Wiese *et al.*, 2009).

APPLICATION OF AR AND BF AS DRINKING WATER (PRE-)TREATMENT WORLDWIDE

A review of the available technical information of 194 drinking water facilities integrating aquifer recharge and bank filtration was performed (Wiese *et al.*, 2009). The review should not be considered as exhaustive (since many other facilities are extracting water from alluvial aquifer, without referring to aquifer recharge), but it provides a first overview on the current situation of AR practices worldwide. The bulk of field sites identified is situated in Europe (70%) followed by North America (23%), Asia (4%), Australia (2%) and Africa (1%). Within Europe, most information was found on field sites in Germany, Netherlands, France and Hungary.

Among the 79 sites for which the type of recharge was given, bank filtration, artificial recharge via infiltration ponds and sprinkling/infiltration ditches accounted for 74%, 18% and 8% respectively. Other techniques like infiltration meadows or floodplains are of minor relevance.

The type of infiltrating surface water was identified for 178 sites. The majority of source water comes from river water (91%) and only 9% is derived from lake water. Sites using lake water are – with the exception of one site each in Australia and India – all located in Central and Northern Europe.

Out of the 62 AR/BF sites for which the recharged or abstracted volume is known, 23 are situated in Germany. The capacity of the field sites varies widely (Figure 2) between 200 m^3/d (pilot site at Hong Phong commune, Bac Binh District, Vietnam) to 600.000 m^3/d (Budapest, Szentendre Island, Hungary) with a median value of 55.000 m^3/d. The water works in Berlin, namely Spandau, Friedrichshagen, Beelitzhof and Tegel, are all well above this median value with capacities ranging from 160.000 to 250.000 m^3/d.

At the 81 field sites for which information on the abstraction method was available, the recharged water is usually extracted by wells (91%). In 9% of the analyzed cases, long horizontal subsurface tubes (9%) are used for abstraction. Among the sites for which the information on well type was available, 52% use vertical wells, 34% horizontal wells and 14% both. The type of well is closely linked to the average thickness of the aquifer with vertical wells usually being installed in thick aquifers (median value: 37 m) and horizontal wells in shallow aquifers (<22 m). The maximum aquifer thickness from 69 sites ranges between 5 and 360 m (see Figure 2). For the majority of sites (72%), the aquifer thickness is less than 50 m. It is remarkable that the wells at Csespel Island site (Budapest, Hungary) are situated in an aquifer of 15 m maximum thickness for its water production of 400 000 m^3/d. The maximum hydraulic conductivities from 62 sites are given in Figure 3 (median: 2×10^{-3} m/s). In Berlin, the maximum values (5×10^{-4} m/s) are distinctively lower.

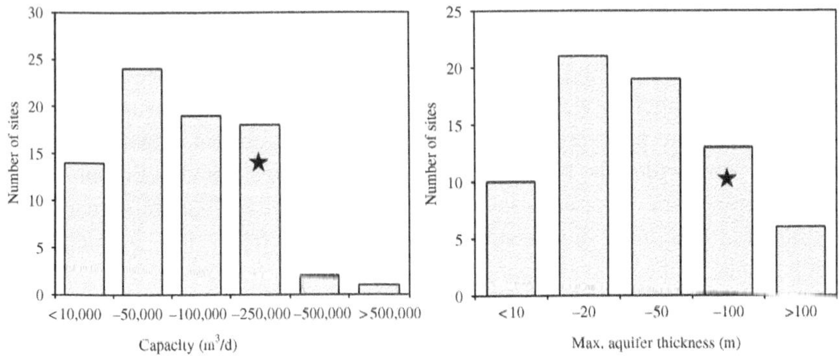

Figure 2. Capacity (left) and max. aquifer thickness (right) of 62 and 69 AR/BF sites worldwide (Wiese *et al.*, 2009). The water works in Berlin fall into the star-labeled categories.

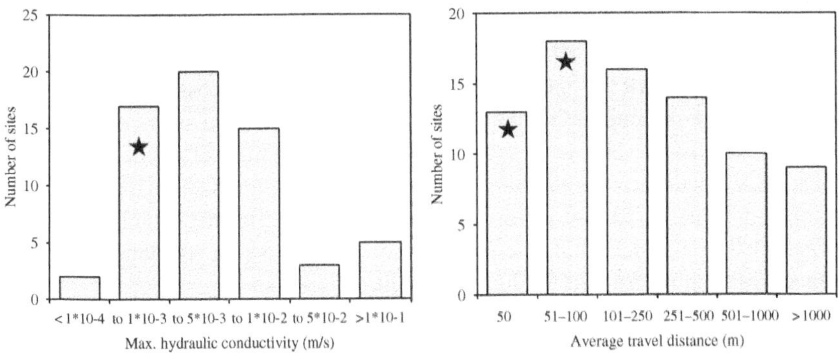

Figure 3. Maximum hydraulic conductivities (left) and average travel distance (right) of 62 and 80 AR/BF sites worldwide (Wiese *et al.*, 2009). The water works in Berlin fall into the star-labeled categories.

Geological information was given for 34 sites and 4 different types could be distinguished:

- sand-gravel sediments (61%),
- a mainly sandy ground (18%),
- mainly gravel (9%),
- and lime karst grounds of which all are located in the USA (12%).

For 72 field sites, information on the distance from abstraction well to bank was available. Where distances were given as range, the arithmetic mean was calculated. Out of the 72 field sites, 4% abstracted water after <20 m subsurface passage, 31% between 50 m and 100 m and 65% kept at least a distance of 100 m to the bank. Especially those field sites located in the Netherlands, Poland, Slovakia and Finland operate BF systems with a long distance to the bank. In the Netherlands, 81% of the sites keep at least a distance of 300 m. The Krajkowo site in Poland is located 565 m from the bank and in Slovakia the distance ranges from 250 m to 1750 m. The lowest travel distances were found for Hungary and the USA. The Hungarian river sites, which use Danube river water, have a distance between 30 m and 101 m and the sites in the USA have a distance between 23 and 131 m from the river bank. The sites in Berlin lie well below the median of 175 m.

Travel times are available for 39 field sites and range from 1 to 4000 days, with most travel times being between 10 and 50 days. Long travel times are especially found at the river Danube with 300 to 700 days in Slovakia and 5 to 1000 days in Hungary. No significant difference was observed between lake and river sites.

CHALLENGES IN NEWLY-INDUSTRIALIZED AND DEVELOPING COUNTRIES

Due to the easy implementation and little maintenance necessary, BF has been suggested as valuable drinking water treatment for developing and newly-industrialized countries. These countries are often located in tropical, semi-arid or arid climate zones being subject to weather extremes such as monsoon, floods and droughts. Scarce freshwater resources are further stressed by urbanization, intense land use and the rapid industrial growth seen in newly-industrialized countries.

It is well known that BF is particularly efficient to reduce pathogens, total suspended solids (TSS) and algal toxins (Table 1). These water quality parameters are of high relevance to developing and newly-industrialized countries where (i) riverine systems carry the bulk of suspended solids worldwide and (ii) eutrophic waters as well as warm temperatures promote the proliferation of certain pathogens and harmful algal blooms. In newly-industrialized countries, the presence of water contaminants originating from agricultural and industrial activities (pesticides, aromatic and chlorinated hydrocarbons, anthropogenic heavy metals) is an emerging concern. The efficiency of BF to remove organic and inorganic micropollutants depends on the sorption behaviour of the substance, redox conditions in the subsurface, microflora, nutrients and co-metabolites present. Oxic conditions are often found to be crucial for removal, e.g. for organic matter degradation or the sorption of heavy metals to Fe and Mn (hydroxy-)oxides. In comparison, anoxic conditions are less efficient and may render long retention times necessary. It is assumed that in developing and newly-industrialized countries, anoxic/ anaerobic groundwater conditions prevail due to warm temperatures and high organic load in source water, whereby the degradation of organic matter under oxygen consumption is promoted.

Table 1. Relevance of contaminants, BF removal potential and limitations (Hülshoff *et al.*, 2009).

Parameter	Relevance for developing countries	Relevance for newly-industrialized countries	BF removal/ possible quality limitation by
Suspended solids	high	high	good/clogging
Pathogens	high	high	good/short travel times
Algal Toxins	high	high	good/coarse aquifer
Nutrients (C,N,P)	high	high	good to moderate/anoxic or anaerobic passage only
Anions & Cations	low	low	poor/salinity, anoxic or anaerobic passage only
Heavy metals (geogenic)	high	high	site and substance specific
Heavy metals (anthropogenic)	low to moderate	high	site-specific/anoxic or anaerobic passage only
Pesticides	low to moderate	high	site and substance specific
Chlorinated hydrocarbons	low	high	moderate to poor/anoxic or anaerobic conditions
Aromatic hydrocarbons	low	high	moderate/long attenuation especially anoxic cond.
Endocrine disrupting chemicals (except pesticides)	low	high	good to moderate/anoxic or anaerobic passage only
Pharmaceuticals	low	low to moderate	site and substance specific
Disinfection by-products	low	moderate to high	site and substance specific/oxic (!) passage only

For BF operation, raw water quality and geochemical contaminants (e.g. arsenic) are to be taken into account. Certain substances present in surface water, which originate from sewage (e.g. ammonia), are not efficiently removed by subsurface passage. Despite indications that some conditions met in developing and newly-industrialized countries may impair cleaning capacity (e.g. clogging, anoxic/anaerobic conditions), riverbank filtration is regarded useful in such countries to remove contaminants from surface water and offer storage capacity in areas of high variations in precipitation and run-off.

CASE STUDY NEW DELHI (INDIA)

In order to assess the benefits and limitations of bank filtration under environmental conditions different from Central Europe, field investigations at three potential BF sites in Delhi (India) were carried out within the TECHNEAU project (Figure 4). In the course of a year, samples were taken monthly from newly installed observation wells arranged in a transsect perpendicular to the river (Yamuna river, Najafgarh drain). In addition to the analysis of drinking water relevant quality parameters, hydraulic investigations on hydraulic conductivities of the aquifer and the bank/river bed were carried out (Pekdeger *et al.*, 2008).

The suitability of BF as drinking water treatment (or pre-treatment) was shown to be very site-specific since the geology, hydrochemistry of the groundwater, and surface water quality vary to great extent. Factors identified that may locally limit the applicability of BF were:

(i) surface water quality: due to the high impact of poorly treated sewage and fertilizers on the Yamuna river at the Nizamuddin field site, ammonia concentrations reach values of up to 35 mg/L in river water. They are not removed during subsurface passage (median value in shallow groundwater: 16 mg/L),

(ii) groundwater quality: saline groundwater at Najafgarh field site with chloride concentrations of up to 3015 mg/L (for shallow groundwater) could render the bank filtrate objectionable for drinking water purpose,

(iii) aquifer geochemistry: mobilization of iron, manganese and arsenic was observed at field site Nizzamuddin making post-treatment necessary.

However, there are indications that pathogens and trace organics are removed efficiently, even after short travel times (5-log removal of bacteriophages after 5 m of travel distance, Sprenger *et al.*, 2008).

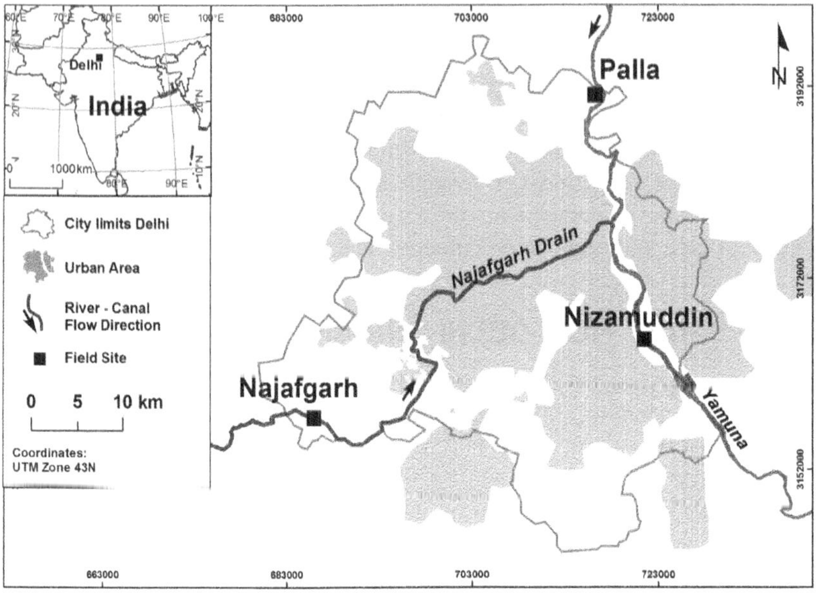

Figure 4. Location of the three investigated field sites in Delhi (Pekdeger *et al.*, 2008).

The identified limitations could be overcome by an optimized design of the bank filtration sites (e.g. achieving high shares of bank filtrate to avoid influence of highly saline groundwater) or by applying an adapted post-treatment.

VULNERABILITY OF BF TO CLIMATE CHANGE

Climate change is regarded an emerging hazard to water supply and yet, literature on the vulnerability of bank filtration (BF) is scarce. For Central Europe where the majority of BF sites is located, an increase in seasonal floods and droughts has been projected and is already experienced. Factors identified that determine BF operation and will be affected by climate change are temperature, raw water quality, river discharge and run-off as well as precipitation and groundwater recharge (Hülshoff *et al.*, 2009).

In a presumed 'dry' climate change scenario of high evaporation rates, low run-off and river discharge, low precipitation and groundwater recharge, the surface water quality will be stronger affected by point-source pollution (e.g. municipal effluent). Pollutants are likely to accumulate in streams due to a lowered dilution potential and water temperature will increase. In combination with low river discharge and a higher influence of municipal effluents, (potentially harmful) algal growth is promoted. Oxygen will not only be depleted in surface waters but also in the subsurface environment altering redox conditions. In a presumed 'wet' scenario of low evaporation rates, high run-off and river discharge, high precipitation and groundwater recharge, the surface water will be stronger affected by diffuse-source pollution (e.g. from agricultural, urban and industrial run-off). Further, sediments, suspended solids and associated

Table 2. Projected climate change effect on contaminant load in source water and BF removal efficiency under two major climate change scenarios (from Hülshoff *et al.*, 2009).

Substances	Removal efficiency of BF systems	a) Transport load in source water two major climate change scenarios		b) Assumed trend of BF removal efficiency under two major climate change scenarios	
scenario		wet[1]	dry[1]	wet[1]	dry[1]
pathogens	good	+		o	
suspended solids	good	+			
algal toxins	good	o	+		
DOC	good to moderate	+	+	o	o
NO_3^-	good to moderate	+	o		+
NH_4^+	good to moderate	+	+		o
anions & cations	poor		+		o
heavy metals	inconsistent	+	+		o
pesticides	inconsistent	+	o	o	
chlorinated hydrocarbons	moderate to poor	+		o	
aromatic hydrocarbons	moderate	+		o	o
endocrine disrupting chemicals	good to moderate	+		o	o
pharmaceuticals	inconsistent	+	+		
disinfection by-products	good			o	+

[1]symbols: 'o' = decreased, '+' increased, 'blank' = no change or uncertain

pathogens will be mobilized into the water phase to a larger extent. In case of flooding water production wells may become inundated, thus raising the chance of potential fecal contamination. An advantage of the 'wet' scenario is that high flow turbulence counteracts algal growth and oxygen depletion. The above-mentioned factors do influence the BF conditions such as redox potential and travel times, which are two major parameters determining removal capacity of BF systems. The majority of water contaminants is better removed (biodegraded and sorbed) under aerobic conditions and by long travel times. Droughts were seen to promote anaerobic BF conditions, while flood events can drastically shorten travel time and result in the breakthrough of pathogens. Sites using bank filtration can compensate weather extremes due to their storage capacity better than sites relying on surface or groundwater abstraction only. However, in both climate change scenarios, BF performance becomes impaired. It is furthermore assumed and has been illustrated by a case study for the BF site in Berlin that the anthropogenic influence on available freshwater quantity and quality is likely to have a larger impact than climate change.

SUMMARY AND CONCLUSIONS

Within the TECHNEAU project, investigations were carried out to analyze the function and relevance of MAR with special focus on bank filtration to enable safe drinking water production especially in newly-industrialized and developing countries.

A review of available information from water works producing drinking water from bank filtrate or aquifer recharge yielded high variations in capacities and natural settings. Based on median values for the respective parameters a typical AR/BF site would be situated in Central Europe along a river and is characterized by:

- 55.000 m^3/d capacity,
- vertical wells,
- 30 m maximum aquifer thickness,
- sand to gravel aquifer with hydraulic conductivities in the range of 2×10^{-3} m/s,
- 175 m distance between well and bank and a travel time of 70 days.

But also for newly-industrialized and developing countries, which are often situated in tropical, semi-arid or arid climate zones, BF may represent an option for safe drinking water production. This is mainly due to its potential to efficiently remove pathogens from surface water, which pose a major threat especially in less developed regions with little or no sewage treatment. Even though chlorination will most probably be necessary to avoid regrowth (in contrast to Berlin, where bank filtered water is not disinfected prior to use), the effort in terms of chemical consumption and/or energy will decrease significantly. On the other hand, certain substances present in surface water, which originate from sewage (e.g. ammonia), are not efficiently removed by subsurface passage. In addition, mobilization of iron and manganese as well as of certain trace metals (e.g. arsenic) needs to be taken into account under anoxic conditions. Finally, the majority of organic trace compounds are more efficiently removed under aerobic conditions. Anoxic/anaerobic conditions, which are believed to occur more frequently in warmer climate zones, may limit the applicability of this technique. Yet, redox conditions depend on a variety of other parameters (e.g. source water DOC, aquifer org. C), so that a system's cleaning capacity has to be evaluated site-specifically. In addition, an adjusted post-treatment can be designed to eliminate problematic substances. The combined system of BF/AR and adjusted post-treatment will follow the multi-barrier approach recommended by the WHO and increase the safety of drinking water supply.

In the course of climate change, source water quality is also believed to deteriorate at existing BF/AR sites in Central Europe. This is especially true for parameters related to increased surface run-off following extreme rain events (e.g. pathogens, suspended solids) but also for parameters that could increase during droughts (e.g. algal toxins). Although subsurface passage is particularly efficient in removing the mentioned major threats under the current conditions, their removal efficiency might also be affected by climate change (e.g. rising temperatures leading to prevailing anoxic conditions). However, it could be shown for an investigated case study (Berlin) that the variations resulting from demographic or land-use changes (though possibly indirectly a result of climate change) will impact drinking water production and the performance of AR/BF systems by far more severely than climatic changes.

ACKNOWLEDGEMENTS

The research work conducted within TECHNEAU at the KWB is co-funded by the European Commission under the Sixth Framework Programme, Sustainable Development, Global Change and Ecosystems Thematic Priority Area (contract number 018320) and Veolia Water. The project "IC-NASRI" benefits from the funding of Veolia Water. The Berlin Centre of Competence for Water and its partners are grateful for this financial support.

REFERENCES

BMI (1985). Künstliche Grundwasseranreicherung – Stand der Technik und des Wissens in der Bundesrepublik Deutschland. Berlin, Erich Schmidt Verlag GmbH.

Grischek, T., Schönheinz, D., Worch, E. and Hiscock, K. (2002). Bank filtration in Europe – An overview of aquifer conditions and hydraulic controls. In: *Management of Aquifer Recharge for Sustainability*, P.J. Dillon (ed.), Swets & Zeitlinger, Lisse.

Hülshoff, I., Greskowiak, J., Wiese, B. and Grützmacher, G. (2009). Relevance and opportunities of RBF systems in developing and newly-industrialised countries. TECHNEAU deliverable 5.2.9.

Hülshoff, I., Greskowiak, J. and Grützmacher, G. (2009). Analysis of the vulnerability of riverbank filtration systems to climate change by comparing their effectiveness under varying environmental conditions. TECHNEAU deliverable 5.2.3.

Pekdeger, A., Lorenzen, G. and Sprenger, C. (2008). Preliminary report on data of all inorganic substances and physicochemical parameters listed in the Indian and German Drinking Water Standards from surface water and groundwater at the 3 (+1) field sites. TECHNEAU deliverable 5.2.2.

Sprenger, C., Lorenzen, G., Grunert, A., Ronghang, M., Mittal, A., Selinka, H.-C., Girones, R. and Lopez-Pila, J.M. (2008). The Potential of River Bank Filtration For Reducing Chemical Pollutants and Pathogens from River Water in Mega-Cities: The New Delhi Experience. Paper submitted to *1st IWA Development Congress, Mexico City*, 06–09 September 2009.

Weiss, W.J., Bouwer, E.J., Aboytes, R., LeChevallier, M.W., O'Melia, C.R., Le, B.T. and Schwab, K.J. (2005). Riverbank filtration for control of microorganisms: results from field monitoring. *Water Research*, **39**, 1990–2001.

Wiese, B., Hülshoff, I., Orlikowski, D., Hoa, E., Grützmacher, G. and Yann, M.-L. (2009). Final report: Analysis of the role and sustainability of BF and AR. *IC-NASRI*.

Sensors and Monitoring

State-of-the-art in drinking water monitoring

F. Sacher and B. Hambsch

DVGW-Technologiezentrum Wasser (TZW), Karlsruher Strasse 84, 76139 Karlsruhe, Germany (E-mail: sacher@tzw.de)

Abstract This paper gives an overview about recent developments of analytical instrumentation and monitoring technologies and their implications for drinking water surveillance from the authors' perspective. Several trends have been identified which are briefly discussed such as increasing automation, the introduction of high-end technologies like the coupling of liquid chromatography and mass spectrometry or the use of high resolution mass spectrometers for the identification of unknown contaminants or transformation products. With respect to microbiological parameters, culture-based methods are still the golden standard in drinking water analysis and novel techniques such as polymer-chain reaction haven't yet overcome research status. Due to the ever increasing number of chemical contaminants, effect-related methods become more and more important. Modern bioassays allow for the detection of adverse effects in water samples like e.g. cell toxicity, genotoxicity, or estrogenic activity on a molecular or cellular level. With respect to continuous monitoring of raw and drinking waters and to operational control of treatment processes, on-line technologies have already entered the market. Such on-line technologies include on one hand the continuous determination of parameters like conductivity or UV absorption and on the other hand the surveillance of water quality by biological test systems like fish biomonitors or mussel tests.

Keywords analytical technologies, chemical analysis, drinking water monitoring, effect-related analysis, emerging pollutants, microbiological parameters, on-line monitoring

INTRODUCTION

Fast, sensitive and reliable monitoring and control technologies are indispensable for the production of safe drinking water. They allow for the surveillance of source water quality and the detection of biological and chemical threats, thus defining the boundary conditions for the subsequent treatment and providing early-warning in case of unexpected contaminations. They are mandatory for the permanent surveillance of the overall treatment process and the efficacy of each single treatment step, and they safeguard the high quality of the finished water. Furthermore, appropriate analytical techniques are indispensable for the detection of changes in water quality during distribution and for monitoring drinking water quality at consumers' tap. In addition to providing valuable information for

 TECHNEAU: Safe Drinking Water from Source to Tap. Edited by Theo van den Hoven and Christian Kazner. ISBN: 9781843392750. Published by IWA Publishing, London, UK.

water suppliers and health authorities, reliable monitoring technologies contribute to a large extent to consumers' trust in a high drinking water quality.

Today, requirements for state-of-the-art monitoring technologies are enormous: They should offer the ability to detect a large number of biological and chemical contaminants whereby the list of target analytes is increasing day-to-day. They should give immediate answer to any disturbance in source water quality as well as to any malfunction within the treatment process. Early-warning systems may be used by water suppliers to detect sudden changes in source water quality and may provide rapid information necessary to implement appropriate responses such as closing intakes or adjusting treatment parameters. Especially river waters are susceptible to rapid changes in quality as a result of accidental, intentional, or natural contamination. To protect consumers from potentially harmful contaminants, to avoid disruptions in the treatment process, and to ensure compliance with regulations, water suppliers must respond rapidly to spills and sudden pollution events and make appropriate adjustments in drinking water treatment and operations. In addition to this, monitoring and control technologies should be easy to use, should require a low level of maintenance but for all that should be robust and sensitive...and cheap. Requirements for appropriate monitoring technologies however cannot be generalised but have to be defined on a case-to-case basis taking into account the special situation of a water supply including vulnerability of raw water sources, size of the water supply, and status of the treatment train and the distribution system.

Looking at the scientific developments of analytical instrumentation and monitoring technologies and their implications for drinking water surveillance during the last decade, several trends can be identified which will be briefly discussed in the following chapters.

CHEMICAL PARAMETERS

With respect to the analysis of basic chemical water quality parameters like temperature, pH, conductivity, or alkalinity, but also for routine analysis of classical cations and anions like calcium and magnesium, chloride, nitrate and sulphate, a clear trend towards automation can be observed and instruments like flow-injection analysers or robot-like systems have found their way into modern water laboratories. These systems allow for the rapid analysis of a large number of samples with low levels of manual work. In many cases, the results of these automated analysers are directly fed into a laboratory information system where further data manipulation takes place. Thus, involvement of laboratory technicians into the analytical procedure for the determination of these basic parameters is reduced to a minimum.

With respect to conventional techniques for chemical analysis of target compounds like gas chromatography or liquid chromatography it can be stated that the introduction of mass spectrometric detection methods into water laboratories is almost complete, resulting in a significant improvement of reliability of analytical data. Especially the coupling between liquid chromatography and mass spectrometry (HPLC-MS) was a remarkable step forward, opening the analytical window for the trace-level determination of a multitude of new analytes which exhibit rather polar chemical structures and thus are not amenable to other analytical technologies. Due to their high polarity, however, these compounds might pose a particular challenge to drinking water utilities as removal can not easily be achieved. Without the possibilities offered by HPLC-MS coupling the recent developments in the emerging field of pharmaceuticals in the environment (including source and drinking waters) would have been unimaginable. The increasing sensitivity of the high-end versions of these systems offers on the one hand the possibility to analyse water samples without sample pre-treatment and thus allows for the determination of analytes for which no appropriate pre-concentration technique is available and on the other hand enables the detection of ever lower concentrations of single compounds. For some analytes, limits of quantification in the sub ng/L range are feasible already now. Although it has to be critically discussed whether such low detection limits are really needed for all types of compounds, there are definitely questions that can only be answered by application of such elevated analytical instrumentation. Looking e.g. at the Environmental Quality Standards for some of the priority substances defined in the EU Water Framework Directive or at the no-effect levels of some endocrine disrupting compounds, a need for such low limits of quantification can be seen.

Although recent developments in gas chromatography (GC) have not been as spectacular as in LC-MS, several improvements could be observed. The possibilities of large-volume injection has lead to an increasing sensitivity also in GC-MS analysis which e.g. for the carcinogenic nitrosamines is definitely required. Novel extraction technologies like solid-phase micro-extraction (SPME) (Pawliszyn, 1997; Keller *et al.*, 2000) or stir bar sorptive extraction (SBSE) (Sandra *et al.*, 2000) on one hand allow for a more sensitive determination of a multitude of analytes and on the other hand can rather easily be linked to conventional GC-MS systems and thus can be used for further automation of the analytical procedures.

In addition to monitoring raw and drinking water for chemical target compounds, a new analytical discipline has recently approached on the horizon: The identification of new (or yet unknown) analytes by enhanced analytical tools like high-resolution mass spectrometers. Innovative new instruments like time-of-flight mass spectrometers (Cotter, 1994) or the orbitrap system (Hu *et al.*, 2005) offer possibilities to accurately determine the mass of an unknown compound and thus facilitate its identification. It can be expected that in the near future such techniques, although rather expensive, will be applied more and more in modern water laboratories, e.g. for the identification of transformation products formed during treatment (e.g. biological treatment, oxidative treatment, disinfection).

MICROBIOLOGICAL PARAMETERS

Since decades, the determination of microbiological parameters in drinking water focuses on the detection of pathogens by indicator organisms. Most water-borne pathogens are transmitted by drinking water via the faecal-oral route. For these faecal contaminations the indicator principle has proven to be very effective. Therefore the EC Drinking Water Directive defines the indicator bacteria *E. coli* and *intestinal enterococci* as microbiological parameters which must not be present in 100 mL. For the monitoring of these microbiological parameters in drinking water, the classical culture-based methods are still the golden standard. Cultural analysis is usually completed within one or two days and although it is not an on-line method it can provide a quasi-continuous overview of water quality when analysed with a high frequency. In the presence of free chlorine, however, the use of *E. coli* and *enterococci* as indicators for pathogenic microorganisms is limited as some of those are more resistant to chlorine, e.g. cryptosporidium, giardia or some viruses.

During the last few years, there is a growing demand for faster results of microbiological analysis. In some cases, molecular techniques like those based on polymer chain reaction (PCR) can be used instead of cultural methods. So far, however, there are limitations for their routine use in drinking water monitoring, such as lack of sensitivity on one hand and the detection of dead cells (which are not relevant with respect to drinking water quality) on the other. Moreover, these molecular methods are far from being standardized. These techniques are currently used in research and as screening methods for some parameters which are extremely time-consuming to be done in culture (e.g. Legionella). Additionally, PCR-based methods are used in such cases, where no cultural methods exist, e.g. for the detection of certain viruses.

Further improvements and developments in monitoring of microbiological parameters can be expected during the next few years. Flow cytometry e.g. might become a future tool in drinking water microbiology, since the method is fast and quantitative. It can be used for the direct enumeration of total cell concentrations, the detection of specific cellular features like viability, or the enumeration of specifically targeted cells after staining with fluorescence-labelled antibodies or oligonucleotides (Amann and Fuchs, 2008; Hammes *et al.*, 2008).

For special needs, for instance the detection of specific pathogens, future analysis can be done by molecular methods. However, as cultural methods are sensitive and robust, it can be assumed that culture-based methods and modern culture-independent methods will exist in parallel.

EFFECT-RELATED ANALYSIS

Conventional analytical techniques allow for the analysis of an ever increasing number of chemical compounds. But even by monitoring thousands of single chemicals it's impossible to completely exclude that a water sample causes any harmful effect in humans or wild-life. That's where effect-related monitoring technologies come into play. These biological test methods allow for the detection of adverse effects in water samples like e.g. cell toxicity, genotoxicity, or estrogenic activity without determining the chemical composition of the water under investigation.

In-vitro bioassays are becoming increasingly popular as screening tools in environmental monitoring because they can provide complementary information to chemical analysis. Such bioassays can detect biologically-active compounds based on their effect at the molecular or cellular level and do not require prior knowledge of the chemical structure of the contaminant (Leusch, 2008). However while bioassays have undoubtedly proven their worth in pharmaceutical and laboratory conditions to test the biological activity of individual compounds, there are still many questions about their usefulness with environmental samples. Bioassay results may be affected by other unrelated toxic effects from complex mixtures, and there are still questions about the reliability and robustness of bioassays in such conditions (Leusch, 2008).

A practical example of recent application of bioassays can be found in water monitoring for a variety of endocrine disrupting compounds (van der Linden *et al.*, 2008). In this study bioassays have been particularly useful since they enable to measure the combined endocrine activity of compounds, even when present in a complex environmental mixture and the identity of the compounds of interest is unknown. So far, most of the attention in endocrine disruption research has been paid towards male and female sex steroid hormones, the androgens and estrogens. It has been found that the receptors of these steroid hormones can also be easily activated or inactivated by a variety of unnatural chemicals like drugs and several industrial chemicals. While other steroid hormones, like progestins and glucocorticoids also are of major importance in regulating normal physiological functions, these have yet received little attention in the context of endocrine disruption. One reason for this is the paucity of good biological detection systems for compounds interfering with these hormonal systems. Within the EU funded TECHNEAU project, the existing panel of bioassays for the detection of estrogens and androgens was expanded with assays responding to other classes of hormones, including glucocorticoids, progestins and thyroid hormones. A panel of bioassays was applied to detect multiple types of hormone-like activity in a range of water samples, including surface water, drinking water and several types of wastewater (van der Linden *et al.*, 2008). The results show that the levels of estrogens are in the normal range for surface waters and effluents, but also that all other types of hormonal activity can be detected in these water samples. In particular, the surprisingly high levels of glucocorticoids found compared to the other types of activity are of importance.

ON-LINE MONITORING

In addition to the above mentioned analytical technologies which are most often laboratory-based and provide information only after some time of sample transport, sample analysis, and data evaluation, the application of on-line technologies can provide fast, almost in-time water quality information. Such immediate response is especially needed for early-warning purposes, e.g. to instantly detect sudden contaminations of source waters or malfunctions of a treatment step within the drinking water production process.

In the last years several approaches for on-line monitoring systems have entered the market. Most of them monitor rather basic water quality parameters such as conductivity or UV absorption which might be linked by intelligent software to more elaborated water quality parameters. A recent example for such an on-line monitoring system is the application of a set of two UV/Vis-spectrometers for the measurement of dissolved ozone and assimible organic carbon (AOC) in a drinking water treatment plant (van der Broeke *et al.*, 2008). On-line surveillance of treatment steps is also in-line with the WHO Water Safety Plan approach which asks for a continuous process control as complement to monitoring the quality of the finished drinking water.

With respect to the continuous surveillance of raw and drinking waters for toxic compounds biological test systems can be used in addition to chemical analysis. On-line biomonitors are able to rapidly detect a large variety of toxic compounds and thus might give immediate response to any deterioration in water quality as a result of accidental, intentional, or natural contamination. Biological test systems make use of cells or whole organisms that might be affected by toxic compounds. Systems currently used include tests with bacteria, algae, daphnia, mussels, or fish (Sagner *et al.*, 2008). Parameters monitored are either markers of biological activity such as oxygen consumption, carbon dioxide production or bioluminescence, or sudden changes in the behaviour of the organism. Many of these systems have been used for years for real-time surveillance of wastewaters or river waters (von Danwitz *et al.*, 1992). When transferring the systems to drinking water utilities, however, it has to be taken into account that the ambient conditions are different and that biological systems might be very sensitive to these differences. The limited availability of substrate in drinking waters e.g. might easily affect the results of a biological test. Furthermore, as most of these test systems exhibit different sensitivities towards different chemicals, the application of a whole test battery that consists of different biotests might be recommendable.

Regarding the practical application of biological test systems for drinking water monitoring, there are still some doubts about the sensitivity and the reliability of the existing systems. Recent experiences however prove that especially the reliability has been significantly improved and that false positive results can almost be excluded. Detection limits of most biomonitors are in the μg/L to mg/L range, depending on the test organisms and the specific mode of action of the compound under investigation (Sagner *et al.*, 2008). This might be sufficient to detect an intentional contamination of drinking waters; for routine monitoring and especially for comparison to chemical analyses, however, a higher sensitivity is desirable.

Although possibilities for on-line monitoring of organic micro-pollutants are still limited several attempts have been undertaken to develop a water-monitoring device, which is robust, cost-effective, automated and able to measure several tens of organic pollutants at low nanogram per litre level in a short time, preferably without any time-consuming sample concentration step and prior sample pre-treatment. In the EU funded project AWACSS (Automated Water Analyser Computer Supported System), a multi-analyte immunoassay-based system has been constructed and successfully tested for analysis of aqueous samples (Tschmelak *et al.*, 2004a and 2004b; Proll *et al.*, 2005). Two prototypes of the biosensor system were designed and constructed within the project and over 20 antibodies and their analyte derivatives were developed. Analytical performance of the systems was tested in comparison to a wide range of conventional chromatography-based analytical techniques. The results showed that the biosensor system is fully comparable to conventional analytical techniques in terms of accuracy, repeatability and reproducibility. Detection limits of all tested analytes were in the low nanogram per litre range, while selectivity allowed for trace analysis even in complex matrices such as sediment extracts. Time of a single analytical run was less than 18 min and during the system validation more than 70 analyses were performed within a day in a fully automated regime.

CONCLUSIONS

State-of-the-art drinking water monitoring is always affected by recent results of scientific research on the one hand and by latest developments in analytical instrumentation on the other. In many cases, progress in scientific research and improvements in analytical instrumentation are closely linked to each other. The detection of "new" or emerging contaminants triggers the development of new analytical instruments while improvements in analytical instrumentation lead to the detection of new compounds. It can be foreseen that the development of more sensitive and more reliable instruments will continue in the coming years. Key-drivers for this process will be the competition between manufacturers and between different research groups. It will be interesting to see how regulation deals with the detection of ever new compounds and the evaluation of the relevance of these compounds, especially with regard to human and eco toxicity will be a challenging task.

One solution to this problem might be the increasing application of effect-related analytical methods which allow for the detection of adverse effects of water samples. This would ask however for more reliable and more sensitive test systems which can be used on a routine basis.

Another development which can be foreseen is the increasing demand for on-line control systems, either for simple instruments which can be used for continuous process control or for elaborated instrumentation which allow for the real-time surveillance of drinking water quality and thus contribute to the protection of the consumer.

REFERENCES

Amann, R. and Fuchs, B.M. (2008). Single cell identification in microbial communities by improved fluorescence in situ hybridization techniques. *Nature Reviews Microbiol.*, **6**(5), 339–348.

Cotter, R.J. (1994). *Time-of-Flight Mass Spectrometry*. American Chemical Society, Columbus, OH.

Hammes, F., Berney, M., Yingying, W., Vital, M., Köster, O. and Egli, T. (2008). Flow-cytometric total bacterial cell counts as a descriptive microbiological parameter for drinking water treatment processes. *Water Research*, **42**(1–2), 269–277.

Hu, O., Noll, R.J., Li, H., Makarov, A., Hardman, M. and Cooks, R.G. (2005). The Orbitrap: A new mass spectrometer. *Journal of Mass Spectrometry*, **40**(4), 430–443.

Keller, B., Merkel, T., Sacher, F., Maier, M. and Maier, D. (2000). Einsatz der Festphasenmikroextraktion zur online-Überwachung der Trinkwasserqualität (Solid Phase Micro Extraction for the Online Monitoring of Drinking Water Quality). *Vom Wasser*, **94**, 213–227.

Leusch, F.D.L. (2008). *Tools to detect estrogenic activity in environmental waters.* Report of the Global Water Reseach Coalition (GWRC), London, UK.

Pawliszyn, J. (1997). *Solid Phase Microextraction – Theory and Practice*. Wiley-VCH, New York.

Proll, G., Tschmelak, J. and Gauglitz, G. (2005). Fully automated biosensors for water analysis. *Analy. Bioanalyti. Chem.*, **381**, 61–63.

Sagner, A., Zawadsky, C. and Tiehm, A. (2008). *Toxizitätstests zur Überwachung von Trinkwasser* (*Toxicity tests for drinking water monitoring*). Final report of DVGW research project W7/01/06, Karlsruhe, Germany.

Sandra, P., Baltussen, E., David, F. and Hoffmann, A. (2000). *Stir Bar Sorptive Extraction* (*SBSE*) applied to Environmental Aqueous Samples. Gerstel Application Note 2/2000, Mülheim, Germany.

Tschmelak, J., Proll, G., Kaiser, J., Kraemmer, P., Bárzaga, L., Wilkinson, J., Hua P., Hole, P., Nudd, R., Jackson, M., Abuknesha, R., Barceló, D., Rodriguez, S., Sacher, F., Stien, J., Slobodník, J., Oswald, P., Kozmenko, H., Korenková, E., Tóthová, L., Krascsenitcs, Z. and Gauglitz, G. (2004a). Automated water analyser computer supported system (AWACSS) – Part I: Project objectives, basic technology, immunoassay development, software design & networking. *Biosensors and Bioelectronics*, **20**(8), 1499–1508.

Tschmelak, J., Proll, G., Kaiser, J., Kraemmer, P., Bárzaga, L., Wilkinson, J., Hua, P., Hole, P., Nudd, R., Jackson, M., Abuknesha, R., Barceló, D., Rodriguez, S., Sacher, F., Stien, J., Slobodník, J., Oswald, P., Kozmenko, H., Korenková, E., Tóthová, L., Krascsenitcs, Z. and Gauglitz, G. (2004b). Automated water analyser computer supported system (AWACSS) – Part II: Intelligent, remote-controlled, cost-effective, on-line, water-monitoring measurement system. *Biosensors and Bioelectronics*, **20**(8), 1509–1519.

van der Broeke, J., Ross, P.S., van der Helm, A.W.C., Baars, E.T. and Rietveld, L.C. (2008). Use of on-line UV/Vis-spectrometry in the measurement of dissolved ozone and AOC concentrations in drinking water treatment. *Water Science and Technology*, **57**(8), 1169–1175.

van der Linden, S.C., Heringa, M.B., Man, H.-Y., Sonneveld, E., Puijker, L.M., Brouwer, A. and van der Burg, B. (2008). Detection of multiple hormonal activities in wastewater effluents and surface water, using a panel of steroid receptor CALUX bioassays. *Environmental Science and Technology*, **42**(15), 5814–5820.

von Danwitz, B., Merschhemke, C., Löbbel, H. and Stein, P. (1992). Zeitnahe Gewässerüberwachung mit kontinuierlichen Biotests (Real-time river water monitoring with continuous biotests). *Gewässerschutz Wasser Abwasser*, **134**, 55–65.

Comparison between conventional and comprehensive GC for the search of 58 target compounds and screening of undesirable pollutants in an environmental complex matrix

G. Semard[a,b], A. Bruchet[a], P. Cardinaël[b], V. Peulon[b] and J.-P. Bouillon[b]

[a]CIRSEE (Centre International de Recherche Sur l'Eau et l'Environnement), Suez-Environnement, 38, rue président Wilson, 78230 Le Pecq, France
(E-mail: auguste.bruchet@suez-env.com)
[b]Laboratoire Sciences et Méthodes Séparatives (EA 3233), Université de Rouen, IRCOF, F-76821 Mont Saint Aignan cedex, France

Abstract This study reports on the separation of 58 priority target compounds in a raw waste water extract. After comparison of a standard mixture separation by conventional GC and comprehensive GC (GC×GC), the optimization of the analytical conditions has been carried out by GC×GC-FID. Then a raw waste water sample collected at the inlet of a urban waste water treatment plant has been analysed by GC/MS and GC×GC-TOFMS. GC×GC has proven superior to GC both for the identification of target compounds and the screening of non target pollutants of health concern present in this extract. GC×GC appears as the method of choice for the broad screening of a wide array of target and non target compounds in complex environmental matrices.

Keywords complex environmental matrix, comprehensive GC, GC/MS, GC×GC-FID, GC×GC-TOFMS, raw waste water

INTRODUCTION

The European legislation aims by 2015 at restoring a «good ecological state» of natural waters. In order to reach this objective, dangerous priority substances in these waters must be eliminated. For this reason, a list of 33 priority substances has been published in annex X of the European Water Framework Directive (2000) (EC, 2000). A follow up directive published environmental Quality Standards (EQS) for these 33 plus 8 new priority substances (EC, 2008). The European Parliament has also published a regulation which has set the threshold of rejection in waters, air and soils for 71 pollutants (EC, 2006). By combining the three legislations, a list of 58 target compounds was established. These compounds have to be quantified in waters at low levels (ng/L). Table 1 presents the list of these 58 target compounds

Table 1. List of 58 target compounds which have to be measured in water.

Family	Compound	Environmental quality standards (EC, 2008)
Pesticides		
Organochlorinated insecticides	Aldrin	SUM = 0.01 μg/L
	Dieldrin	
	Endrin	
	Isodrin	
	α-Endosulfan	0.005 μg/L
	β-Endosulfan	0.005 μg/L
	2,4′-DDT	SUM = 0.025
	4,4′-DDT	4,4′-DDT = 0.010
	Heptachlor	1 kg/year
	α-hexachlorocyclohexane	0.1 μg/L
	β-hexachlorocyclohexane	
	γ-hexachlorocyclohexane (lindane)	
	δ-hexachlorocyclohexane	0.02 μg/L
	ε-hexachlorocyclohexane	0.02 μg/L
	Chlordane	1 kg/year
	Chlordecone	1 kg/year
	Mirex	1 kg/year
Organophosphorus insecticides	Chlorfenvinphos	0.1 μg/L
	Ethyl Chlorpyrifos	0.03 μg/L
	Methyl Chlorpyrifos	0.03 μg/L
Herbicides	Alachlor	0.3 μg/L
	Atrazine	0.6 μg/L
	Simazine	1 μg/L
	Trifluraline	0.03 μg/L
Fungicides	Hexachlorobenzene	0.01 μg/L

Table 1. Continued

Family	Compound	Environmental quality standards (EC, 2008)
Industrials Pollutants		
Chlorobenzenes	1,2,3-trichlorobenzene	0.4 μg/L
	1,2,4-trichlorobenzene	0.4 μg/L
	1,3,5-trichlorobenzene	0.4 μg/L
	Pentachlorobenzene	0.007 μg/L
Phenol	4-tertiobutylphenol	
Organochlorinated	Hexachlorobutadiene	0.1 μg/L
Phthalate	Phthalic acid, bis-2-ethylhexyl (DEHP)	1.3 μg/L
Heterocyclic compound	Benzothiazole	
Phosphate	Tributylphosphate	
PAH	Naphtalene	2.4 μg/L
	Anthracene	0.1 μg/L
	Fluoranthene	0.1 μg/L
	Benzo(b)fluoranthene	SUM = 0.03 μg/L
	Benzo(k)fluoranthene	
	Benzo(a)pyrene	0.05 μg/L
	Indeno(1,2,3-cd)pyrene	SUM = 0.02 μg/L
	Benzo(g,h,l)perylene	
Polybrominated diphenyl ethers (PBDE)	BDE-28	0.0005 μg/L
	BDE-47	0.0005 μg/L
	BDE-99	0.0005 μg/L
	BDE-100	0.0005 μg/L
	BDE-153	0.0005 μg/L
	BDE-154	0.0005 μg/L
	BDE-183	0.0005 μg/L
	BDE-205	0.0005 μg/L
	BDE-209	0.0005 μg/L

(*continued*)

Table 1. Continued

Family	Compound	Environmental quality standards (EC, 2008)
Polychlorinated biphenyls (PCB)	PCB-28	0.1 kg/year
	PCB-47	0.1 kg/year
	PCB-101	0.1 kg/year
	PCB-118	0.1 kg/year
	PCB-138	0.1 kg/year
	PCB-153	0.1 kg/year
	PCB-180	0.1 kg/year

with their respective environmental quality standards. The list contains 8 polynuclear aromatic hydrocarbons (PAH), pesticides, 9 polybrominated diphenylethers (PBDE) and 7 polychlorinated biphenyls (PCB). Some of these components are considered as carcinogenic by the World Health Organization (WHO) and the European Union (Danielsson *et al.*, 2005). Additional compounds considered as relevant including tributylphosphate, 4-tert-butylphenol and benzothiazole which are commonly found in waste waters, were also included. The combined analysis of these substances near their limits of detection by conventional chromatographic methods (7.0 ng/L for pentachlorobenzene, 5.0 ng/L for endosulfan, 10 ng/L for the sum of endrin, aldrin, dieldrin and isodrin, between 10 and 100 ng/L for the other substances) remains a challenge. The separation of these compounds in a complex matrix which may contain more than 1000 other organic compounds, overtakes the resolution of the currently used 30 to 60 m capillary gas chromatographic columns. Hence comprehensive two-dimensional gas chromatography (GC × GC) which provides a high peak capacity could be considered as an alternative technique to overcome these difficulties.

Conventional GC has been widely used during the 3 past decades with clean-up steps to analyse environmental matrices. Nevertheless, this technique is limited by its low peak capacity for complex mixtures. Multi-dimensional gas chromatography which allowed separation improvements is not suitable when numerous coelutions are observed because only a small part of the sample can be diverted to a second separation column (Phillips and Xu, 1995). A better alternative is to subject the sample to a GC × GC separation. GC × GC was developed by Liu and Phillips in 1991 (Liu and Phillips, 1991). It involves two columns connected in series separated by a modulator which focuses the analytes eluted from the first column and reinjects them into the second one. This process is repeated every few seconds, thus the entire sample can be separated on two columns which possess different geometrical parameters and selectivities. The increase in sensitivity and the high peak capacity of GC × GC enable to improve the separation of complex matrices and hence the detection of target compounds into these matrices. In the environment field, GC × GC has already been applied to separate fractions of BTEX (benzene toluene ethylbenzene xylene) and MTBE (methyl tert-butyl ether) in water (Gaines *et al.*, 1998), PAH and PCB in sediments (Hyötiläinen *et al.*, 2002) and nonylphenols in river samples (Ieda *et al.*, 2005).

The aim of this study is to show the potential of GC × GC for the analysis of complex environmental matrices. The first part of the study deals with the comparison of separations of a 58 target compounds mixture (standard mixture) by GC-FID and GC × GC-FID. For GC × GC-FID both chromatographic conditions and modulation time were optimized. The second part reports on the search for the 58 target compounds in a urban waste water collected at the inlet of a waste water treatment plant. A screening of other pollutants present in this extract has also been carried out. Chemical identification was also completed both by GC/MS and GC × GC-TOFMS for comparison purposes.

EXPERIMENTAL

Samples

Pure reference standards of aldrin, dieldrin, 2,4′-DDT (dichlorodiphenyltrichloroethane), 4,4′-DDT, DEHP (bis-2-ethylhexyl phtalic acid), α-endosulfan, β-endosulfan, endrin, heptachlor, technical grade hexachlorocyclohexane (HCH), α-HCH, β-HCH, lindane, δ-HCH, technical grade chlordane, chlordecone hydrate, isodrin, mirex, chlorfenvinphos, ethyl chlorpyrifos, methyl chlorpyrifos, alachlor, atrazine, simazine, trifluralin, hexachlorobenzene, hexachlorobutadiene, 1,2,3-trichlorobenzene, 1,2,4-trichlorobenzene, 1,3,5-trichlorobenzene, pentachlorobenzene, naphtalene, anthracene, fluoranthene, benzo(b)fluoranthene, benzo(k)fluoranthene, benzo(a)pyrene, indeno(1,2,3-cd)pyrene, benzo(g,h,i)perylene were supplied by Cluzeau Info Labo (Ste Foy la Grande, France). Nine PBDEs and seven PCBs were purchased from Interchim (Montluçon, France). Pure standards of benzothiazole, 4-tert-butylphenol and tributylphosphate were supplied by Sigma-Aldrich (St Quentin Fallavier, France). The methylene chloride solvent was Pestipur purity grade from SDS (Val de Reuil, France).

The standard mixture containing the 58 target compounds was prepared in methylene chloride at 1ng/μL and stored at −18°C.

For the analysis of waste water by GC/MS and GC × GC-TOFMS, 500 mL of raw waste water have been sampled at the inlet of a urban waste water treatment plant. The non filtered sample was extracted at basic pH (pH = 11) with methylene chloride (2×100 mL) in order to recover the base-neutral fraction. Then, the extract was concentrated to 1 mL under a nitrogen stream to be directly injected in GC and in GC×GC. No purification steps have been carried out on the sample extract to avoid losing traces of compounds.

Instruments

Conventional and comprehensive two-dimensional GC analyses of the standard mixture were performed on a Thermo Fischer Scientific Trace 2DGC Ultra instrument equipped with a FID detector and a dual-jet CO_2 modulator for GC×GC analysis. A Triplus autosampler was used for the injection. The data were acquired at a 100 Hz frequency and processed with HyperChrom software (version 2.3.3). The columns were connected by means of two siltite unions and were installed in the same oven.

Conventional GC analysis of the raw waste water sample extract was carried out on a Varian 4000 Ion Trap GC/MS using 70ev electron impact ionization. The instrument was equipped with a Combi-Pal autosampler (CTC Analytics, Terce, France). The data were acquired and reprocessed with MS Workstation software (version 6.6).

The raw waste water extract was also analysed by GC×GC-TOFMS using a Pegasus IVD from Leco Corporation (St Joseph, MI, USA) equipped with a quad-jets nitrogen modulator (two cold jets and two hot jets). The columns were connected by means of a press-fit connector. The second column was installed in a second oven. Injections were carried out with a Combi-Pal autosampler (CTC Analytics, Terce, France). The data were acquired and reprocessed with ChromaTOF software (version 3.21).

Analytical conditions

Table 2 presents the analytical conditions for the analysis of the standard mixture and of the raw waste water extract.

RESULTS AND DISCUSSION

Separation of 58 target compounds: GC versus GC×GC

The theoretical peak capacity of conventional gas chromatography for a 30 m column is generally estimated at 500–1000. However practical experience is very different and GC often shows coelutions during the analysis of a few dozen or hundred compounds. For example congeners of 209 PCBs (Focant *et al.*, 2004) and

Table 2. Analytical conditions for analysis of standard mixture by GC-FID and GC×GC-FID and for analysis of raw waste water extract by GC/MS and GC×GC-TOFMS.

Separation of 58 target compounds mixture		
Device	Trace **GC-FID** (Thermo Fischer Scientific)	Trace **GC×GC-FID** **(Thermo Fischer Scientific)**
Injector	280°C splitless (1 min)	280°C splitless (1 min)
Column(s)	BPX-5 (5% phenyl): 15 m×0.32 mm×0.25 μm (SGE)	BPX-5: 15 m×0.32 mm×0.25 μm (SGE) DB-5: 0.1 m×0.1 mm×0.4 μm in modulator (Agilent) BPX-50 (50% phenylene): 1 m×0.1 mm×0.1 μm (SGE)
Flow	Helium at 99.999% Constant flow 1 mL/min	Helium at 99.999% Constant flow 1 mL/min
Oven	60°C (1 min) at 340°C (1 min) at 3°C/min	60°C (1 min) at 340°C (1 min) at 3°C/min
Modulator		Dual jet of CO_2 Modulation time: 6 sec
Detector	FID: 100Hz, 350°C	
Analysis of raw waste water sample		
Device	GC/MS 4000 (Varian)	Pegasus IVD GC×GC-TOFMS (Leco)
Injector	280°C split 10:1	PTV split 10:1
Column(s)	Factor 4 (5% phenyl): 30 m×0.25 mm×0.25 μm (Varian)	VF-5ms (5% phenyl): 30 m×0.25 mm×0.25 μm (Varian) VF-17 ms (50% phenyl): 2 m×0.1 mm×0.1 μm (Varian)
Flow	Helium at 99.999% Constant flow 1.5 mL/min	Helium at 99.999% Constant flow 1.5 mL/min
Oven	40°C (2 min) at 190°C at 20°C/min 190°C at 230°C à 2.5°C/min 230°C at 320°C (10 min) at 15°C/min	40°C (2 min) at 190°C at 20°C/ min 190°C at 230°C à 2.5°C/min 230°C at 290°C (20 min) at 15°C/min
Modulator		Quad jet of N_2 Modulation time: 10 sec
Detector	Ion Trap Ion trap temperature: 220°C Source temperature: 200°C Acquisition range: 40–1000 m/z Acquisition rate: 0.75 scans/sec	Time of Flight Source temperature: 200°C Transfer Line temperature: 250°C Acquisition range: 50–1000 m/z Acquisition rate: 200 spectra/sec

PBDEs (Korytar *et al.*, 2005) cannot be resolved using only one column. For this reason, it is interesting to compare the separation of the standard mixture by GC-FID and GC × GC-FID (Figure 1).

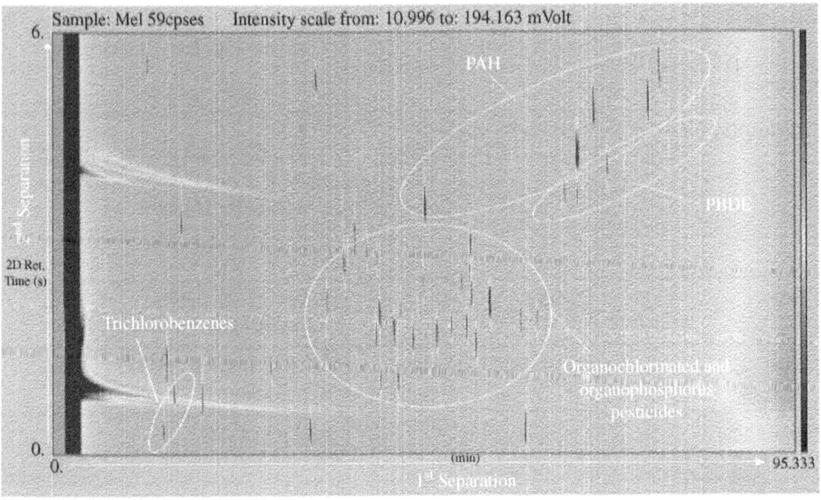

Figure 1. 2D chromatogram of 56 target compounds mixture by GC × GC-FID compensated by 4.2 sec.

Eight partial coelutions were observed in the analysis of the 58 target compounds mixture by conventional GC between naphthalene and hexachlorobutadiene, hexachlorobenzene and simazine, and two isomers of PAH (benzo(b) and benzo(k)fluoranthene). The separation capacity of GC × GC for this mixture is depicted in Figure 1. The reader should be aware that some of the detected compounds appear in light yellow colour and are not visible in Figure 1. Overall, 56 of the 58 target compounds were identified and were separated. Indeed BDE-209 which was not detected in the standard mixture, is thermally unstable and known to be degraded at high temperature (Korytar *et al.*, 2005). Moreover 2 HCH isomers have not been resolved in these conditions. Compounds which are eluted in the middle of the chromatogram are all separated (organochlorinated and organophosphorus pesticides). In conventional GC in the same part of the chromatogram, coelutions are observed between β-endosulfan, 4,4-DDT and BDE-28 as well as between PCB-28 and δ-hexachlorocyclohexane. On the other hand, only the two isomers of PAH are still coeluted in GC × GC. The use of more polar stationary phase and/or chiral stationary phase in the second dimension should be necessary to separe these compounds.

Thanks to the ability of GC×GC providing "structured" chromatograms, it was possible to gather several groups of compounds in the retention space which gave precious additional information for the identification of the target compounds. The volatile trichlorobenzenes are eluted at the begining of the chromatogram, while the semi-volatile organochlorinated and organophosphorus pesticides are grouped in the middle. Finally, less volatile compounds such as PAH and PBDE are eluted at the end of the chromatogram along two distinct isovolatility lines (Figure 1).

To find the best separation for the standard mixture, 2 parameters were optimized: oven temperature and modulation time. Various temperature rates between 3 and 5°C/min were tested. The modulation time has also been optimized for the analysis of the 58 target compounds mixture. The modulator is responsible for collecting, focusing and transfer of the entire sample from the first column to the second one. The modulation time, which is a key factor in GC×GC, is the period corresponding to the sampling of one fraction of analytes on the two jets. It is equal to the time between two successive injections into the second column. There are two rules to choose the modulation time. First, to ensure a good sampling of one analyte, it is advised to cut it 3 or 4 times (Dallüge *et al.*, 2003). Secondly, it is important to avoid "wraparound peaks" which result from a second separation longer than the modulation time. Three modulation times have been investigated: 3, 6 and 8 sec. With a 3 sec modulation a lot of "wraparound" has been noticed preventing separation of compounds of a similar chemical family. This phenomenon was not encountered with a 8 sec modulation time. However, most peaks were sampled only twice and some of them only one. The best compromise was a 6 sec modulation time in which all peaks were sampled three or four times without "wraparound" (Figure 1). The optimized conditions for the separation of the 58 compounds mixture by GC×GC-FID are summarized in Table 2.

Analysis of an environmental complex matrix: GC/MS versus GC×GC-TOFMS

The second part of this study was devoted to the detection of the target compounds in a waste water sample. A comparison of GC/MS and GC×GC-TOFMS was carried out. A screening of other pollutants of potential health concern present in this extract was also completed. Due to its high peak capacity, GC×GC allows reduction of sample preparation and clean-up steps. To avoid losing traces of compounds, the sample was just subjected to a liquid-liquid extraction at basic pH (pH = 11). The complexity of raw waste water extracts prevents detection of the target compounds by FID detection. Consequently, the analysis of waste water sample was done at the same time by GC/MS and GC×GC-TOFMS. The deconvolution software used for GC×GC-TOFMS experiments allowed extraction of compounds from the complex matrix.

The analysis by GC×GC-TOFMS was carried out on a Leco GC×GC system. The analytical conditions (Table 2) and program temperature rates (10, 15 and 20°C/min) have been chosen to test the focusing capacity. First the standard mixture was analysed by GC×GC-TOFMS to estimate the retention time and limit of detection of the device. The limit of detection established for GC×GC-TOFMS analysis (2.0 ng/L for pentachlorobenzene, 3.0 ng/L for isodrin, 8.0 ng/L for endosulfan and endrin, and between 2 and 50 ng/L for the other substances) are close to the environmental quality standards recommended by the legislators (EC, 2008). Figure 2 presents the chromatograms of the raw waste water analysis by GC/MS and GC×GC-TOFMS.

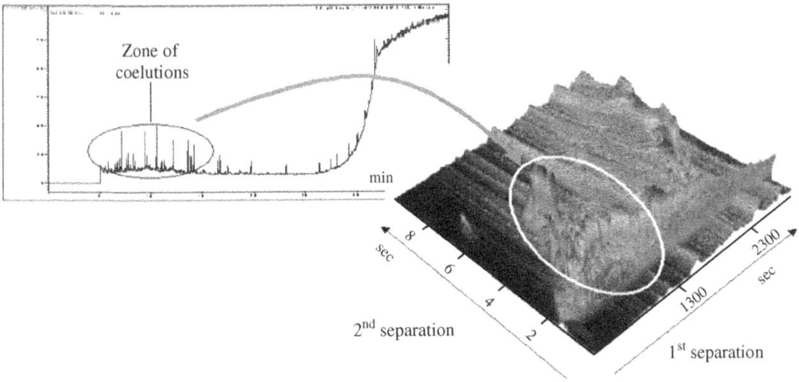

Figure 2. Chromatograms of raw waste water extracted at basic pH by GC/MS (top left) and GC×GC-TOFMS (bottom right).

The chromatogram of the raw waste water at basic pH by GC/MS is presented on the left side, whereas the right side shows the analysis of the same extract by GC×GC-TOFMS. Saturation was noticed on the two chromatograms. A higher temperature rate (20°C/min) can be used with the Leco GC×GC chromatograph due to the use of quad jet focusing. Indeed no peak splitting has been observed. However a lower temperature rate should further improve the separation. Figure 3 shows the complexity of this matrix which was not observed in GC/MS due to its lower resolving power and sensitivity. A zone of coelution was partially resolved in GC×GC. Indeed the second column allows spreading this zone of coelution. Further improvements are currently under investigation.

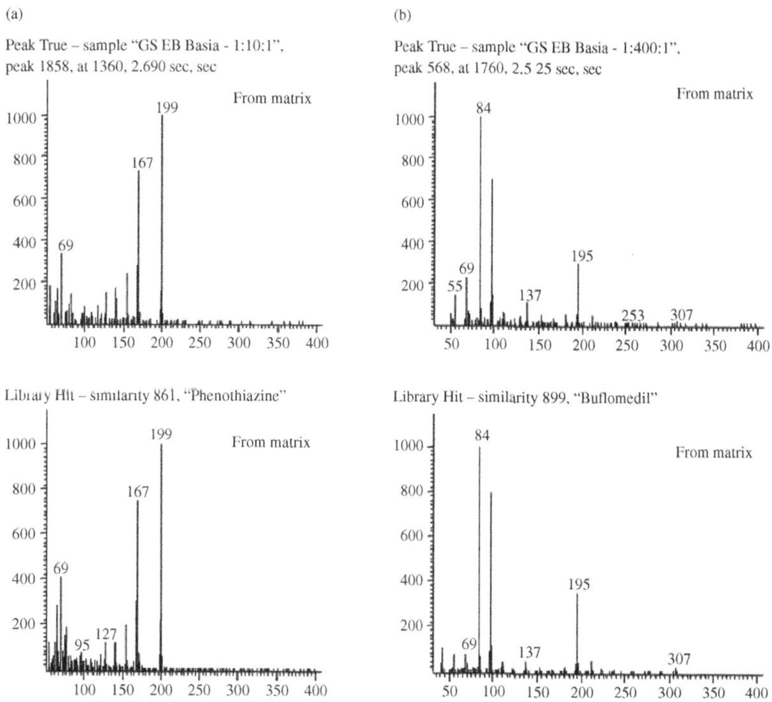

Figure 3. (a) Mass spectra for phenothiazine extracted from the matrix (on the top) and from the library (below). (b) Mass spectra for buflomedil extracted from the matrix (on the top) and from the library (below).

GC/MS detected 400 compounds whereas GC × GC detected more than 1000 compounds. Among the 56 detectable target compounds, only 4-tert-butylphenol has been identified by GC/MS with a low similarity index. On the other hand, GC × GC allowed the identification of 6 target compounds including 4-tert-butylphenol. For the screening of other pollutants present in the matrix, only 5 compounds were tentatively identified using GC/MS (similarity index higher than 70%) including 4-methylphenol, menthol, o-hydroxybiphenyl, toluene, aspirin. When applying a temperature rate of 3°C/min to improve separation of the target compounds, 20 more compounds were identified. During the tests on Leco GC × GC-TOFMS, more than 1000 other compounds were observed. The mass spectra of 200 compounds showing the highest similarity were examined. 50 compounds including a few pesticides which were not listed as priority pollutants in European regulations (for example phenothiazine and permethrin) were tentatively identified. Many drugs have also been found in this extract including mostly antidepressants (codeine, venlafaxine, piracetam and buflomedil). Some aromatics (eucalyptol, eugenol...) and organic compounds (aniline, caprolactam...) have also been tentatively identified. GC×GC-TOFMS has also found various compounds such as caffeine, cholesterol and tocopherol which are often found in waste water samples. In Figure 3, the mass spectra for phenothiazine and buflomedil extracted from the matrix are in good agreement with those from the library. In general, GC×GC-TOFMS yields good quality mass spectra because coelutions are limited; furthermore, unlike quadrupoles and ion-traps, a Time-of-Flight mass spectrometer is not a scanning instrument and therefore yields mass spectra with no skewing effect as the compound concentration quickly changes in the ion source during peak elution In summary, GC×GC-TOFMS has enabled better identification of target compounds and a much better screening of non target organic pollutants in the raw waste water extract. Confirmation of identifications with authentic standards is currently underway and the results for a range of urban wastewaters will be published in the near future.

CONCLUSION

The preliminary results reported in this paper indicate that comprehensive two-dimensional gas chromatography is a method of choice for the detection of target compounds and for the screening of other, non target organic pollutants in a complex matrix such as a urban wastewater. GC×GC-FID allowed to resolve seven coelutions for a standard mixture (58 target compounds) compared with GC-FID. GC×GC-TOFMS (50 compounds tentatively identified from manual examination of 200 mass spectra) was much more powerful than GC/MS (5 compounds tentatively identified) to analyse a complex environmental matrix (urban wastewater). Moreover, the detection limits of the GC×GC-TOFMS device enable the simplification of clean-up steps. Although the raw waste water extract did contain a few of the selected target chemicals, more than 1000 other natural and anthropogenic compounds were detected. Among them were many drugs, aromatic products used in household products, food additives, and pesticides which are not listed in the recent environmental European regulations. GC×GC will lead to a much better understanding of the efficiency of waste water treatment processes, due to its superior resolving power. Thus not only GC×GC will be useful to comply with new regulations but it will also allow to anticipate future legislations.

ACKNOWLEDGEMENT

G. Semard would like to thank ANRT and CIRSEE for CIFRE fellowship.

REFERENCES

Dallüge, J., Beens, J. and Brinkman, U.A.Th. (2003). Comprehensive two-dimensional gas chromatography: a powerful and versatile analytical tool. *J Chromatogr. A*, **1000**, 69.

Danielsson, C., Wiberg, K., Korytar, P., Bergek, S., Brinkman, U.A.Th. and Haglund, P. (2005). Trace analysis of polychlorinated dibenzo-p-dioxins, dibenzofurans and WHO polychlorinated biphenyls in food using comprehensive two-dimensional gas chromatography with electron-capture detection. *J. Chromatogr. A*, **1086**, 61–70.

EC – European Communities (2000). Directive 2000/60/CE. Official Journal of the European Communities, 22.12.2000, L327–73.

EC – European Communities (2006). Regulation 2006/166/CE. Official Journal of the European Communities, 18.1.2006, L 33/1–15.

EC – European Communities (2008). Directive 2008/105/CE. Official Journal of the European Communities, 16.12.2008.

Focant, J.-F., Sjödin, A. and Patterson Jr., D.G. (2004). Improved separation of the 209 polychlorinated biphenyl congeners using comprehensive two-dimensional gas chromatography–time-of-flight mass spectrometry. *J. Chromatogr. A*, **1040**, 227.

Gaines, R.B., Ledford, E.B. and Stuart, J.D. (1998). Analysis of water samples for trace levels of oxygenate and aromatic compounds using headspace solid-phase micro-extraction and comprehensive two-dimensional gas chromatography. *J. Microcol. Sep.*, **10**, 597–604.

Hyötyläinen, T., Kallio, M., Hartonen, K., Jussila, M., Palonen, S. and Riekkola, M.-L. (2002). Modulator design for comprehensive two-dimensional gas chromatography: quantitative analysis of polyaromatic hydrocarbons and polychlorinated biphenyls. *Anal. Chem.*, **74**, 4441.

Ieda, T., Horii, Y., Petrick, G., Yamashita, N., Ochiai, N. and Kannan, K. (2005). Analysis of nonylphenol isomers in a technical mixture and in water by comprehensive two-dimensional gas chromatography–mass spectrometry. *Environ. Sci. Technol.*, **39**, 7202–7207.

Korytar, P., Covaci, A., Leonards, P.E.G., de Boer, J. and Brinkman, U.A.Th. (2005). Comprehensive two-dimensional gas chromatography of polybrominated diphenyl ethers. *J. Chromatogr. A*, **1100**, 200–207.

Liu, Z. and Phillips, J.B. (1991). Comprehensive two-dimensional gas chromatography using an on-column thermal modulator interface. *J. Chromatogr. Sci.*, **29**, 227–231.

Phillips, J.B. and Xu, J. (1995). Comprehensive multi-dimensional gas chromatography. *J. Chromatogr. A*, **703**, 327–334.

Advances in microbiological methods for drinking water analysis: flow cytometry, assimilable organic carbon and pathogen growth potential

F. Hammes, M. Vital, D. Stucki, K. Lautenschlager and T. Egli

Eawag, Swiss Federal Institute of Aquatic Science and Technology, Überlandstrasse 133, CH-8600 Dübendorf, Switzerland (E-mail: frederik.hammes@eawag.ch)

Abstract Flow cytometry (FCM) offers immense potential for both drinking water research and routine monitoring. Total cell counts and basic viability assays allow rapid assessment of disinfection efficacy and analysis of bacterial growth in biofilters and during distribution. During the Techneau project we have optimised and tested several FCM methods in comparison to conventional methods on pilot scale and full-scale treatment and distribution systems. In addition, we have used the FCM total cell count method to establish an alternative assimilable organic carbon (AOC) assay where an indigenous microbial community is used as inoculum. The advantage of the new AOC method is that it is fast and allows simultaneous processing of a larger number of samples. We have furthermore expanded this AOC assay to test specifically for the growth potential of pathogenic bacteria (pathogen growth potential (PGP)) assay. These methods offer new tools to assess biological stability of drinking water.

Keywords assimilable organic carbon (AOC), drinking water, flow cytometry, pathogen growth

INTRODUCTION

Bacteria are indigenous to drinking water treatment and distribution systems in concentrations typically between 10^4–10^6 cells/mL (suspended cells) and comprising a wide variety of species (Berry *et al.*, 2006; Hammes *et al.*, 2008). The presence of these bacteria is usually not a result of exogenous contamination, but rather due to micro-organisms that proliferate in these systems under the prevailing low nutrient conditions. On the one hand, bacteria perform an essential functional role in the biofilters during drinking water treatment; their removal of biologically available organic carbon is the key process in the production of biologically stable drinking water. On the other hand, bacterial growth in the distribution network is opportunistic and unwanted, potentially leading to a deterioration of water quality (Van der Kooij, 2000). In both cases it is important to understand and quantify the basic principles and parameters of growth (i.e. biomass production and organic carbon removal).

 TECHNEAU: Safe Drinking Water from Source to Tap. Edited by Theo van den Hoven and Christian Kazner. ISBN: 9781843392750. Published by IWA Publishing, London, UK.

Fundamental research and routine monitoring both require methodologies that are fast, accurate and reproducible. However, the introduction of new microbiological methods for drinking water analysis has been limited. This is at least partially because the conventional heterotrophic plate count (HPC) method has been ingrained through a century of use in the water industry. However, to form a proper understanding of bacterial growth and inactivation (disinfection), HPC is of limited value, simply because this method fails to detect the majority of bacteria living in drinking water (Van der Kooij *et al.*, 1995; Siebel *et al.*, 2008).

In this paper we discuss the use of flow cytometry as an advanced cell counting tool that is fast, accurate and reproducible. We have focused specifically on applications that are straightforward and which can be used readily for routine monitoring and research tools. In addition, we have established an alternative assimilable organic carbon (AOC) assay and pathogen growth potential (PGP) assays to be used in combination with FCM for the assessment of biological stability of drinking water.

FLOW CYTOMETRY: RAPID ANALYSIS OF BACTERIAL CELLS

How does flow cytometry work?

A flow cytometer is essentially a high precision cell counter that can be combined with a wide variety of fluorescent dyes to gain information on the cells that are measured, e.g. nucleic acid content, cell size, viability, activity, type, etc. In short, bacterial cells are labelled with a fluorescent dye, and the stained cells are pumped through a thin glass capillary that is crossed by a laser beam. The laser-cell interaction causes light scattering (including fluorescence) from the bacterial cell, which is quantified on detectors. Analysis of a single sample takes about 3–5 minutes, the quantification limit is around 200 cells/mL and typical throughput is about 1000 events/second. Figure 1 shows typical FCM dot-plot data, in which two fluorescence signals are acquired for each particle (dot). The bacteria are separated from the background signals with electronic gates based on appropriate controls. FCM is an ideal tool for drinking water assessment. The reasons are that drinking water as a matrix is extremely clean from particles that might interfere with the analysis, and the bacteria are usually in concentrations that require no dilution. This renders analysis fast and uncomplicated. In addition, unlike the conventional HPC method, FCM allows detection of the entire microbial community in the water sample, irrespective of cultivability.

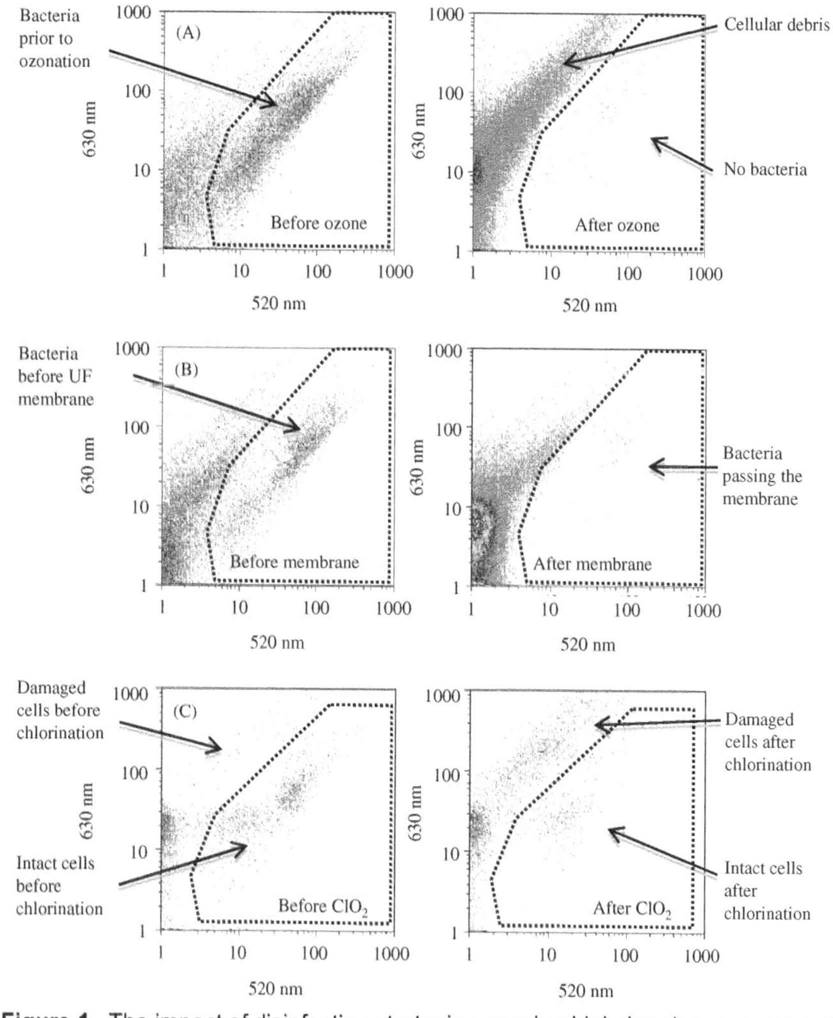

Figure 1. The impact of disinfection strategies on microbial abundance measured with flow cytometry. (A) *Ozonation*: samples from full-scale treatment of lake water (Zurich, CH), cells stained with SYBR Green I; (B) *membrane filtration*: samples from a pilot scale ultra-filtration system, cells stained with SYBR Green I; (C) *chlorination*: samples from a full-scale system treating groundwater, cells stained with SYBR Green I and propidium iodide. Dotted lines show the electronic gate around the bacterial cells prior to the treatment.

Straightforward applications in drinking water research

Analysis of disinfection processes during treatment

Ozonation is an aggressive oxidation process, which is often used as the first step in a drinking water treatment train. Ozone, depending on the dose, destroys bacterial cells completely (Hammes *et al.*, 2008). This can be easily observed and quantified with flow cytometry and a simple total cell count stain such as SYBR Green I (Figure 1(A)). In case the ozone dose is not sufficient to destroy the cells completely, the addition of a membrane permeable stain (e.g. propidium iodide) can be used to assess damage to the bacteria cells (see chlorination section below).

Membrane filtration is often used as a final or even a single treatment step in case of high quality source water. The principle of membrane filtration is the exclusion of particles of a certain size. Therefore, in case the membrane cut-off is below the size of bacterial cells (which it typically is), membrane filtration efficacy can simply be measured with a total cell count stain such as SYBR Green I (Figure 1(B)).

Chlorination (chlorine, chlorine dioxide or mono-chloramine) is used worldwide for drinking water disinfection. Chlorination results in oxidative damage to cells, which means that cell membranes are damaged/destroyed in the process. As a result, a membrane permeable fluorescent dye such as propidium iodide can provide quick and clear information on the process (Figure 1(C)).

Quantification of bacterial growth during biofiltration processes

Biofiltration (activated carbon filters, rapid sand filters or slow sand filters) are used commonly in drinking water treatment. In such filters, bacteria grow as fixed biomass on the filtration material. These cells consume biodegradable organic matter and in turn produce new biomass. A major part of this growth can be measured as planktonic cells released in the effluent of the filter. Since the target is bacteria that have grown in the biofilter, a simple total cell count stain such as SYBR Green I is sufficient (Figure 2).

Quantification of growth in distribution networks and household installations

Bacteria grow opportunistic in distribution networks and household installations. A part of the additional growth ends up in the bulk phase and can be measured relatively easily as an increase in suspended cell concentrations between any two points (Figure 3). This is particularly important because such data can be used to assess the biological stability of drinking water, the impact of household conditions (stagnation, temperature), as well as any problems in the network.

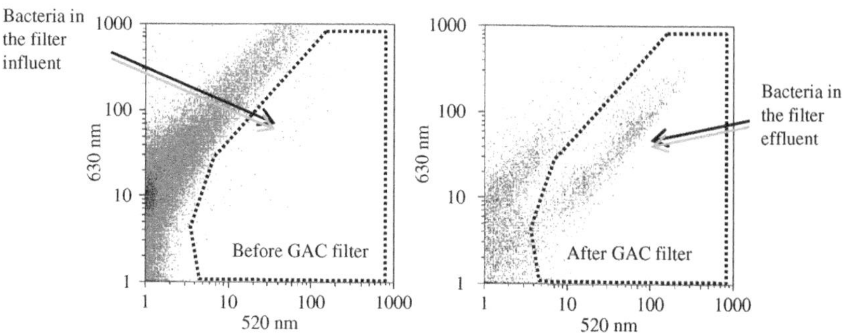

Figure 2. Bacterial growth during biofiltration measured with SYBR Green I staining. Data from a full-scale GAC filter (Zürich, CH) treating ozonated surface water.

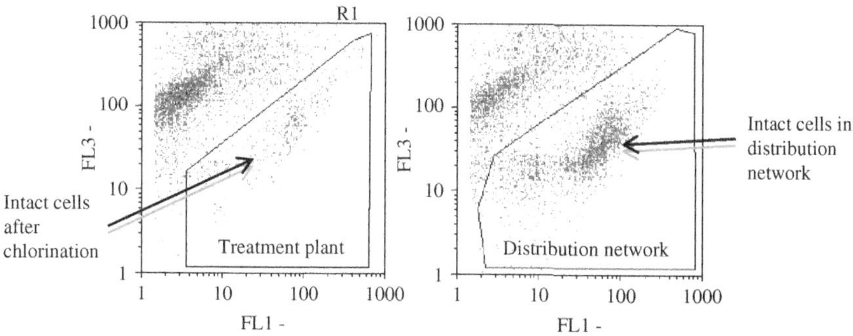

Figure 3. Bacterial growth in drinking water after chlorination, measured as the difference between intact cells immediately following treatment and at a hydrant 40 h after distribution (data from the full-scale network of Riga, LV). Due to the use of chlorine, these samples were measured with SYBR Green I and propidium iodide.

Data from a full-scale drinking water system

Figure 4 shows data from the drinking water system of Zürich (CH), which was assessed in the frame of the Techneau project. This system uses no final chlorination and the assessment was done using only total cell counts (SYBR Green I staining). It is clear that both of the disinfective ozonation processes could be clearly quantified, while regrowth in the biofilters could be seen as well. During distribution, only a small increase in the cell concentration occurred, which suggests a high level of biological stability of the water.

However, on household level a considerable increase in cell concentration occurred (note that the latter was only in the 1st litre after overnight stagnation).

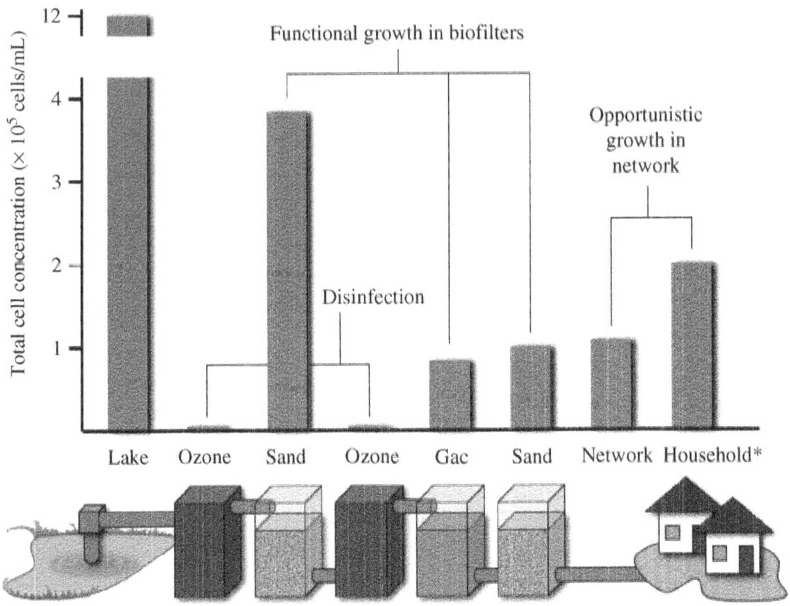

Figure 4. Flow-cytometric quantification of bacterial abundance from source to tap in the Zürich drinking water system. GAC = granular active carbon filter; sand = rapid sand and slow sand filters; network = 48 h residence time; household* = overnight stagnation.

Correlation with other microbiological parameters

The relation between FCM data and alternative microbiological parameters was investigated in several studies related to the Techneau project (Berney *et al.*, 2008; Hammes *et al.*, 2008; Siebel *et al.*, 2008). These included samples from treatment systems (Zurich, Basel), samples from distribution networks (Zurich, Riga, Basel) and samples from household installations. In none of the samples did we ever observe a meaningful correlation between conventional HPC data and FCM data (data not shown). This is not surprising because not many bacteria are cultivable on conventional media, while FCM can detect all bacteria, irrespective of cultivability (Siebel *et al.*, 2008). However, good correlations were observed between FCM data and the concentration of bacterial adenosine tri-phosphate (ATP) in water samples (Figure 5) (see Berney *et al.* (2008) for methodological details). This is significant because ATP is a molecule associated only with living cells, suggesting that the FCM data collected correlated well with the viable bacteria in the water. We calculated an average ATP-per-cell value of 8×10^{-17} g/cell (n = 410). The correlation between these two individual parameters highlights the value of cultivation-independent methods to assess the general microbial quality of water samples.

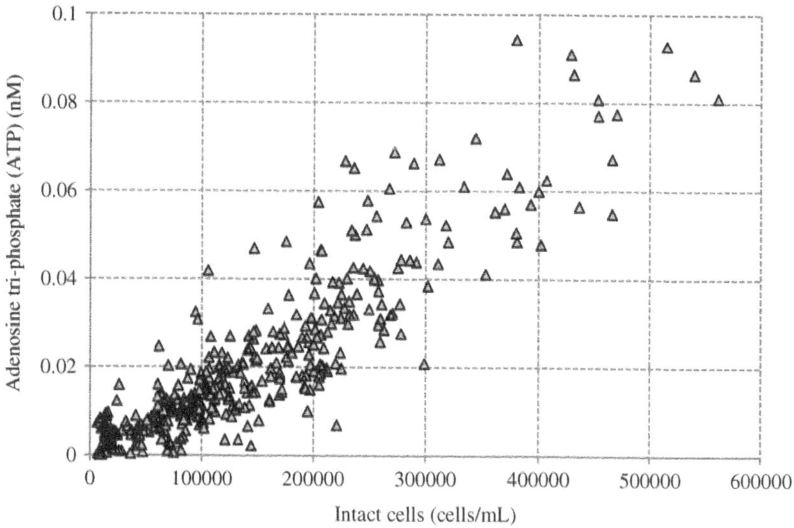

Figure 5. Correlation between the concentration of intact bacterial cells and ATP in drinking water samples collected throughout the Techneau project ($R^2 = 0.8$; n = 410).

Future of FCM: challenges and cautions

A couple of clear challenges exist for FCM method development in the field of drinking water. A key question in drinking water systems is pathogen detection. While the detection limit of FCM suffices for quantifying the indigenous communities, pathogens are usually present in significantly lower concentrations. Much work has already been done in the field of sample concentration (e.g. immunomagnetic separation), but it is clear that for the time being there is no FCM method that is better than existing plating methods for routine pathogen or indicator organism detection. Virus detection is another point that has received only limited attention in drinking water research. Numerous studies are looking into the use of FCM for virus analysis (Brussaard, 2004), but none of these methods are currently on a level where they can be used routinely for drinking water monitoring. UV disinfection is emerging as a key treatment technology across Europe. While FCM methods for disinfection with UV-A light (sunlight) has been investigated in detail (Berney *et al.*, 2006), at this moment there exist no simple FCM methods that can be used to assess UV-C damage to bacterial cells accurately.

A note of caution: Many fluorescent dyes for FCM are sold commercially as ready-to-use kits. However, these stains/protocols are usually developed using laboratory pure cultures of bacteria, which differ considerably from indigenous drinking water communities with regards to cell size and physiology. Therefore, FCM dyes and protocols have to be optimised and standardised for each water sample and application (Berney *et al.*, 2008). In addition, one has to be careful with the interpretation of FCM data. An example is staining with the well-known Live/Dead kit: too often such data are reported as actually meaning "live" and "dead", which is absolutely not the case. This leads to wild conclusions about the research or the value of the assay. The fact is that this method measures only membrane damage in cells. While it is fair to say that membrane damage is usually irreversible, thus meaning dead cells, it is more than possible that cells can be dead without showing membrane damage (Berney *et al.*, 2008).

AN ALTERNATIVE ASSIMILABLE ORGANIC CARBON (AOC) ASSAY

Principle of the AOC method

The purpose of measuring AOC in drinking water is to gain information on the growth-supporting organic carbon fraction in the water, which in turn could help predict the biological stability of the water. AOC assays typically measure the concentration of suspended cells that can grow in a sterilised water sample in batch culture under defined laboratory conditions. The amount of grown cells is then converted to a concentration of consumed carbon (or carbon equivalents). The principle has been introduced through the seminal work of Dick van der Kooij (Kiwa/KWR, The Netherlands) during the last three decades. The original AOC assay utilised the growth of two pure cultures (*Pseudomonas fluorescence* P17 and *Aquaspirillum* NOX) (Van der Kooij, 1992), while several adaptations or alternative methods have been proposed (Werner, 1985; Hammes and Egli, 2005; Van den Broeke *et al.*, 2008).

Adopting a new AOC method

During the course of the Techneau project, Eawag has optimised and tested an alternative AOC assay. This approach differs from the original assay in that it uses a natural microbial community as inoculum, incubation at 30°C for faster growth, and detection/enumeration with fluorescent staining and flow cytometry (Figure 6) (Hammes and Egli, 2005; Hammes *et al.*, 2006; Hammes *et al.*, 2007; Vital *et al.*, 2007). The main advantages of this approach are that it is fast (3 days to results), it detects a broad range of carbon compounds (due to the use of the microbial community), and the use of flow cytometry allows for high throughput of samples. A first comparison within the Techneau project with two previously established AOC methods (Van der Kooij method (Van der Kooij, 1992) and Werner-Hambsch method (Werner, 1985) have suggested that similar results are obtained by all three methods (data not shown).

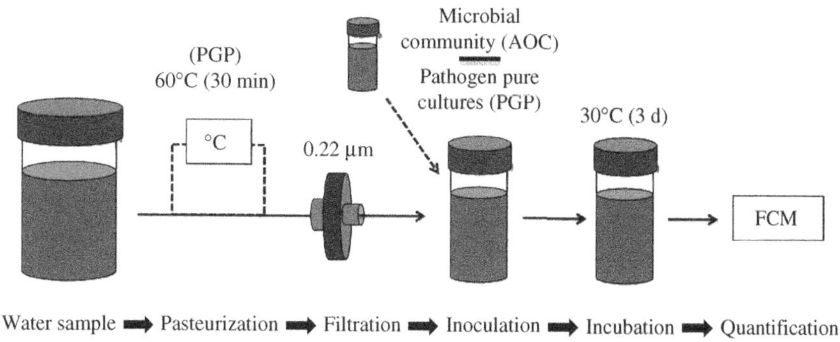

Figure 6. Layout of the AOC and PGP assays. The PGP assay includes an additional pasteurization pre-treatment step. Inoculation is around 1 × 10^3 cells/mL. For the PGP assay, inoculation includes pure cultures of *E. coli O157, V. cholera, P. aeruginosa.*

Applications of AOC analysis

AOC analysis is used to study the impact of specific treatment processes such as ozonation (Hammes *et al.*, 2006, 2007), chlorination and biofiltration on the biodegradability of organic carbon (Figure 7), and it is seen as a measurement of biological stability of drinking water (Van der Kooij, 1992; Hammes *et al.*, 2006, 2007). However, AOC data should also be treated with a level of caution – it is well known that AOC methods detect only a fraction of the actual biodegradable carbon in a water sample (Volk and LeChevallier, 1999). Hence, it is possible that relying on only AOC data, one may underestimate biodegradability and overestimate biological stability of the water.

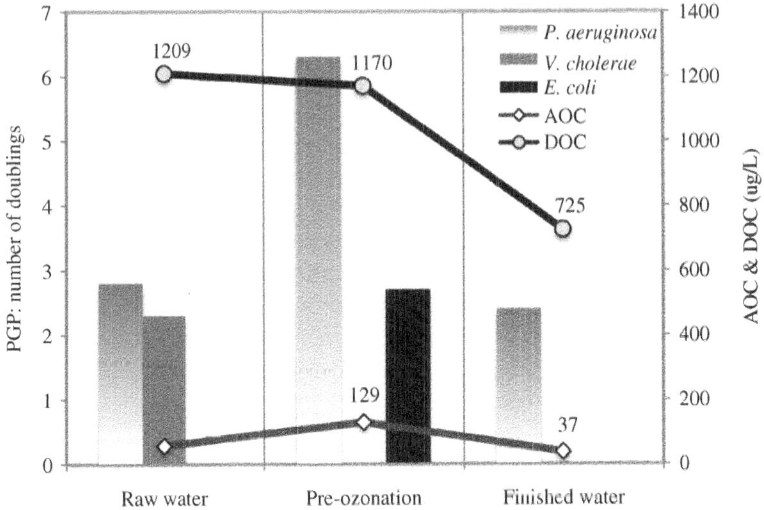

Figure 7. Assessment of a full-scale drinking water treatment plant (Zurich, CH) with AOC and PGP assays in relation to DOC data. PGP is expressed as the number of times that the concentration of the individual pathogens doubled in the water samples.

PATHOGEN GROWTH POTENTIAL (PGP) ASSAY

Principle of the PGP assay

In the context of biological stability of drinking water, it is most important to produce/distribute water that does not support the growth of pathogenic bacteria. Both faecal and environmental pathogens can grow in freshwater at low carbon concentrations (Vital *et al.*, 2007, 2008; Van der Kooij, 1982). The PGP assay is essentially an advanced AOC assay that measures specifically the potential of a water sample to support the growth of pathogenic bacteria in relation to the concentration of AOC (as measured above). The assay procedure is similar to the AOC assay (Figure 6), with the exception that the water sample is pasteurised in addition to filtration. The test organisms that are used include: *Vibrio cholerae, Escherichia coli O157 and Pseudomonas aeruginosa*. Typical growth of such organisms in sterile freshwater, together with the main laboratory procedures, has been described previously (Vital *et al.*, 2007, 2008). Note that due to the nature of these bacteria, the PGP assay requires a Safety Level 2 laboratory and skilled researchers. The main use of the PGP assays would be (1) to assess different treatment technologies on their impact specifically with regards to pathogen growth, (2) to test materials used for installation for the leaching of compounds favouring pathogen growth, (3) to test finished waters including drinking water, wastewater effluent, process water and swimming pools for pathogen growth potential.

Example: the PGP assay on samples from full-scale water treatment

As a part of the Techneau project, the PGP assay was tested on water from several full-scale treatment plants. Figure 7 shows a data set derived from one such test. The key findings from this experiment are that *P. aeruginosa* was capable of growing in all water samples and that growth of this organism correlated well with the AOC concentrations in the water. Neither *V. cholerae* nor *E. coli O157* could grow in the finished water, which is probably a result of substrate thresholds demonstrated previously for these organisms (Vital *et al.*, 2007, 2008). However, whereas *E. coli* responded positively to increased AOC concentrations after ozonation, *V. cholerae* could not grow in ozonated water. This phenomenon was observed in multiple data sets (data not shown). The importance of this data is that while a low available substrate (AOC) concentration is a main step towards ensuring the absence of pathogenic growth, one cannot use AOC or DOC values alone to estimate pathogen growth potential. Selective usage of specific carbon substrates, or selective inhibition by certain compounds, can alter the growth pattern of specific pathogenic strains.

CONCLUSIONS

Bacterial growth in drinking water can be beneficial/functional (bioreactors in the treatment train), opportunistic (distribution network), and severely problematic (biofouling and/or pathogen proliferation). Heterotrophic growth is a matter of substrate utilisation and resulting biomass production, which are two quantifiable variables. In between lies a more complex range of physiological, stoichiometric and kinetic parameters (e.g. yield, substrate thresholds, substrate preference, specific growth rates), which dictates the relationship between these parameters. However, this can only be understood and described if the means exist with which to measure growth and substrate utilisation accurately. FCM provides a tool which is fast, sensitive, accurate and easy to use, and of which the most basic applications are already useful and directly applicable for drinking water analysis. The combination of FCM analysis with AOC assays and advanced PGP assays provides the opportunity to assess drinking water quality and biological stability in more detail.

ACKNOWLEDGEMENTS

The authors appreciate the technical and scientific support from the Zurich Waterworks (WVZ), Basel Waterworks (IWB), Talis Juhna (RTU), Janis Rubulis (RTU), Hansueli Weilenmann (Eawag), as well as financial support from the 6th Framework European project TECHNEAU (018320).

REFERENCES

Berney, M., Vital, M., Hülshoff, I., Weilenmann, H.-U., Egli, T. and Hammes F. (2008). Rapid, cultivation-independent assessment of microbial viability in drinking water. *Water Research*, **42**(14), 4010–4018.

Berney, M., Weilenmann, H.-U. and Egli, T. (2006). Flow-cytometric study of vital cellular functions in *Escherichia coli* during solar disinfection (SODIS). *Microbiology*, **152**, 1719–1729.

Berry, D., Xi, C. and Raskin, L. (2006). Microbial ecology of drinking water distribution systems. *Current Opinion Biotechnology*, **17**(3), 297–302.

Brussaard, C.P.D. (2004). Optimization of procedures for counting viruses by flow cytometry. *Applied Environmental Microbiology*, **70**, 1506–1513.

Hammes, F., Berney, M., Wang, Y., Vital, M., Köster, O. and Egli, T. (2008). Flow-cytometric total bacterial cell counts as a descriptive microbiological parameter for drinking water treatment processes. *Water Research*, **42**(1–2), 269–277.

Hammes, F. and Egli, T. (2005). New method for assimilable organic carbon determination using flow-cytometric enumeration and a natural microbial consortium as inoculum. *Environmental Science and Technology*, **39**, 3289–3294.

Hammes, F., Meylan, S., Salhi, E., Köster, O., Egli, T. and von Gunten, U. (2007). Formation of assimilable organic carbon (AOC) and specific natural organic matter (NOM) fractions during ozonation of phytoplankton. *Water Research*, **41**, 1447–1454.

Hammes, F., Salhi, E., Köster, O., Kaiser, H.P., Egli, T. and von Gunten, U. (2006). Mechanistic and kinetic evaluation of organic disinfection by-product and assimilable organic carbon (AOC) formation during the ozonation of drinking water. *Water Research*, **40**, 2275–2286.

Siebel, E., Wang, Y., Egli, T. and Hammes, F. (2008). Correlations between total cell concentration, total adenosine tri-phosphate concentration and heterotrophic plate counts during microbial monitoring of drinking water. *Drinking Water Engineering and Science*, **1**, 1–6.

Van den Broeke, J., Ross, P.S., Rietveld, L.C., van der Helm, A.W.C. and Baars, E.T. (2008). Use of on-line UV/Vis-spectrometry in the measurement of dissolved ozone and AOC concentrations in drinking water treatment. *Water Science and Technology*, **57**(8), 1169–1175.

Van der Kooij, D. (1992). Assimilable organic carbon as indicator for bacterial regrowth. *Journal of the American Water Works Association*, **84**, 57–65.

Van der Kooij, D. (2000). Biological stability: a multidimensional quality aspect of treated water. *Water, Air, and Soil Pollution*, **123**, 25–34.

Van der Kooij, D., Harm, R.V., Baars-Lorist, C., Van der Klift, D.W. and Drost, Y.C. (1995). Biofilm formation on surfaces of glass and Teflon exposed to treated water. *Water Research*, **29**(7), 1655–1662.

Van der Kooij, D., Oranje, J.P. and Hijnen, W.A. (1982). Growth of *Pseudomonas aeruginosa* in tap water in relation to utilization of substrates at concentrations of a few micrograms per liter. *Applied and Environmental Microbiology*, **44**(5), 1086–1095.

Vital, M., Fuchslin, H.P., Hammes, F. and Egli, T. (2007). Growth of *Vibrio cholerae* O1 Ogawa Eltor in freshwater. *Microbiology*, **153**, 1993–2001.

Vital, M., Hammes, F. and Egli, T. (2008). *E. coli* O157 can grow in natural freshwater at low carbon concentrations. *Environmental Microbiology*, **10**, 2387–2396.

Volk, C.J. and LeChevallier, M.W. (1999). Impacts of the reduction of nutrient levels on bacterial water quality in distribution systems. *Applied and Environmental Microbiology*, **65**(11), 4957–4966.

Werner, P. (1985). Eine Methode zur Bestimmung der Verkeimungsneigung von Trinkwasser. *Vom Wasser*, **65**, 257–270.

Use of effect-directed assays in assessing the quality of drinking water and its sources

M.B. Heringa[a], S.C. van der Linden[b], M. Schriks[a], B. van der Burg[b] and A.P. van Wezel[a]

[a]KWR Watercycle Research Institute, PO Box 1072, 3430 BB Nieuwegein, The Netherlands (E-mail: minne.heringa@kwrwater.nl; merijn.schriks@kwrwater.nl; annemarie.van.wezel@kwrwater.nl)
[b]BioDetection Systems BV, Science Park 406, 1098 XH, Amsterdam, The Netherlands (E-mail: sander.van.der.linden@bds.nl; bart.van.der.burg@bds.nl)

Abstract Current monitoring programs for drinking water or its sources only measure a limited amount of often well-known chemicals. To detect all harmful compounds present, and also the total effect of all compounds, *in vitro* effect-directed assays are a valuable addition. Relevant effects to analyze in drinking water and its sources are genotoxicity, endocrine disruption, developmental and teratogenic effects and possibly neurotoxicity and immunotoxicity, as these are effects that can occur at relatively low exposures already. A short overview is given of the various *in vitro* assays available for the mentioned effects. Application of these assays to various water types in the Netherlands has shown the presence of possibly harmful chemicals which were not detected by the current monitoring programs. In order to identify when a response of these tests is unacceptable, effect-directed water quality limits are necessary. A first idea on how to develop these, is by basing them on the ADI of a reference compound, assuming a worst case scenario for absorption, distribution, metabolism and elimination. Further work on this topic is considered necessary.

Keywords effect directed bioassays, endocrine disruption, genotoxicity

INTRODUCTION

In the sources for drinking water, thousands of industrial chemicals can be present (Schwarzenbach *et al.*, 2006). In the EU, 30,000 to 70,000 chemicals are in daily use (EINECS database), varying from industrial chemicals (such as solvents, petrochemicals), consumer chemicals (such as pharmaceuticals, personal-care products) to biocides. Due to global trends such as a growing and older population, increasing prosperity, urbanization and more extreme events (increasing transport rates between environmental compartments) (Diamond and Hodge, 2007), there is an increasing chemical pressure on drinking water sources. For example, only about 10% of European river water samples could be classified as "very clean" (Loos *et al.*, 2009). Pharmaceuticals, perfluorinated compounds, personal care products, detergents and endocrine disruptors could be found throughout Europe up to high ng/L median concentrations (Loos *et al.*,

 TECHNEAU: Safe Drinking Water from Source to Tap. Edited by Theo van den Hoven and Christian Kazner. ISBN: 9781843392750. Published by IWA Publishing, London, UK.

2009), many entering the water cycle via wastewater (Reemstma *et al.*, 2006). This occurrence of chemicals in sources of drinking water is, depending on the treatment processes, sometimes reflected in (much lower) occurrence of these chemicals in finished drinking water (Benotti *et al.*, 2009).

Only a very limited part of the chemicals which are found in drinking water sources are currently regulated via (drinking) water quality regulations. Additionally, current monitoring programs only measure a limited amount of often well-known chemicals. Especially the presence of so-called "emerging chemicals" such as pharmaceuticals, personal-care products, drugs-of-abuse, endocrine disruptors and nanochemicals may lead to consumers concern (e.g. Turgeon *et al.*, 2004). Emerging chemicals can be defined as newly detected chemicals, which are not covered by existing water quality legislation, and for which relatively little information is available on their environmental behaviour and (eco)toxicological properties.

Despite rapidly evolving chemical analytical techniques (Richardson, 2008), it is impossible to analyze and identify all different chemicals present in the aqueous environment. In addition, for many identified chemicals no statutory health-based (drinking) water quality guideline values have been established and no or scarce toxicological information is available to estimate human health risks (e.g. van Genderen *et al.*, 2000). Finally, our understanding of the combined mixture effects in these complex mixtures of many different chemicals occurring in low concentrations is still limited (McCarty and Borgert, 2006). The use of *in vitro* effect-directed assays can be an additional tool to interpret health risks of complex mixtures in drinking water and its sources. *In vitro* effect-directed bioassays do not determine the presence of (a group of) compounds directly, but determine their collective effect in a biological system such as cultured cells. Chemicals that cannot be revealed by analytical techniques but do attribute to the toxicological effects are included in the effect assays, and thus, the assays give a clue of the toxicity of the total mixture of chemicals present in the sample.

This paper discusses which effect assays are useful for assessing the quality of drinking water and its sources, and gives some examples on their application.

SELECTION OF RELEVANT *IN VITRO* EFFECT-DIRECTED ASSAYS

For quality assessment of the drinking water production chain, relevant effect-directed assays are related to carcinogenicity and genotoxicity, to hormonal disruption, or to developmental and teratogenic effects. These end-points are relevant human health effects which can occur after chronic exposure to relatively low concentrations. Toxicity assays that focus on acute effects are considered as less relevant; human health effects after acute exposure are not expected given the drinking water quality in developed countries.

Hormonal (or endocrine) disruption occurs when chemicals disturb the delicate balance of the levels of the various hormones. Hormones are very potent chemical messengers that travel from one organ to another, specific target organ, to regulate a variety of physiological processes in the human body. For example, the estrogenic hormones regulate the various female reproductive processes, as the menstrual cycle and breast development. The androgenic hormones regulate the typical male processes, such as body hair growth and sperm production. Progestagens play a pivotal role during pregnancy and thyroid hormones are important for fetal development and energy metabolism. Glucocorticoid hormones play a role in inflammation and in the availability of glucose (hence the name glucocoritcoids) for the energy metabolism. When a chemical compound structurally resembles one of these hormones sufficiently to bind to the specific receptor for this hormone, the action of this hormone is unnecessarily increased. Alternatively, a chemical can block the action of a hormone by occupying its receptor without leading to action. Both processes can lead to a disruption of the finely regulated functions of the hormones. Presently, several effect-directed bioassays are available to measure estrogen, androgen, progesteron, glucocorticoid and thyroid activity of contaminants in human and yeast cells (e.g. Jin, 1997; Sonneveld, 2005; Marchesini, 2006).

Genotoxic chemicals are compounds that damage DNA. They can for example bind to the DNA, preventing it from being copied or read. When certain genes on the DNA can not be read, certain essential cellular components cannot be produced anymore, which may be lethal to the cell. The cell therefore has several repair systems, to cut out the piece of DNA with the chemical bound to it and repair the DNA. Some of these repair systems can make errors, leading to changes (mutations) in the DNA. These can be very small changes, such as the change of a nucleotide (i.e. gene mutations), or changes on a larger scale, such as chromosomal breaks (i.e. chromosomal aberrations). Such changes in the DNA can cumulatively lead to a distorted function of a cell, thereby becoming a cancer cell. Thus, genotoxic chemicals can ultimately cause cancer. Mutations in germ cells can lead to changes in the offspring, which can cause congenital diseases.

For detection of genotoxicity, various tests are available. Genotoxicity can be tested by measuring the formation of DNA damage itself (gene mutations or chromosomal aberrations) or by measuring the induction of the various DNA repair enzymes. As DNA damage can occur through different mechanisms, a battery of tests is necessary in both cases. However, for the repair enzymes, no full battery covering all repair types for the different types of damage is available.

Therefore, a battery consisting of a gene mutation test and a chromosomal aberration test is recommended (Heringa, 2005; Pfuhler *et al.*, 2007).

Finally, developmental and teratogenic effects are relevant effects of drinking water contaminants. Teratogenic effects are structural birth defects caused during pregnancy by chemicals passing from the mother to the foetus. A well known example of such an effect are the limb malformations caused by the sedative Softenon in the 1960s. Developmental effects can occur during the whole lifestage from conception to adult, and can range from subtle effects on cognitive function (for example the loss of IQ by lead exposure) to gross effects such as growth retardation. There are several *in vivo* assays to cover adverse developmental (and teratogenic) effects using *Daphnia*, fish or tadpoles as model systems (e.g. Abe *et al.*, 2000; Berry *et al.*, 2007; Gutleb *et al.*, 2007). These are not used in screening of drinking water quality at the moment, but seem promising and relevant tools also as model systems for humans (Berry *et al.*, 2007).

Other toxicological end-points that might be relevant to study in future work are assays related to immunotoxicity and neurotoxicity.

COMPARING ROBUSTNESS AND SENSITIVITY OF VARIOUS *IN VITRO* EFFECT ASSAYS

Effect assays for endocrine disruption

Recently, Leusch (2008) evaluated the performance of five *in vitro* effect-directed assays to assess estrogenic activity in the same water samples; the yeast estrogen screen (YES), ER-CALUX, MELN, T47D-KBluc and E-screen. These assays were selected from an initial broader set of 24 bioassays. Both ER-CALUX and E-screen were shown to be robust and able to sensitively detect estrogenic activity in water, at concentrations of 0.1 to 320 ng/L EEq (equivalents of the most potent estrogenic chemical). The YES and MELN assay gave respectively less sensitive and less quantitative results and results also depended on the matrices. Data were too limited to draw firm conclusions for the KBluc assay.

Similar comparative studies have not yet been performed for other effect assays related to hormone disruption. However, based on 41 literature references the sensitivity, robustness, time to result and operational specifications were evaluated for 20 available effect assays for hormonal disruption in the context of the TECHNEAU project (Table 1; Mons, 2008). The assays are based on mammalian or yeast cells, and are reporter gene assays or cell proliferation assays. As especially for recently developed assays available literature is still scarce, this evaluation should be viewed as indicative. For more detailed information we refer to Annex I of Mons, 2008.

Table 1. Indicative evaluation of different effect assays for hormonal disruption (Mons, 2008).

	Sensitivity		Robustness		Time to result	Operational specifications	
	Sour-ces	Drink-ing water	Operational robustness	Selec-tivity		Ease-of-use	Maintenance requirements
Bioassays for estrogenic activity							
ER CALUX®	4	4	5	3	3	4	3
MVLN and MELN	2	2	2	2–3	2		
T47D-KBluc	3	3			3		
YES	1	1	4	3	2–3	3	3
E-screen	4	4	4	2	1	3	3
Bioassays for androgenic activity							
AR CALUX®	3	3	3	5	3	4	3
MDA-bk2	3	3		1	3		
PALM	3	3	3	3	3		
YAS	2	1	3	3	3		
A-screen	3	3		2	1		
Bioassays for progestagenic activity							
PR CALUX®	3	3	3	5	3	4	3
TM-Luc	3	3		1	3		
Yeast based assays	2	1		4	3		
Bioassays for glucocorticoid activity							
GR CALUX®	3	3	3	5	3	4	3
TGRM-Luc	3	3		1	3		
MDA-kb2	3	3		1	3		
Bioassays for thyroid activity							
TRβ CALUX®			3		3	4	3
T-Screen					1		
PC-DR-Luc							
xL58-TRE-Luc							

1 = very poor, 2 = poor, 3 = average, 4 = good, 5 = very good; when insufficient information was available, the box was left empty.

Effect assays for genotoxicity

Similarly, high-throughput effect assays for genotoxicity were evaluated in the context of the TECHNEAU project (Mons, 2008). Here, 13 different bioassays were evaluated based upon 49 individual literature studies (Table 2). Existing assays are gene mutation assays, chromosomal aberration tests or DNA repair assays. Often the enzyme mix S9 is added to simulate metabolic processes that may lead to the formation of genotoxic metabolic products. As Table 2 shows, information is far from complete.

Table 2. Indicative evaluation of different high-throughput effect assays for genotoxicity (Mons, 2008).

	Sensitivity		Robustness		Time to result	Operational specifications	
	Sources	Drinking water	Operational robustness	Selectivity		Ease-of-use	Maintenance requirements
Gene mutation assays							
Ames test	2	2	4	4	3	4	3
Vibrio harveyi test	2	2			3		
Chromosomal aberration tests							
Comet assay	3	3	4		3		
Micronucleus assay					2–3		
Alkaline elution assay				2	2–3		
Polymerase inhibition assay					3–4		
DNA repair assays							
UMU test	4	4	3–4	3	3	3	
SOS Chromotest					4		
Vitotox®					4		
Mutatox™	3	3	4	3	4		
Greenscreen®					3		
Greenscreen HC					3		
MCF-7-p53R2					3		

1 = very poor, 2 = poor, 3 = average, 4 = good, 5 = very good; when insufficient information was available, the box was left empty.

As stated before, a battery consisting of a gene mutation test and a chromosomal aberration test is recommended (Heringa, 2005; Pfuhler *et al.*, 2007). Such a combination has been shown to have good sensitivity (i.e. a high degree of correct positives) for rodent carcinogenicity, but seems not to be able to escape a poor specificity (i.e. a high degree of false positives) (Kirkland *et al.*, 2005). To assess genotoxicity in surface water, a combination of the Comet assay with human lymphocytes next to the Ames or umu-test has been proven useful (Grummt and Erprobung, 2001; Minnear and Plewa, 2003). If one of the tests is positive in such a scheme, a third assay should be performed, preferably in mammalian cells. If a genotoxic sample remains genotoxic after conventional water treatment, additional research, including unravelling of responsible chemicals and risk assessment, will be necessary

APPLICATION IN DRINKING WATER AND ITS SOURCES AND UNRAVELLING RESPONSIBLE COMPOUNDS

Often sample preparation and concentration are needed before aqueous samples can be tested in effect assays, because of the detection limits of the assays and because water samples need to be cleaned of excess dirt and microorganisms to allow the cells of the test to function properly. During this sample preparation, the original chemical mixture should preferably not be modified due to e.g. volatilization or sorption. Detection limits of the *in vitro* effect-directed bioassays, including sample concentration, should be sufficiently low to detect water concentrations that are relevant for human health risks.

The identity of the compounds responsible for the toxicological effect remains unknown, unless the effect directed bioassays are combined with analytical chemical techniques in a toxicity identification evaluation (Houtman *et al.*, 2006). Especially accurate mass screening and identification using LC–LTQ FT Orbitrap MS has proven useful in elucidating unknown compounds responsible for observed effects in *in vitro* effect assays (Hogenboom *et al.*, 2009). This asks for sample preparation techniques that are suitable for both chemical analysis and evaluation in effect assays. Using solid phase extraction (SPE) with modern, broad-spectrum extracting columns, this can be accomplished.

Van der Linden *et al.* (2008) applied the various CALUX assays for hormonal disruption to various effluents, surface waters and drinking waters in the Netherlands. Estrogen, androgen, progesterone and glucocorticoid activities were shown in effluents and surface waters, but not in drinking water. They showed that androgen and progesterone activities were very low, but especially glucocorticoid activities were high compared to the other hormonal effects.

Schriks *et al.* (in preparation) identified and quantified the compounds responsible for the observed effects in these samples, which were mostly immunosuppressive pharmaceuticals used against diseases such as asthma and eczema. These compounds were not included in any regular monitoring programme thus far.

Heringa *et al.* (in preparation) have applied the Ames II, Comet and micronucleus assays to a sewage water treatment plant effluent, several surface waters and several contaminated ground waters in the Netherlands. They found genotoxic activity in the effluent, one surface water and one ground water in the Ames II and/or Comet assays, but no responses were measured with the micronucleus assay. The responsible chemicals have not been elucidated, but the occurrence of detectable levels of genotoxic compounds in these water samples was not known so far. It remains to be investigated whether the responsible genotoxic compounds pass the treatment steps in drinking water preparation.

The aforementioned examples show that a combination of using sensitive *in vitro* effect assays and subsequent accurate mass screening to identify and quantify responsible chemicals is a powerful combination to improve the assessment of the drinking water quality.

INTERPRETATION OF EFFECT ASSAYS IN TERM OF HUMAN HEALTH RISKS

Currently, the significance of test results in the effect-directed bioassays in terms of human health risks is still in debate, as *in vitro* cell cultures are clearly very different from the whole of the human body. Current human-health based drinking water quality limits are generally based on Acceptable Daily Intake values (ADIs), the quantity of a compound (expressed per kg body weight per day) that can be ingested without adverse effects. This ADI is mostly based upon *in vivo* mammalian toxicological data, taking into account the exposure to one compound only. In *in vivo* assays adsorption, distribution, metabolism and elimination (ADME) of the compound will influence its toxicity. These factors will not always be well mimicked in *in vitro* assays. In order to come to effect-based quality limits for complex mixtures, it is proposed to base these on the Acceptable Daily Intake value (ADI) of a highly potent reference compound, and to translate this value into effect-based limits using worst case assumptions for ADME (Van der Oost, 2008). This method has the disadvantage of probably leading to overprotective limits in most cases, but seems the only way to effectively apply effect-directed assays in water quality assessment. Further work is necessary to develop such water quality limits expressed in terms of acceptable effects in *in vitro* assays.

CONCLUSIONS

- *In vitro* effect assays are useful in the assessment of (drinking) water quality as they detect the presence of harmful chemicals that analytical chemical methods might not pick up, and they show the cumulative effect of all chemicals present in the water.
- Relevant effects to analyze in drinking water and its sources are genotoxicity, endocrine disruption, developmental and teratogenic effects and possibly neurotoxicity and immunotoxicity, as these are effects that can occur at relatively low exposures already.
- There are various assays available for the mentioned effects, differing in sensitivity, specificity and robustness.
- Application of these assays to various water types in the Netherlands has shown the presence of possibly harmful chemicals which were not detected by the current monitoring programme.
- There is a first idea on how to develop water quality limits expressed in terms of acceptable effects in *in vitro* effect assays, by basing them on the ADI of a reference compound. Further work on this topic is necessary.

REFERENCES

Abe, T., Saito, H., Niikura, Y., Shigeoka, T. and Nakano, Y. (2000). Embryonic development assay with Daphnia magna: Application to toxicity of chlorophenols. *Water Science and Technology*, **42**, 297–304.

Benotti, M.J., Trenholm, R.A., Vanderford, B.J., Holady, J.C., Stanford, B.D. and Snyder, S.A. (2009). Pharmaceuticals and endocrine disrupting compounds in U.S. drinking water. *Environmental Science and Technology*, **43**(3), 597–603.

Berry, J.P., Gantar, M., Gibbs, P.D.L. and Schmale, M.C. (2007). The zebrafish (Danio rerio) embryo as a model system for identification and characterization of developmental toxins from marine and freshwater microalgae. *Comparative Biochemistry and Physiology C*, **145**, 61–72.

Diamond, M.L. and Hodge, E. (2007). Urban contaminant dynamics: From source to effect. *Environmental Science and Technology*, **41**(11), 3796–3800.

Grummt, T. and Erprobung (2001). Vergleich, Weiterentwicklung and Beurteilung von Genotoxizitätstests für Oberflächenwasser. *Wasser Abwasser*, **142**(5), 346–355.

Gutleb, A.C., Schriks, M., Mossink, L., Van den Berg, J.H.J. and Murk, A.J. (2007). A synchronized amphibian metamorphosis assay as an improved tool to detect thyroid hormone disturbance by endocrine disruptors and apolar sediment extracts. *Chemosphere*, **70**, 93–100.

Heringa, M. (2005). *Approach for Assessment of Carcinogenic Activity in Water – Part 1*. BTO report 2005.022. KWR Watercycle Research Institute, Nieuwegein, The Netherlands.

Hogenboom, A.C., Van Leerdam, J.A. and De Voogt, P. (2009). Accurate mass screening and identification of emerging contaminants in environmental samples by liquid chromatography-hybrid linear ion trap Orbitrap mass spectrometry. *Journal of Chromatography A*, **1216**, 510–519.

Houtman, C.J., Booij, P., Jover, E., Pascual del Rio, D., Swart, K., Van Velzen, M., Vreuls, J.J., Legler, J., Brouwer, A. and Lamoree, M.H. (2006). Estrogenic and dioxin-like compounds in sediment from Zierikzee harbour identified with CALUX assay-directed fractionation combined with one and two dimensional gas chromatography analyses. *Chemosphere*, **65**, 2244–2255.

Jin, L., Tran, D.Q., Ide, C.F., McLachlan, J.A. and Arnold, S.F. (1997). Several synthetic chemicals inhibit progesterone receptor-mediated transactivation in yeast. *Biochim. Biophys. Res. Commun.*, **233**(1), 139–146.

Kirkland, D., Aardema, M., Henderson, L. and Müller, L. (2005). Evaluation of the ability of a battery of three in vitro genotoxicity tests to discriminate rodent carcinogens and non-carcinogens. I. Sensitivity, specificity and relative predictivity. *Mutation Research*, **584**, 1–256.

Leusch, F.D.L. (2008). *Tools to Detect Estrogenic Activity in Environmental Waters*. GWRC, London, UK.

Loos, R., Gawlik, B.M., Locoro, G., Rimaviciute, E., Contini, S. and Bidoglio, G. (2009). EU-wide survey of polar organic persistant pollutants in European river waters. *Environmental Pollution*, **157**, 561–568.

Marchesini, G.R., Meulenberg, E., Haasnoot, W., Mizuguchi, M. and Irth, H. (2006). Biosensor recognition of thyroid-disrupting chemicals using transport proteins. *Analytical Chemistry*, **78**, 1107–1114.

McCarty, L.S. and Borgert, C.J. (2006). Review of the toxicity of chemical mixtures: Theory, policy, and regulatory practice. *Regulatory Toxicology and Pharmacology*, **45**, 119–143.

Minear, R.A. and Plewa, M.J. (2003). *Comparative Genotoxicity Assessment of DBPs in Drinking Water*; Awwa Research Foundation and American Water Works Association, report number 90939.

Mons, M. (2008). *Monitoring and Control of Drinking Water Quality – Inventory and Evaluation of Monitoring Technologies for Key-parameters*. TECHNEAU report D.3.1.3.

Pfuhler, S., Albertini, S., Fautz, R., Herbold, B., Madle, S., Utesch, D. and Poth, A. (2007). Genetic toxicity assessment: employing the best science for human safety evaluation part IV: Recommendation of a working group of the Gesellschaft fuer Umwelt-Mutationsforschung (GUM) for a simple and straightforward approach to genotoxicity testing. *Toxicological Sciences*, **97**(2), 237–240.

Reemtsma, T., Weiss, S., Mueller, J., Petrovic, M., Gonzalez, S., Barcelo, D., Ventura, F. and Knepper, T.P. (2006). Polar pollutants entry into the water cycle by municipal wastewater: A European perspective. *Environmental Science and Technology*, **40**, 5451–5458.

Richardson, S.D. (2008). Environmental mass spectrometry: Emerging contaminants and current issues. *Analytical Chemistry*, **80**, 4373–4402.

Schwarzenbach, R.P., Escher, B.I., Fenner, K., Hofstetter, T.B., Johnson, C.A., Von Gunten, U. and Wehrli, B. (2006). The challenge of micropollutants in aquatic systems. *Science*, **313**, 1072–1077.

Sonneveld, E., Jansen, H.J., Riteco, J.A., Brouwer, A. and van der Burg, B. (2005). Development of androgen- and estrogen-responsive bioassays, members of a panel of human cell line-based highly selective steroid-responsive bioassays. *Toxicological Sciences*, **83**(1), 136–148.

Turgeon, S., Rodriguez, M.J., Theriault, M. and Levallois, P. (2004). Perception of drinking water in the Quebe region (Canada): The influence of water quality and consumer location in the drinking water system. *Journal of Environmental Management*, **70**, 363–373.

Van der Linden, S.C., Heringa, M.B., Man, H.Y., Sonneveld, E., Puijker, L.M., Brouwer, A. and Van der Burg, B. (2008). Detection of multiple hormonal activities in wastewater effluents and surface water, using a panel of steroid receptor CALUX bioassays. *Environmental Science and Technology*, **42**, 5814–5820.

Van der Oost, R. (2008). *Vision Upon Mixture Toxicity in Drinking Water*. BTO report 2008.009. KWR Watercycle Research Institute, Nieuwegein, the Netherlands (in Dutch).

van Genderen, J., Mons, M.N. and van Leerdam, J.A. (2000) *Inventory and Toxicological Evaluation of Organic Micropollutants, revision 1999*. Report of the Association of River Waterworks – RIWA, Amsterdam.

The versatility of online UV/Vis-spectrometry – an overview

J. van den Broeke[a], P.S. Ross[b], F. Edthofer[a], A.W.C. van der Helm[b,c], A. Weingartner[a] and L.C. Rietveld[b]

[a]scan Messtechnik GmbH, Brigittagasse 22–24, A-1200 Vienna, Austria
(E-mail: jvandenbroeke@s-can.at)
[b]Delft University of Technology, PO BOX 5048, 2600 GA Delft, The Netherlands
(E-mail: p.s.ross@tudelft.nl)
[c]Waternet, PO Box 8169, 1005 AD Amsterdam, the Netherlands

Abstract On-line optical instruments for water quality measurements can be found in all types of applications, from drinking water to wastewater. Traditionally, these instruments measure one or two wavelengths in the visible or UV range of the spectrum, but more advanced instruments that measure the full range of UV and visible light with high resolution have now established themselves on the market. These advanced spectrometer probes can measure multiple parameters, such as turbidity, nitrate and sum organics, simultaneously and in real time with a high frequency. In the TECHNEAU project, such instruments are in use to monitor all aspects of the drinking water production and distribution chain. New developments and applications using this technology that have been triggered by the TECHNEAU project are described in this contribution.

Keywords alarm, AOC, drinking water, multi-parameter, online, spectrometry, UV/Vis

INTRODUCTION

During the eighties, the introduction of miniature diode array detectors, combined with powerful microprocessors and state-of-the-art mathematical tools, led to a renaissance of UV-Vis spectrometry and enabled the rapid spread of compact, relatively low-cost yet still powerful laboratory UV-Vis machines. During the nineties, the technology moved from the lab into the field, but still in the form of relatively large, complicated and expensive off-line analysers. However, the market for in-situ optical sensors/probes still was – and still is – dominated by relatively simple photometers, which are able to measure only one or two wavelengths at a time. These instruments are thus limited to measuring one parameter only, and they generally employ quite crude and unstable methods to compensate for cross-sensitivities to variations, for example in a water matrix. Examples of widely used applications of UV/Vis-spectroscopy in water quality monitoring are the measurement of nitrate, turbidity and organics (SAC254). A lot of additional information, however, as well as reduced cross sensitivity, can be

 TECHNEAU: Safe Drinking Water from Source to Tap. Edited by Theo van den Hoven and Christian Kazner. ISBN: 9781843392750. Published by IWA Publishing, London, UK.

obtained when using the entire absorption spectrum instead of single wavelengths. The developments in electronics and optics in the recent years have enabled the marriage between full spectrum UV/Vis spectroscopy and small scale instruments (van den Broeke *et al.*, 2006). Moreover, these developments have allowed the design of robust spectroscopic analysers that can function in harsh environments, such as inside sewer and industrial process streams, with little or no maintenance because neither chemicals nor moving parts for cleaning are necessary for their functioning. The first compact on-line instruments for use in the field have been on the market since the start of the decennium and have since established themselves in all imaginable types of water quality monitoring applications (Figure 1). Within the TECHNEAU project, such online spectrophotometer instruments have been employed in drinking water monitoring applications. Furthermore, additional capabilities and features have been developed which enhance the capabilities of systems based on such instruments.

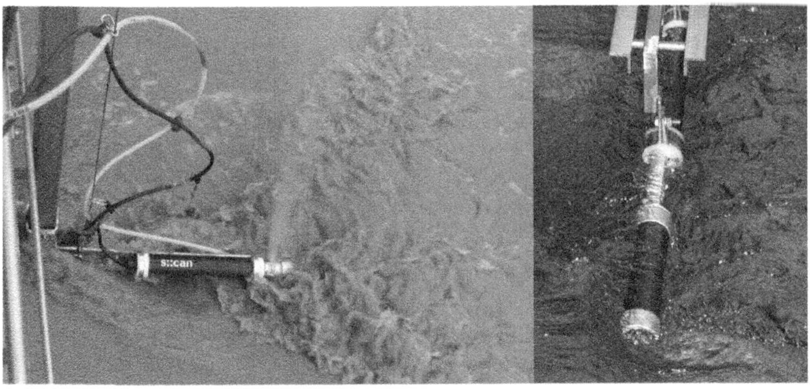

Figure 1. Examples of in-situ applications of UV/Vis-spectroscopy; process control in waste water treatment (left) and drinking water monitoring (right).

UV/VIS SPECTROMETER

The spectro::lyser™, from scan Messtechnik, was used in all measurement campaigns described herein. The commercially available version of this instrument was used, and it was not developed or changed in the course of the project. The instrument is a fully submersible on-line two beam UV/Vis spectrophotometer. The probes employed were equipped with 100 and 35 mm optical measuring cells, and measured absorbance between 200 and 750 nm, with a resolution of 2.5 nm (Figure 2).

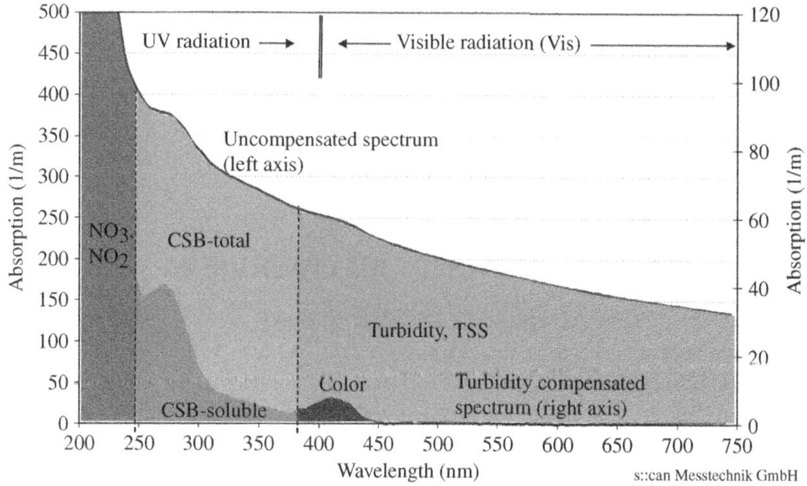

Figure 2. UV/Vis spectrum and examples of parameters derived out of this spectrum together with their characteristic absorbance profiles.

The instruments are built as a compact submersible sensor enabling measurement of UV/Vis spectra directly in liquid media with a detector precision of 0.004 extinction units, which corresponds to 0.035 and 0.10 abs/m for the 35 and 100 mm instruments respectively. Signal drift is <0.01 extinction units in 30 days both in water and in air. The linear range of the instrument in absorption measurements is between 0–70 abs/m (35mm) or 0–25 abs/m. (100 mm).

DEVELOPMENT OF SPECTRAL ALGORITHMS

The spectro::lyser™ is equipped with calibration algorithms that allow the measurement of turbidity, nitrate and TOC and also perform automatic turbidity compensation of the spectrum. The latter allows the subsequent calculation of DOC. The wavelengths used for the determination of all these parameters have been selected using principal component analysis (PCA) and partial least square regression (PLS) on hundreds of datasets containing both UV/Vis spectra and reference values of these parameters (Langergraber *et al.*, 2003), the latter being determined using established laboratory techniques. The use of such multi-wavelength algorithms allows much higher specificity than can be achieved with conventional, single or dual wavelength, photometry (Rieger *et al.*, 2004). As absorption peaks of substances often overlap, no distinction between such substances would be possible with the more primitive systems, whereas multi-wavelength spectroscopy can do so as a matter of course.

The applied PCA and PLS models combine full spectral analysis with various validation procedures and outlier tests, to reach both high correlation quality and robustness (Otto, 1999), even in complicated mixtures of substances. For development of new measurement parameters in the work described here an identical approach was employed.

TECHNEAU APPLICATIONS, AN OVERVIEW

Within the TECHNEAU project, the spectro::lyser™ was used both as a monitoring tool as well as a development tool. The aim was the broadening of the applicability of the technology for use in drinking water. This was done through research and development triggered by experience gathered in the installations performed in the framework of TECHNEAU. Up until the time of writing, it has been used in:

- Vienna Waterworks: installation at a mountain spring, where water quality events were monitored by measuring turbidity, TOC, SAC254 and spectral alarms parameters.
- Waternet/TU Delft – Amsterdam: measurement at Weesperkarspel pilot plant, focused on analysis of spectral changes during ozonation, pellet softening and activated carbon filtration. Two new parameters were developed based on these results.
- NTNU – Trondheim: use as a research tool to monitor and control coagulation as pretreatment step for the oxidation – biofiltration – membrane process.
- SINTEF – Trondheim & Bergen: colour and turbidity monitoring as control parameter for coagulation.
- RTU – Riga: monitoring of compositional change of water throughout the distribution network using spectral data as well as turbidity, nitrate and DOC measurements.

Several additional tests are due in 2009 and 2010.

NEW MEASUREMENT PARAMETERS – OZONE

Ozone, a very powerful oxygen agent, is used in drinking water treatment as a disinfectant and the oxidation of organic material present in the raw water, both for organoleptic and public health purposes. Measurement of the effective ozone dose is important to determine the efficacy of the treatment, however, the very short half life time of ozone in water makes concentration measurements, especially using grab sampling, difficult. The use of an online and in-situ tool that is simple to use and needs little maintenance is therefore highly interesting. As ozone is a strong absorber of UV-light (Molina, 1986), the use of UV/Vis spectroscopy was expected to be suitable for this purpose.

During the research performed at Waternet, the spectrometer probes were used to measure samples taken from 36 sampling points in the ozone reactors available at the Weesperkarspel pilot plant (van den Broeke *et al.*, 2008). Using existing spectral information about ozone as well as the results collected during this study, a spectral algorithm for the measurement of dissolved ozone was developed. Very high correlations between results from the spectrometer and analytical reference technologies were obtained ($R^2 > 0.97$). Furthermore, the algorithm was also applied in measurement of communal waste water. This was done in Regensdorf (Switzerland), where ozone oxidation was used for tertiary waste water treatment in a pilot study performed under the framework of the EC research project NEPTUNE. In this application again very good results were obtained, and a comparison of the spectral ozone measurement with an electrochemical analyser at this site showed similar performance for both instruments, but with lower measurement noise on the spectral measurement (Figure 3).

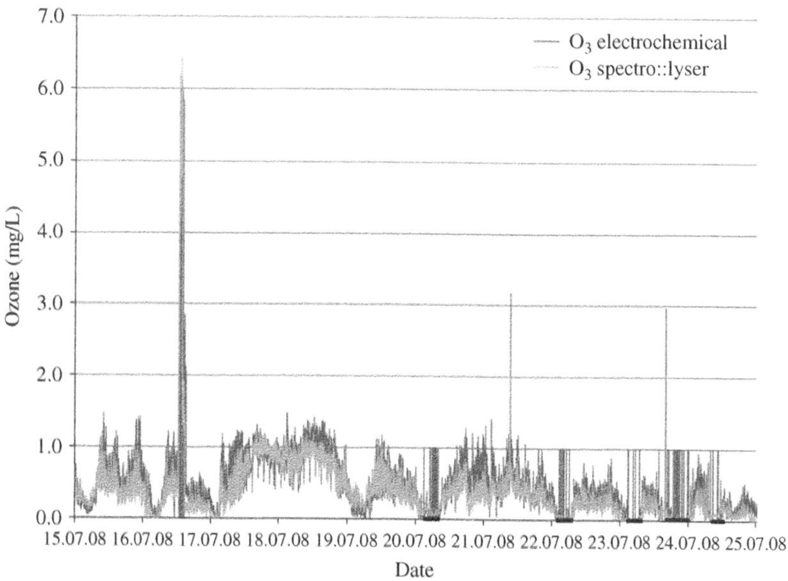

Figure 3. Overlaid measurement results of spectral and electrochemical ozone measurement in waste water.

New measurement parameters – assimilable organic carbon

Besides measuring the ozone concentration profile across the reactor, the spectrometer probe also allowed the measurement of the change in composition of the organics during oxidation. During oxidation of the natural organic substances in the source water large molecules, such as humic and fulvic acids, are cracked into smaller ones (Hammes *et al.*, 2006). A side-effect of this procedure is the increased availability of microbiologically assimilable organic carbon (AOC). This is an important parameter especially in drinking water networks where no residual disinfectant is present in the water. The measurement of AOC is performed using cell cultures (van der Kooij, 1992), which is a lengthy procedure. A more rapid assessment of the AOC levels produced during treatment would be very valuable, as it can allow the optimisation of treatment for the actual water quality. During oxidation the total amount of organics in the water does not change significantly. As a result, classical spectroscopically measurable sum parameters such as DOC are not very well suited to assess AOC concentrations. For estimation of AOC a new spectral algorithm was required.

Online UV/Vis spectroscopy has proven itself as a tool that allows the collection of specific information on the removal efficiency and subsequent concentrations of (organic) substances in water. For example, the parameters total suspended solids, chemical oxygen demand (COD) and nitrate (NO_3-N) can be readily monitored with UV/Vis spectral measurements (Langergraber *et al.*, 2004). Other examples are the prediction of disinfection by-product formation by chlorination (Korshin *et al.*, 1997) and chlorine demand (Chow *et al.*, 2006). The use of two on-line spectrometer instruments, placed before and after a treatment step, and the calculation of the differential spectrum between these two sites could open up a further area in water quality monitoring and process control as it allows calculation and prediction of water quality parameters previously unavailable. In the case of DBP formation, the differential spectral information is used as an indicator parameter (i.e. as soft sensor); the concentration level of the substances of interest is too low to be measured directly, but relationships between spectral change and the concentration levels of the substances do allow reliable calculation of their concentration (Korshin *et al.*, 1997). A similar approach allowed the measurement of changes in AOC concentrations during ozonation and during subsequent pellet softening and activated carbon filtration. Laboratory analyses of AOC levels, using cell culturing methods, and UV/Vis spectra were combined to develop spectral algorithms that predict the change in AOC from the spectral changes incurred in a treatment process (Figure 4). The relation was determined for the entire treatment process, but also visible, but less strong, for the three individual treatment steps evaluated.

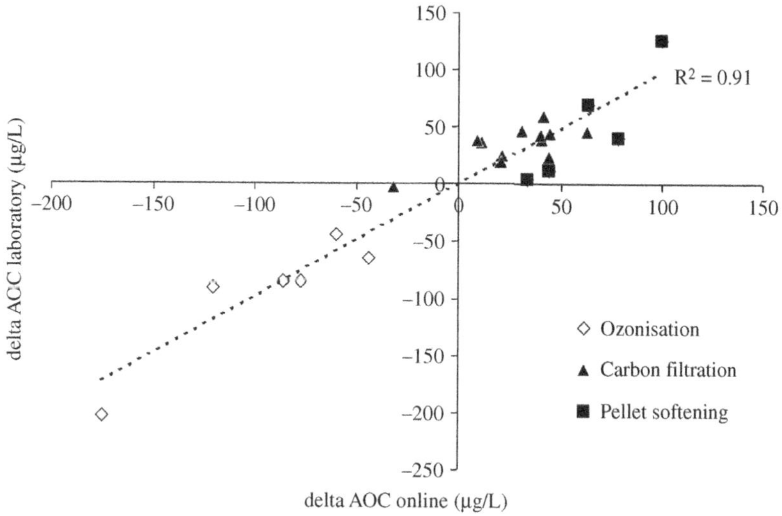

Figure 4. Correlation between AOC changes over individual process steps and predicted AOC change based on differential spectra.

Changes in the character of NOM throughout the year result in changes in the UV-spectra during the treatment steps. Due to these changes the relation between the delta AOC and the differential spectrum should be refined, using data from a longer period. Then, on the basis of the on-line measurement of the UV-spectrum of the raw water the operation of the different treatment steps can be adjusted to optimal settings.

DATA EVALUATION AND WATER QUALITY ALARMS

Besides looking at parameters for particular substances or sum parameters, developments in data interpretation and event detection were undertaken within TECHNEAU. It is well known that no water quality sensor is perfect, and issues like measurement noise, drift and measurement outliers lead to erroneous measurements. Such issues lead to untrustworthiness of sensors and, particularly when used in an automated detection system, trigger "false alarms". Experience shows that false alarms typically result from feeding non-validated data from various sources into alarm algorithms. All alarm algorithms principally have a hard time distinguishing between sensor-related anomalies and water-related anomalies ("events"). Events typically disappear in the noise of non-validated online data, although they theoretically could be detectable if the baseline data were validated, noise-reduced and clean. Alarm algorithms can not even be evaluated with regards to their performance as long as they are fed by such noisy and unreliable data. Both the issues of generating validated data and building an alarm tool have been addressed.

The validation tool that has been developed combines multiple functionalities. It cleans the data through automatic outlier removal and noise correction. It also flags data when steps and or drift are recognised, so that the subsequent use of these results in the alarm system can be considered. The validation builds upon software for station management that allows tracking of all station properties and activities that are used as input for the validation as well. In this way maintenance and calibration that, respectively, affects water quality at the station or leads to a sudden change in sensor outputs, do not lead to a false alarm. This software is intentionally designed as an open platform that can accept data from various sensors, regardless of the sensor type or manufacturer.

Anomaly detection is performed using validated data only. The detection consists of two different modules, one that works with spectral data only and one that works with all types of data. The spectral alarm software "learns" the shape and features of the "normal" spectrum of the water body in a defined range. The initial definition of the reference – or baseline – spectra for later identification of extraordinary composition is done fully automatic, and can be continuously improved / re-trained. This functionality is able to detect operational deviations as well as many contaminations, and has been in use for several years already (Langergraber *et al.*, 2004).

To this the analysis of patterns between all available parameters, including non-spectral parameters as well as the above mentioned alarm parameters, has been added. All collected and validated parameters from any sensor (including e.g. biological sensors) can be fed into a module for baseline definition and anomaly detection. Functionalities of this module include: static alarm thresholds, dynamic alarm thresholds (time-gradients), pattern recognition based alarm parameter and composite alarms. In this method, linear combinations are composed of (a selection of) the parameters measured and these linear combinations are then classified. When the results for different measurements give the same or similar outcomes, the measurements are allocated to the same class. When the results are significantly different, they are allocated to different classes. This approach was tested during the monitoring campaign at Vienna Waterworks, where it was found that different states of the water composition could be identified. This spring water, which is sometimes affected by heavy rainfall, showed a constant ratio between NO_3-N, DOC and turbidity when no rainfall event occurred. For all these measurements a linear combination of these parameters will provide similar results. During the rainfall events, the ratios change, with NO_3-N remaining constant, but turbidity going up very strongly. This changing ratio makes possible the separation of rainfall events into a separate class. During training of a monitoring system different classes can be identified and the appropriate causes can be allocated (different water sources, operational changes, etc). Such a trained system will then only respond to anomalies that have not been seen before. Figures 5 and 6 show an example of the classification performed. In this example, from the city of Glendale (Arizona, USA) who have already implemented this system (Thompson *et al.*, 2009), a combination of 8 parameters was used to determine the linear combination. These parameters were Turbidity, TOC, DOC, nitrate (all measured with the spectrometer probe) as well as pH, conductivity, chlorine residual and temperature.

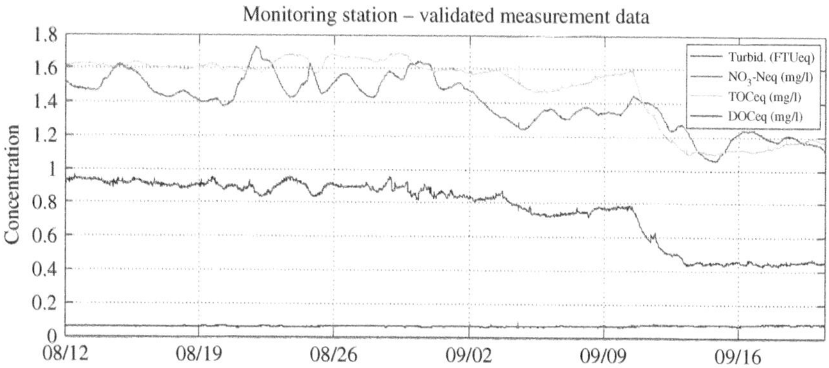

Figure 5. Variation of water quality parameters as a result of a change in water source.

CONCLUSIONS

This paper has presented a brief overview of the developments in the field of online water quality measurement in general, and online UV/Vis spectroscopy in general, that have been triggered by the TECHNEAU project. UV/Vis spectrometry, by presenting a rich and multidimensional measurement signal, has been used to measure a broad variety of water quality parameters. The development of a new algorithm to directly measure dissolved ozone and an algorithm that functions as an indicator parameters for changes in AOC levels

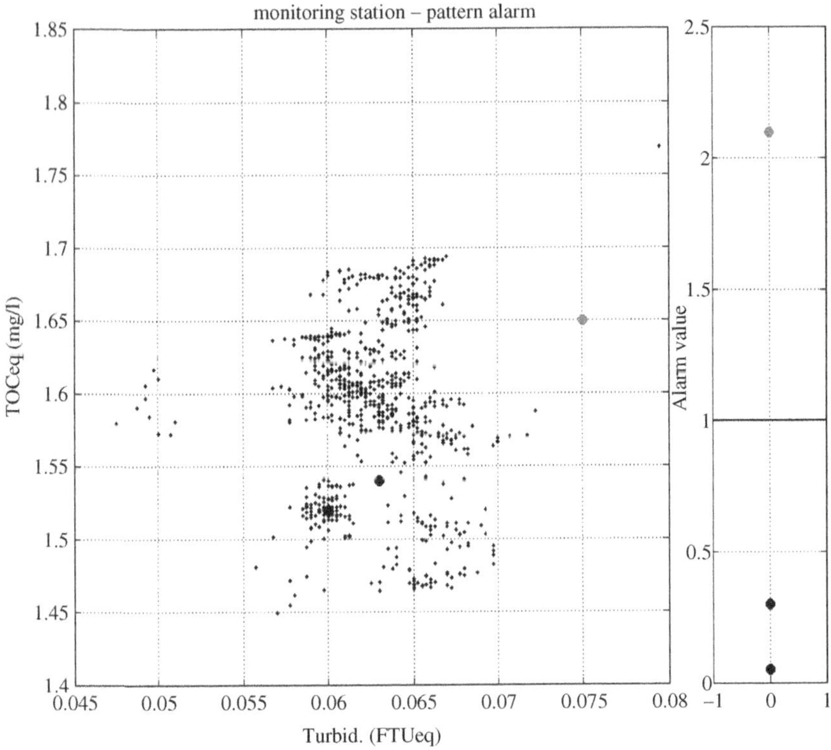

Figure 6. A visualisation of the pattern recognition and classification tool. Displayed are 2 of the 8 dimensions used in the analysis of this example (see Figure 5). On the x-axis Turbidity is plotted and on the y-axis TOC. On the right hand side of the figure the alarm response of 3 patterns is shown, after the system had learned the pattern of both water compositions.

during treatment are a result of this. The latter in particular is an example of how a combination of information can be used to determine parameters that can not be measured by single sensors. Such concepts of virtual sensors, also referred to as soft sensors, were also successfully applied to develop a system for detection and classification of changes in water quality.

ACKNOWLEDGEMENTS

This research was financially supported by the European Community through the project TECHNEAU (contract number 018320).

REFERENCES

Chow, C.W.K., Fitzgerald, F., Sutherland-Stacey, L., Dexter, R., Fabris, R. and Drikas, M. (2006). *Applications of UV/Vis spectrometry in Drinking Water Quality Management.* Proceedings of the Australian Water Association '2nd Annual Water Industry Engineers and Operators (SA) Conference, Adelaide (AUS), April 4, 2006.

Hammes, F., Salhi, E., Köster, O., Kaiser, H.-P., Egli, T. and von Gunten, U. (2006). Mechanistic and kinetic evaluation of organic disinfection by-product and assimilable organic carbon (AOC) formation during ozonation of drinking water. *Water Research*, **40**(12), 2275–2286.

Korshin, G.V., Li, C.-W. and Benjamin, M.M. (1997). The decrease of UV-absorbance as an indicator of TOX formation. *Water Research*, **31**(4), 946–949.

Langergraber, G., Fleischmann, N. and Hofstädter, F. (2003). A multivariate calibration procedure for UV/Vis spectrometric quantification of organic matter and nitrate in wastewater. *Water Science and Technology* **47**(2), 63–71.

Langergraber, G., Weingartner, A., Fleischmann, N. (2004). Time resolved delta spectrometry: a method to define alarm parameters from spectral data. *Water Science and Technology*, **50**(11), 13–20.

Molina, L.T. and Molina, M.J. (1986). Absolute absorption cross sections of ozone in the 185–350 nm wavelength region. *Journal of Geophysical Research*, **91**, 14501–14508.

Otto, M. (1999). *Chemometrics: Statistics and Computer Application in Analytical Chemistry.* Wiley VCH Verlag GmbH, Weinheim, Germany.

Rieger, L., Langergraber, G., Thomann, M., Fleischmann, N. and Siegrist, H. (2004). Spectral in-situ analysis of NO_2, NO_3, COD, DOC and TSS in the effluent of WWTP. *Water Science and Technology*, **50**(11), 143–152.

Thompson, K., Scott, R.A., Kadiyala, R., Weingartner, A. and van den Broeke, J. (2009). *Advances in Data Validation, Event Detection, and Communications Structures for a CWS – case study: Glendale, Arizona.* Proceedings of AWWA 2009 Water Security Congress, Washington DC (USA), April 8–10.

van den Broeke, J., Langergraber, G. and Weingartner, A. (2006). Online and insitu UV/vis spectroscopy for multi-parameter measurements: a brief review. *Spectroscopy Europe*, **18**(4), 15–18.

van den Broeke, J., Ross, P.S., van der Helm, A.W.C., Baars, E.T. and Rietveld, L.C. (2008). Use of online UV/Vis-spectrometry in the measurement of dissolved ozone and AOC concentrations in drinking water treatment. *Water Science and Technology*, **57**(8), 1169–1175.

van der Kooij, D. (1992). Assimilable organic carbon as indicator of bacterial regrowth. *Journal AWWA*, **84**(2), 57–65.

Risk Assessment and Risk Management

International outlook on WSP tools and standards

T.J.R. Pettersson

Department of Civil and Environmental Engineering, Chalmers University of Technology, SE-412 96 Göteborg, Sweden (E-mail: thomas.pettersson@chalmers.se)

Keywords risk management, tools, water safety plan

EXTENDED ABSTRACT

The World Health Organisation (WHO, 2004) has in its Guidelines for drinking-water quality concluded that end-product testing, i.e. analyses of water quality in treatment and distribution systems, is not sufficient to guarantee safe drinking water to consumers. The results from the analyses are often received after the water is consumed and they may not provide correct information regarding the health effects of the water. To ensure safe drinking water a holistic risk management approach from catchment to consumer is emphasised. WHO stresses the importance of using the risk-based approach when managing drinking water supplies and in particular through comprehensive Water Safety Plans (WSPs), where all steps in the water supply chain, from catchment to consumer, must be integrated. WSPs are currently being implemented in several countries and are expected to become an increasingly important part of water management in both developed and developing countries. The Bonn Charter strategy (IWA, 2004) also promotes an integrated approach and further specifies the use of WSP in drinking water management.

During 2008 an assessment study of the revision of the European Drinking Water Directive (DWD) was carried out and the scope of the revised DWD suggests a risk-based approach in the management of water supply (Hulsmann, 2009). The risk-based approach is comparable to water safety plans and all water supplies (large, medium, small and individual) have to adopt WSP and risk-based management. It is clear that the large or medium sized (and complicated) supplies need to make a full-scale WSP. Another requirement will be independent audit of the WSP's on a regular basis, suggesting annually audits for large supplies whereas the frequency for the medium and small sized supplies will be on a 2–5 years basis or decided by the Member States. The long-term implication for the European water utilities is an increased need for applicable tools and methods to carry out a WSP, especially for smaller utilities with no or little experiences of risk-based management.

In the literature there are only a few, if any, methods or tools that are directly applicable setting up and implement a WSP for water utilities. On the other hand there are several methods, approaches and tools that individually can assist the water suppliers in assessing separate parts of the supply chain. In a state-of-the-art review article MacGillivray *et al.* (2006) describe a variety of methods and tools for risk analysis for water utilities, including strategic (new technology), programme (asset management) and operational (plant performance) risk analysis. However, there is a need for the utilities to develop their own risk analysis strategy to fit the purpose.

WHO provides general but not detailed information on tools and methods for risk assessment and risk management in WSPs, but provide general descriptions of hazard identification and a method for qualitative (or semi-quantitative) classification of risks (WHO, 2004; Davison *et al.*, 2005). The method provides a useful structure for risk assessment and facilitates a ranking of risks as a basis for prioritizing safety measures. The result is often presented using a risk matrix but a major drawback of risk ranking methods is that it is difficult to properly model complex systems with interactions between components.

In the TECHNEAU project (WA4) a generic framework for integrated risk management has being developed, The framework is aimed at providing a comprehensive structure for integrated risk management from catchment to consumer in WSPs, considering both water quality and water quantity at different levels of complexity and for both quality and quantity risks. A clear definition of the scope of the risk management process is provided.

Guidance reports on risk management methods in WSP are being developed as well. An introduction to risk analysis methods for water supply, from source to tap, is published describing which problems can be addressed by various methods. The capabilities of the methods and typical results of the analyses are presented.

Specific tools for risk assessment and risk managelent that are being developed in the TECHNEAU project include:

- The TECHNEAU Hazard Database (THDB) aimed at supporting end-users in identifying hazards providing a catalogue with potential hazards of generic technical, environmental or human origin in the entire supply system divided into 12 sub-systems (10 physical, one organizational and one on future hazards). The THDB has been tested with water companies in case studies and being constantly updated.
- The TECHNEAU Risk Reduction option Database (TRRDB), provides a catalogue with risk reduction options listed. The objective of the TRRDB

is to help personnel in water supply companies to identify relevant risk reduction measures to mitigate the hazards identified. All options are divided into the same 12 sub-systems, as for the THDB, providing a catalogue with three different types of risk reducing options (control, information/education and physical barrier options).

- The TECHNEAU decision support tool (TDST) developed for risk management is integrating asset management and cost optimisation. In decision-making situations a number of alternative actions may be considered where these are typically a trade-off between risk reduction and economic costs. Cost optimisation will be based on the concept of economic decision analysis, where alternatives for risk reduction are compared to the costs for obtaining them, from simple qualitative ranking methods to more advanced methods such as cost-effectiveness and cost-benefit analysis.
- Good examples on integrated risk assessment from six case studies (Göteborg, Bergen, Amsterdam, Freiburg-Ebnet, Březnice/Czech Republic and Upper Mnyameni/South Africa) were carried out in 2007–2008. The aim was to apply and evaluate the applicability of different methods for risk analysis and to test and update the THDB and TRRDB tools.
- Training sessions on risk assessment and risk management in WSP for water utility personnel in have been developed and arranged in three different parts of the word (North and South Europe and South Africa). Water utility personnel attended a two-days seminar carrying out a risk assessment on a real water supply system using the tools THDB and TRRDB.

All tools and reports within the TECHNEAU project are available on: www.techneau.eu

REFERENCES

Davison, A., Howard, G., Stevens, M., Callan, P., Fewtrell, L., Deere, D. and Bartram, J. (2005). *Water Safety Plans: Managing drinking-water quality from catchment to consumer*. WHO/SDE/WSH/05.06, World Health Organisation, Geneva.

Hulsmann, A. (2009). *Decision time for Europe's drinking water directive revision*, Water 21, February 2009, pp. 42–44.

IWA (2004). *The Bonn Charter for Safe Drinking Water*. International Water Association, London.

MacGillivray, B.H., Hamilton, P.D., Strutt, J.E. and Pollard, S.J.T. (2006). Risk analysis strategies in the water utility sector: An inventory of applications for better and more credible decision making. *Critical Reviews in Environmental Science and Technology*, **36**(2), 85–139.

WHO (2004). *Guidelines for drinking-water quality*. Third Edition, Volume 1 Recommendations, World Health Organization, Geneva.

Quantitative risk assessments of water supply systems from source to tap

A. Lindhe[a], L. Rosén[a], O. Bergstedt[b], T. Norberg[c] and T.J.R. Pettersson[a]

[a]Department of Civil and Environmental Engineering, Chalmers University of Technology, SE-412 96 Göteborg, Sweden (E-mail: lars.rosen@chalmers.se; andreas.lindhe@chalmers.se; thomas.pettersson@chalmers.se)
[b]Göteborg Vatten, Box 123, SE-424 23 Angered, Sweden
(E-mail: olof.bergstedt@vatten.goteborg.se)
[c]Department of Mathematical Sciences, Göteborg University and Chalmers University of Technology, SE-412 96 Göteborg, Sweden (E-mail: tommy@chalmers.se)

Abstract In the 3rd edition of the *Guidelines for Drinking-water Quality*, the World Health Organisation concludes that an integrated management of risks in source waters, treatment systems and distribution networks is the most efficient way to guarantee safe drinking water to consumers. The integrated approach is fundamental to avoid sub-optimisation of risk-reduction efforts. This paper presents an application of an integrated and quantitative risk model for comparing risk-reduction alternatives to support decisions for reaching specified water safety targets. A fault tree approach is used for structuring the risk analysis and for estimating the risk, expressed as Costumer Minutes Lost (CML). Input information is a combination of hard data and expert judgements. Uncertainties in input information are considerable and modelled by a Bayesian statistical approach. The Göteborg drinking water system is used to exemplify model application. Quantitative safety targets have been confirmed at the political level as a basis for long-term planning of investments and reinvestments. Four different risk-reduction alternatives concerning additional raw water supplies and increased treatment capacity were compared. A combination of increased treatment capacity and additional raw water sources was shown to provide the greatest risk reduction. The paper describes how a structured and thorough analysis of risk-reduction options can facilitate transparency and long-term planning of drinking water systems.

Keywords fault tree, risk assessment, risk management, risk reduction, uncertainty, Water Safety Plan

INTRODUCTION

The World Health Organization (WHO, 2004) concludes that end-product testing in treatment and distribution systems is not sufficient to guarantee safe drinking water to consumers. To ensure safe drinking water a holistic risk management approach from catchment to consumer is emphasised. The Framework for Safe Drinking Water presented by Davison *et al.* (2005) includes the preparation of Water Safety Plans (WSPs) in which risks in source

 TECHNEAU: Safe Drinking Water from Source to Tap. Edited by Theo van den Hoven and Christian Kazner. ISBN: 9781843392750. Published by IWA Publishing, London, UK.

waters, treatment systems and distribution networks should be assessed and managed. WSPs are currently being implemented in several countries and are expected to become an increasingly important part of water management in both developed and developing countries. The Bonn Charter strategy presented by the International Water Association (IWA, 2004), further specifies the use of WSP in drinking water management. The guidelines on WSPs (WHO, 2004; Davison *et al.*, 2005) provide general descriptions of hazard identification and a method for qualitative (or semi-quantitative) classification of risks. The method provides a useful structure for risk assessment and facilitates a ranking of risks as a basis for prioritizing safety measures. Qualitative methods for risk ranking are commonly used in many disciplines, e.g. engineering, mining, environmental and industrial management. It relies on qualitative, often subjective, estimates of probabilities and consequences. The result is often presented using a risk matrix. Favourable aspects of the approach are that it is easy to use, provides transparency and facilitates a basis for communication between stakeholders. A major drawback of risk ranking methods is that it is difficult to properly model complex systems with interactions between components. This may cause suboptimal prioritisation of risk-reduction efforts and an inefficient use of available resources. Furthermore, risk ranking methods assume a discrete nature of hazards, cannot provide quantitative estimates that can be directly compared to performance targets, and typically lack procedures for sensitivity and uncertainty analysis (Burgman, 2005).

Drinking water systems are often complex with numerous components that interact in various ways. In this paper, a fault tree model is used to properly account for different types of interactions between system components and to include the entire drinking water system, from source to tap. The approach allows for two very important types of assessments in risk management: (1) quantitative comparison to established water safety targets; and (2) evaluation of the efficiency of risk-reduction options. The fault tree model considers quantity as well as quality failures and is described in detail by Lindhe *et al.* (2009) and Norberg *et al.* (2008).

The purpose of this paper is to illustrate and evaluate the use of the fault tree model for comparing four alternative risk-reduction options for the Göteborg drinking water system. Water safety targets have been quantified and confirmed at the political level as a basis for the long-term planning of investments and reinvestments. The targets include events such as waterborne diseases and major delivery failures. A previous risk analysis of the raw water availability showed high probability of interruption in the supply to one of the two treatment plants due to failures in the raw water supply system (Rosén and Steier, 2006). The fault tree model used in this paper has also been used to analyse the system (Lindhe *et al.*, 2008) and it was shown that failures in the raw water system contributes most to the total risk level, expressed as Customer Minutes Lost (Lindhe *et al.*, 2009). The maximum capacity of each plant is far below the average need and drinking water storage is well below the daily need. An additional analysis of the dominating failure scenario showed that increased treatment capacity would reduce the risk of extensive delivery interruptions to an acceptable level, and that the raw water supply would not be a limiting factor in this scenario. As a result, decisions have been made to increase the treatment capacities. However, there are other events, including pollution of the raw water reservoir, where additional raw water sources would be crucial for facilitating sufficient drinking water production. Several alternative options for additional raw water sources have been identified and there is a need of quantifying their contribution towards meeting the established safety targets.

Four alternative options for risk reduction were analysed in this paper: (1) increase of drinking water treatment capacities (already decided); (2) regulation of existing raw water lake reservoirs in a small watershed adjacent to the drinking water system; (3) change of regulation of the largest lake and watershed in the area; and (4) a combination of alternatives (2) and (3). The total risk for each of the four alternative options was calculated and compared to the risk level for the present day situation.

THE GÖTEBORG DRINKING WATER SYSTEM

Göteborg is the second largest city in Sweden and approximately 500,000 consumers are supplied with drinking water. The system is based solely on surface water and the main raw water source is the Göta Älv river, a moderately polluted river (Westrell, 2003). A schematic description of the Göteborg raw water system, including possible alternative raw water sources, is given in Figure 1. Two smaller lakes (*Main reservoir lakes* in Figure 1) are used for intermediate storage of water and to improve the water quality. Approximately half of the water taken from the river is transferred directly to *Treatment plant no 1*. The other half is transferred via a 12 kilometres rock tunnel to the *Main reservoir lakes*. From the *Main reservoir lakes* water is pumped to *Treatment plant no 2*. Due to the variable quality of the river water, the intake is closed 100 days a year on average (see e.g. Åström *et al.*, 2007). Decisions to close the intake are based on on-line monitoring and reports from operating bodies upstream, e.g. industries and municipalities. Typical parameters monitored on-line are, turbidity, conductivity, redox-potential and pH. In addition, microbial sampling is regularly carried out.

When the intake from the main water source is closed, the *Main reservoir lakes* supply both treatment plants. During long periods (weeks) of closure of the

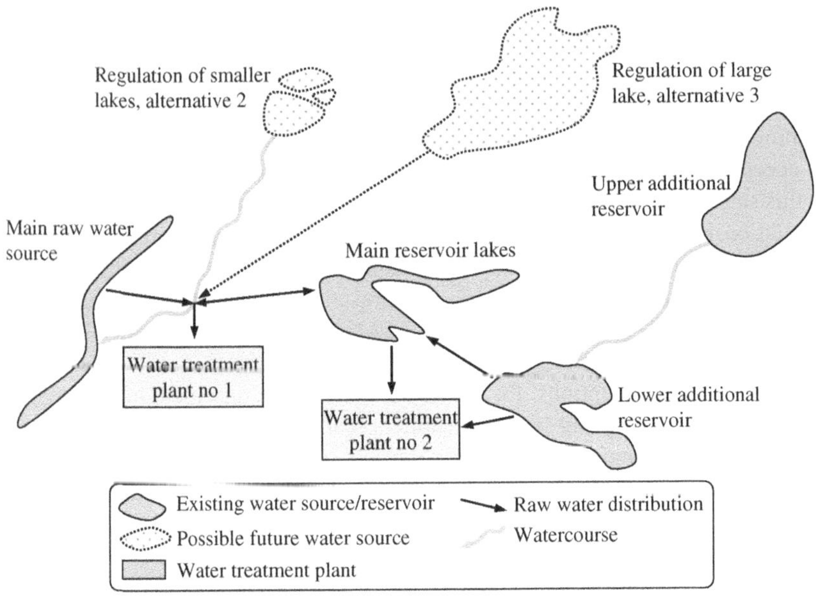

Figure 1. Schematic description of the Göteborg raw water system. Possible future raw water supplies included in the analysis are further described in the "Risk-reduction alternatives" section below.

intake, water from a reservoir in an additional raw water supply system (*Lower additional reservoir* in Figure 1) can be pumped to the *Main reservoir lakes* or directly to *Treatment plant no 2*. However, the situation where *Treatment plant no 1* has no alternative raw water supply of its own makes the system vulnerable. The treatment plants have similar treatment processes, including chemical flocculation, sedimentation, filtration and disinfection. On-line measurements and laboratory analyses are used to monitor the water quality at treatment plants. Both treatment plants contribute in approximately equal parts too meet an average drinking water demand of 165,000 m^3/d. Normally the water demand varies between 120,000 and 210,000 m^3/d. The maximum treatment capacity is 150,000 m^3/d for *Treatment plant no 1* and 120,000 m^3/d for *Treatment plant no 2*. Treatment capacities are normally lower than the maximum due to maintenance work. The distribution network is approximately 1700 km in length. To ensure sufficient pressure in elevated supply zones, the water head is raised through 66 booster stations and 14 service reservoirs are used to meet peaks in the water demand. Monitoring of the water quality in the distribution system is made at different locations, e.g. in private taps and pumping stations.

METHOD

Conceptual model

The overall failure event in the model is supply failure, including: (1) quantity failure, i.e. no water is delivered to the consumer; and (2) quality failure, i.e. water is delivered but does not comply with existing quality standards. Figure 2 shows the two categories of failure and their main causes. Quantity failure may occur due to failure of components, e.g. pumps and pipes, or because the water utility decides to stop the delivery when an unacceptable water quality has been detected. Quality failure may occur when an unacceptable water quality is not detected or when it is detected but no action is taken. The drinking water quality was considered unacceptable when unfit for human consumption, based on the Swedish national quality standards for drinking water (SLVFS 2001:30). Although the applied model can be used to analyse quantity and quality failures, the application in this paper only considered quantity failures.

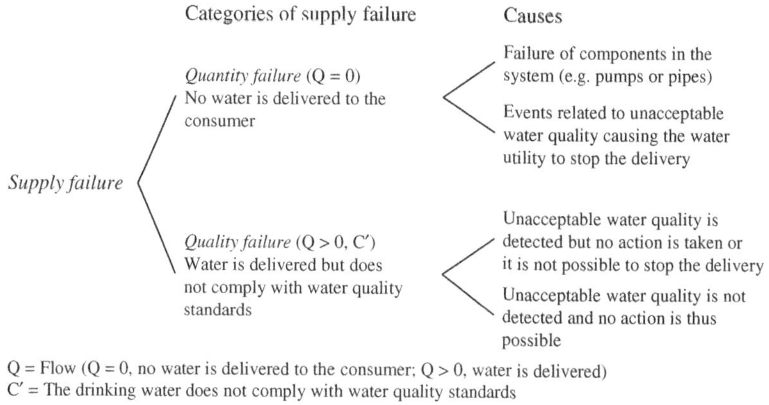

Figure 2. Categories of supply failure and their main causes (Lindhe *et al.*, 2009).

The drinking water system was modelled as a supply chain composed of three sub-systems: raw water source, treatment system and distribution network (Figure 3). Events in any of the sub-systems may cause supply failure, but the sub-systems are also capable of compensating for failure. For example, if one treatment plant fails to produce drinking water, reservoirs in the distribution system and increased production at the other treatment plant may prevent quantity failure. However, the ability to compensate is limited in time, because the reservoir volume is limited.

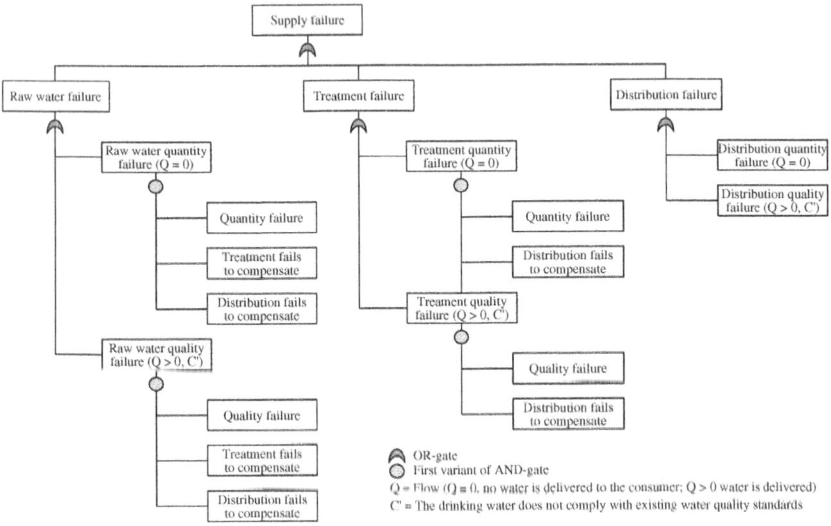

Figure 3. Schematic fault tree including the main events (Lindhe *et al.*, 2008).

The fault tree model

The fault tree method for integrated risk analysis of drinking-water systems is described in detail by Lindhe *et al.* (2009) and Norberg *et al.* (2008). A fault tree analysis is a structured process that identifies potential causes of system failure (Bedford and Cooke, 2001). A fault tree represents interactions between different events by logic gates, and shows how the events may lead to system failure, i.e. the *top event*. The two most common types of logic gates are the OR- and AND-gates, which are used in the present model. Starting with the top event, the tree is developed until a required level of detail is reached. Initiating events are denoted *basic events* and logical combinations of these are denoted *intermediate events*. The top event represents failure of the system while basic events represent failures of components. The capability of representing interactions between critical system components makes fault tree models powerful for analysis of entire drinking water systems. A recent example of a more holistic application of fault trees concerning waterborne outbreaks is presented by Risebro *et al.* (2007). A schematic fault tree illustrating the main type of events included in the analysis is shown in Figure 4. The full model of the Göteborg system includes 116 to 121 basic events, depending on what risk-reduction alternative is analysed.

A Markovian approach is used, which means that the events are modelled using the failure rate λ and mean downtime $1/\mu$ as variables (Rausand and Hoyland, 2004). The mean time to failure is $1/\lambda$, hence the probability of failure

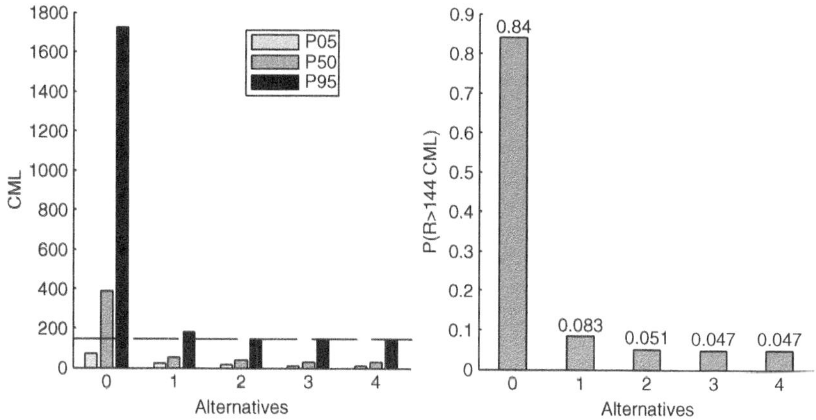

Figure 4. *Left:* Histogram showing the 5th, 50th (median) and 95th percentiles of simulated risk values (CML) for the current system (Alt. 0) and the four risk-reduction alternatives (Alt. 1–4). Target risk level (144 annual CML) indicated by dashed horizontal line. *Right:* The probability of exceeding established water safety target (144 annual CML) for the current system (Alt. 0) and the four risk-reduction alternatives (Alt. 1–4).

is $P_F = \lambda/(\lambda+\mu)$. The main reasons for using the failure rate and mean downtimes instead of direct estimations of probability of failure, is to facilitate relevant elicitation of expert judgements and modelling of the dynamic behaviour of the system. Using uptimes and downtimes of basic events makes it possible to estimate the duration of failures and the time intervals to failures.

The fault tree of the Göteborg system was constructed using four types of logic gates: OR-gates, AND-gates and two new variants of the AND-gate. The OR-gate describes a series system where only one input events has to occur to cause the system to fail. The AND-gate describes a parallel system where all input events have to occur to cause the system to fail. To be able to consider the inherent ability of a system to compensate for failure, Norberg *et al.* (2008) and Lindhe *et al.* (2009) formulated two additional variants to the common type AND-gate. The first variant describes a situation where failure of one component may be compensated for by one or several other components during a limited time period. The second variant represents a situation where a compensating component that has failed may recover and start to compensate again.

Equations for calculating the probability of failure, the failure rate and mean downtime at each intermediate level of the fault tree, are given in Table 1. Here, q_i denotes the probability of *failure on demand* where a compensating component i fails to start compensating when needed, e.g. a reserve pump that cannot be started when the main pump brakes down. Failure during operation is the other type of failure and is represented by the failure rate; e.g. the reserve pump starts but after some time it brakes down.

Table 1. Equations used to calculate the output of the logic gates (Norberg *et al.*, 2008; Lindhe *et al.*, 2009). Variable P_F is the probability of failure, λ_i the mean failure rates, μ_i the mean repair rates ($1/\mu_i$ the mean downtimes) and q_i the probabilities of failure on demand. For the variants of the AND-gate $i = 1$ corresponds to the failure that may be compensated for by events $i = 2, \ldots, n$.

OR-gate	AND-gate
$\lambda = \sum_{i=1}^{n} \lambda_i$	$\mu = \sum_{i=1}^{n} \mu_i$
$\mu = \sum_{i=1}^{n} \lambda_i \cdot \frac{\prod_{i=1}^{n} \mu_i}{\prod_{i=1}^{n} (\lambda_i + \mu_i) - \prod_{i=1}^{n} \mu_i}$	$\lambda = \sum_{i=1}^{n} \mu_i \cdot \frac{\prod_{i=1}^{n} \lambda_i}{\prod_{i=1}^{n} (\lambda_i + \mu_i) - \prod_{i=1}^{n} \lambda_i}$
$P_F = \frac{\lambda}{\lambda + \mu} = 1 - \prod_{i=1}^{n} \frac{\mu_i}{\lambda_i + \mu_i}$	$P_F = \frac{\lambda}{\lambda + \mu} = \prod_{i=1}^{n} \frac{\lambda_i}{\lambda_i + \mu_i}$
1st variant of AND-gate	**2nd variant of AND-gate**
$\mu = \mu_1$	$P_F = \frac{\lambda_1}{\lambda_1 + \mu_1} \cdot \frac{\lambda_2 + q_2(\mu_1 + \mu_2)}{\lambda_2 + \mu_1 + \mu_2}$
$P_F = \frac{\lambda_1}{\lambda_1 + \mu_1} \cdot \prod_{i=2}^{n} \frac{\lambda_i + q_i \mu_1}{\lambda_i + \mu_1}$	$\lambda = \frac{\mu_1 \lambda_1 q_2 (\lambda_2 + \mu_1 + \mu_2) + \lambda_1 \lambda_2 (1 - q_2)(\mu_1 + \mu_2)}{(\lambda_1 + \mu_1)(\lambda_2 + \mu_1 + \mu_2)(1 - P_F)}$
$\lambda = \frac{P_F}{1 - P_F} \cdot \mu$	$\mu = \frac{\mu_1 \lambda_1 q_2 (\lambda_2 + \mu_1 + \mu_2) + \lambda_1 \lambda_2 (1 - q_2)(\mu_1 + \mu_2)}{(\lambda_1 + \mu_1)(\lambda_2 + \mu_1 + \mu_2) P_F}$

Customer Minutes Lost (CML) and risk

The risk (R) was estimated in terms of Customer Minutes Lost (CML). Risk was defined as the expected value of CML. Using the applied fault tree model, CML values for both quantity and quality failures can be calculated. In order to retain transparency these to types of risks should always be presented separately. Since this study only considered quantity failures all CML values refer to quantity failures and are equivalent to the number of minutes per year the average consumer is not supplied with drinking water. A comprehensive description of the measure is presented by Lindhe *et al.* (2009). Blocker *et al.* (2005) describes the use of CML as a performance indicator in the Netherlands. To consider that the system may not fail when in failure mode, it can be shown that the expected value of CML should be calculated as $R = P_F \cdot C$, where P_F is the probability of failure and C is the proportion of all consumers affected (Lindhe *et al.*, 2009). Since the top event in the fault tree, supply failure, is a combination of different types of events, the proportion of consumers affected (C) cannot in a meaningful way be defined for this event. Instead, C should be defined at a lower level in the fault tree for n different types of events. The total risk was thus calculated as

$$R = \sum_{i=1}^{n} P_{Fi} C_i$$

Uncertainty analysis

To enable uncertainty analysis all input parameters were expressed as probability density functions and the calculations were performed by Monte Carlo simulations (10,000 iterations). The parameters λ and μ were modelled as exponential rates using Gamma distributions. The proportion of the consumers affected (C) as well as the probability of failure on demand (q) were modelled by Beta distributions. These distributions facilitate a Bayesian approach, where new information, e.g. monitoring data, can be used for a mathematically formal updating of previous knowledge. The Monte Carlo simulations facilitate two important types of analyses: (1) sensitivity analysis of contributions to the total uncertainty from uncertainties in basic events; and (2) analysis of the probability of not meeting established safety targets for the water utility.

Water safety targets in Göteborg

The City of Göteborg has prepared an action plan which includes performance targets regarding the supply of drinking water (Göteborg, 2006). These targets are politically established and can be considered as acceptable levels of risk. The performance target used in this analysis is: *Duration of interruption in delivery to the average consumer shall, irrespective of the reason, be less than a total of 10 days in 100 years*. For the average consumer this translates to a target value of 144 annual CML.

RISK-REDUCTION ALTERNATIVES

It has been concluded that the current system does not meet the water safety target considered in this paper; this is further described in the results section below. To decrease the risk, possible risk-reduction measures have been identified. The following alternatives for risk reduction were analysed: (1) increased treatment capacity in the two water treatment plants, the increase will make each of the plants capable of producing up to the average total drinking water demand; (2) increased treatment capacity combined with increase in regulation of some fairly pristine lakes to increase the flow in a river for transfer to the drinking water system, these lakes contain relatively large raw water volumes but their watershed is too small for a continuous supply; (3) increased treatment capacity combined with change of regulation of the largest lake and watershed in the area and construction of a pipeline for raw water transfer to the drinking water system, this lake has almost no restrictions in water availability but failures may occur due to dry periods causing water shortage, problems in the transfer of raw water from the lake to the treatment plant or due to pollution events; and (4) combination of alternatives (2) and (3). The alternative raw water supplies (in alternatives 2 and 3) are illustrated in Figure 1 and changes in input variables of the fault tree model are described for each alternative in Table 2.

RESULTS

The results from the simulations are shown in Figure 4. It can be seen that before any risk-reduction measures have been taken (Alt. 0) the acceptable risk level (144 CML) is clearly exceeded for the average consumer (mean value: 621 CML). The probability of exceeding the acceptable level is 0.84. When increasing the treatment capacities in the plants (Alt. 1), the mean value for the risk (79 CML) is below the acceptable level. Due to the uncertainties of input variables in the model, the probability of exceeding the acceptable risk level is 0.083. Adding the regulation of the smaller lakes to the increased treatment capacity (Alt. 2) results in further reduction of the mean risk value to 59 CML and the probability of exceeding the water safety target is 0.051. Adding the changed regulation of the larger lake to the increased treatment capacity (Alt. 3) results in a slightly larger risk reduction; the mean risk value is equal to 52 CML and the probability of exceeding the acceptable risk is 0.047. After the treatment capacity has been increased the risk related to the raw water is dramatically reduced due to an increased ability to compensate for interruptions in the raw water supply. Before increasing the treatment capacity the total risk level is mainly caused by failures in the raw water. After the capacity is increased (Alt. 1) the contribution to the total risk level is similar for the raw water and distribution parts of the system, while the treatment contributes least to the risk. Consequently, when the raw water is no longer the dominating contributor to the risk, measures have to be very efficient to have a large effect on the total risk level. This is the reason the larger lake only provides a slightly larger risk reduction compared to the smaller lakes, although the larger lake is a more reliable water source. Combining increased treatment capacity with regulation of both the smaller lakes and the larger lake (Alt. 4) will provide the same risk reduction as if only the larger lake is included (Alt. 3).

Table 2. Change in input variables for risk-reduction alternatives.

Alternative	Comment and major changes in input variables of the fault tree model
0. No measures taken	–
1. Increased treatment capacity	Increased capacity in each treatment plant to compensate for shortage in raw water supply or treatment failures in the other treatment plant. The increased capacity can compensate up to average demand. Based on statistical data on water demand and estimations regarding the reliability of the treatment plants, the time for compensation (uptime) was estimated to be between 3–120 days (90%-interval) and the probability of failure on demand 0.0025–0.01 (90%-interval).
2. Increased treatment capacity combined with regulation and supply from smaller lakes	Additional raw water source that can be used when other sources are unavailable. The source is limited in volume and availability. If only Treatment plant no 1 needs supply, the source is available (uptime) 25–35 days (90%-interval), whereas if both treatment plants need to be supplied the available time is restricted to 8–18 days (90%-interval). When the lakes are not available, the duration (downtime) is 7–60 days (90%-interval).
3. Increased treatment capacity combined with regulation and supply from larger lake	Additional raw water source that can serve as main source in combination with current main raw water source. Events that can cause the source to become unavailable are shortage of water during dry periods, failures in the transfer of raw water and unacceptable water quality in the lake. The time to failure (uptime) for the possible events is estimated to be 5–15 years (90%-interval) for all events. When failure occurs the duration (downtime) is estimated to be 1–30 days for water shortage, 0.5–2 days for technical failures and 5–30 days for unacceptable quality (all 90%-interavals).
4. Combination of alternatives 2. and 3.	See alternatives 2 and 3.

CONCLUSIONS

The comparison of the risk-reduction alternatives shows that the increased treatment capacity provides a substantial risk reduction. However, if the water safety targets were to be reached with 95% confidence (i.e. the probability of exceeding 144 CML is less than 0.05), additional measures are needed. A 95% confidence is often applied when considering uncertainties in risk evaluation, although other criteria may be used. The analysis shows that the certainty of reaching established water safety target increases when additional water sources are included in the system. The combination of increased treatment capacity and regulation of the larger lake provides the same, and also the largest, risk reduction as when increased treatment capacity is used together with both the larger lake and the smaller lakes as additional raw water sources. However, the risk reduction provided by alternatives 2, 3 and 4 are all similar. It should be noted that the larger lake provides additional benefits not considered in this analysis. These benefits are mainly concerning an enhanced raw water quality. The inclusion of the larger lake in the raw water system is likely to be a rather complicated and long process, for various reasons. Hence, if the larger lake is supposed to be included in the system, it is reasonable to combine increased treatment capacity and regulation of the smaller lakes (Alt. 2) until the larger lake can be installed (Alt. 3). If the larger lake is installed is seems not to be reasonable to also include the smaller lakes. Hence, there seems to be little use of alternative 4. The results presented in this paper are still being analysed and further studies are in progress to also include the economic aspects of the analysed alternatives.

The application of the fault tree model illustrates its possibilities as a decision support tool for structured comparison of alternative risk-reduction options. The risk reduction of alternatives that influence different parts and components of a system can be quantified, taking into consideration the interactions between system components. The model facilitates two very important analyses for decision support in risk management: (1) analysis of the risk reduction provided by a specific action; and (2) analysis of the uncertainty of the risk estimate, which enables an estimation of the probability of not reaching established acceptable risk levels. Furthermore, the model provides transparency and a basis for long-term planning of drinking water systems.

ACKNOWLEDGEMENTS

This study has been carried out within the framework of the TECHNEAU project (Technology Enabled Universal Access to Safe Water), funded by the European Commission (contract number 018320), and with support from the Swedish Water & Wastewater Association and the City of Göteborg. We would like to thank the City of Göteborg for its valuable and fruitful collaboration.

REFERENCES

Åström, J., Pettersson, T.J.R. and Stenström, T.A. (2007). Identification and management of microbial contaminations in a surface drinking water source. *Journal of Water and Health*, **5**(1), 67–80.

Bedford, T. and Cooke, R.M. (2001). *Probabilistic Risk Analysis: Foundations and Methods*. Cambridge University Press, Cambridge.

Blokker, M., Rujig, K. and de Kater, H. (2005). Introduction of a substandard performance indicator. *Water Asset Management International*, **1**(3), 19–22.

Burgman, M. (2005). *Risks and Decisions for Conservation and Environmental Management*. Cambridge University Press, Cambridge.

Davison, A., Howard, G., Stevens, M., Callan, P., Fewtrell, L., Deere, D. and Bartram, J. (2005). *Water Safety Plans: Managing Drinking-water Quality from Catchment to Consumer*. WHO/SDE/WSH/05.06, World Health Organisation, Geneva.

Göteborg, V. (2006). *Action Plan Water: Long-term Goals for the Water Supply in Gothenburg* (In Swedish). City of Göteborg.

IWA (2004). *The Bonn Charter for Safe Drinking Water*. International Water Association, London.

Lindhe, A., Rosén, L., Norberg, T., Pettersson, T.J.R., Bergstedt, O., Åström, J. and Bondelind, M. (2008). Integrated risk analysis from source to tap: Case study Göteborg, *In Proceedings of the 6th Nordic Drinking Water Conference*, Oslo, 9–11 June, pp. 231–241.

Lindhe, A., Rosén, L., Norberg, T. Bergstedt, O. (2009). Fault tree analysis for integrated and probabilistic risk analysis of drinking water systems. *Water Research*, **43**(6), 1641–1653.

Norberg, T., Rosén, L. and Lindhe, A. (2008). Added value in fault tree analyses. In: *Safety, Reliability and Risk Analysis: Theory, Methods and Applications*, Martorell *et al.* (eds.), Taylor & Francis Group, London, pp. 1041–1048.

Rausand, M. and Høyland, A. (2004). *System Reliability Theory: Models, Statistical Methods, and Applications*. 2nd ed., Wiley-Interscience.

Risebro, H.L., Doria, M.F., Andersson, Y., Merdema, G., Osborn, K., Schlosser, O., Hunter, P.R. (2007). Fault tree analysis of the causes of waterborne outbreaks. *Journal of Water and Health*, **5**(1), 1–18.

Rosén, L. and Steier, K. (2006). *Risk Assessment of Water Quantity, Gothenburg* (In Swedish). Consulting report, Contract no. 1310786000, SWECO VIAK AB.

SLVFS 2001:30 *National Food Administration Ordinance on Drinking Water* (In Swedish). Swedish National Food Administration.

WHO (2004). *Guidelines for Drinking-water Quality*. Third Edition, Volume 1 Recommendations, World Health Organization, Geneva.

Westrell, T., Bergstedt, O., Stenström, T.A. and Ashbolt, N.J. (2003). A theoretical approach to assess microbial risk due to failures in drinking water systems. *Inte Journal of Environmental Health Research*, **13**(2), 181–197.

TECHNEAU hazard database and risk reduction option database

R.H.S. Beuken[a] and T.J.R. Pettersson[b]

[a]KWR, P.O. Box 1072, 3430 BB Nieuwegein, The Netherlands
(E-mail: ralph.beuken@kwrwater.nl)
[b]Chalmers University of Technology, Department of Civil and Environmental Engineering, SE-412 96 Göteborg, Sweden (E-mail: thomas.pettersson@chalmers.se)

Abstract One of the first steps to be addressed in risk management is the identification and description of potential hazards for water supply systems. After these hazards have been estimated and evaluated, risk reduction options are to be defined for hazards that are defined as critical. Within the risk assessment/risk management work area in the project tools are developed for identification of hazards (the TECHNEAU Hazard Database – THDB) and for selection of most appropriate risk reduction options (the TECHNEAU Risk Reduction Database – TRRDB). The THDB has been tested with water companies and an updated version is available. The TECHNEAU Risk Reduction Database (TRRDB) is in preparation and is soon to be published.

Keywords checklist, hazard identification, risk management, risk reduction options

INTRODUCTION

The main objective of Work Area (WA4) – Risk assessment and risk management – in the TECHNEAU project, is to integrate risk assessments of the separate parts of a water supply system into a comprehensive decision support framework for cost-efficient risk management in safe and sustainable drinking water supply. The steps in the risk management process are represented in Figure 1. A further description of the risk management process as applied in the TECHNEAU project is provided by Rosén *et al.* (2007). Another objective of WA4 is to develop tools that the water utilities can apply when carrying out risk assessments and risk management work.

One of the first steps to be addressed in risk management is the identification and description of potential hazards for water supply systems. After these hazards have been estimated and evaluated, risk reduction options are to be defined for hazards that are defined as critical. Within WA4 instruments have been set up for these purposes. This paper mainly describes the TECHNEAU Hazard Database (THDB) but also the TECHNEAU Risk Reduction Database (TRRDB) that still is in preparation.

 TECHNEAU: Safe Drinking Water from Source to Tap. Edited by Theo van den Hoven and Christian Kazner. ISBN: 9781843392750. Published by IWA Publishing, London, UK.

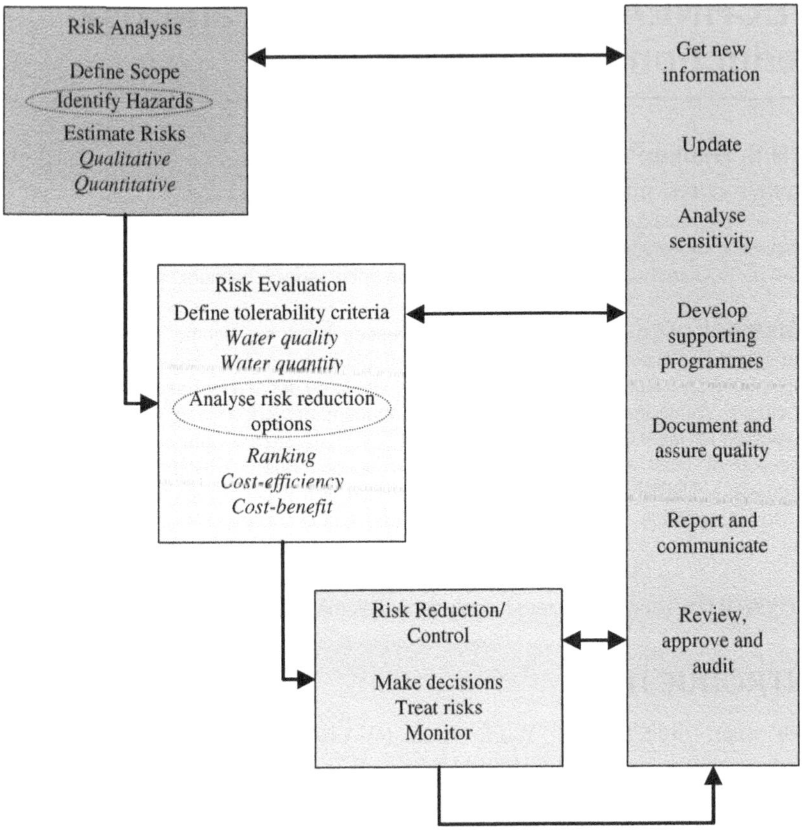

Figure 1. The risk management process and the position of hazard identification (Rosén *et al.*, 2007). The topics of this paper are circled.

A preliminary version of the TECHNEAU Hazard databases has been made in the beginning of 2007 (Beuken, 2007). This version has been tested at various water companies and this has lead to an updated version in 2008 (Beuken, 2008). As the database is continuously improved and updated, the most recent version can be found on the TECHNEAU Knowledge Integrator (TKI) which is available on the TECHNEAU website (www.techneau.eu).

Even though the TECHNEAU Risk Reduction Options Database still is in preparation a short description of this database will be given in this paper.

DESCRIPTION OF THE TECHNEAU HAZARD DATABASE

The objective of the TECHNEAU Hazard Database is to help end-users working at water supply companies with the identification of relevant hazards by providing a catalogue with potential hazards of technical, environmental or human origin for the entire system.

Two different approaches for hazard identification can be applied: a bottom-up approach and a top-down approach. The bottom-up approach has its origin in the recorded experience of the staff directly involved in operation and maintenance. In general, the information gathered is suitable for the assets of the water supply system. It has the disadvantage that focus can be given to well known incidents or that subjective information can be used.

The top-down approach is based on a description of a generic water supply process. As the supply process consists of a wide variety of potential elements, a generic list is extensive.

The approach presented here is a top-down approach, which can be used as a complement to the bottom-up approach.

Structure of the TECHNEAU hazard database

The THDB is related to all physical and organizational structures for the supply of drinking water, stretching out from the catchment area to the customer (from source to tap). Hazards related to the handling of water by consumers at home are covered as well, as these types of hazards can be of importance for the quality of the consumed drinking water, especially for consumers that do not have in-house service connections (Swartz *et al.*, 2009).

The hazards identified in this database are internal as well as external. Internal hazards are mostly related to an incapable supply system or failures of the system. External hazards have their origin outside the system, examples are source water contamination, degradation of mains due to aggressive soils or terrorist actions.

The hazards identified in this report are general hazards of technical, geographical or human origin, having direct effect on the infrastructure for water supply. Other types of hazards, for example of political or economical origin are not assessed specifically in this report. To limit the scope, the database focuses on the primary process of water supply, while supporting activities like buildings, constructions, electrical installations are not addressed.

To manage risks effectively it is important to identify today's hazards as well as future hazards. Based on a literature review, interviews, evaluation work carried out at Chalmers University of Technology, and contributions from WA1 (Rethink the system) partners in the SEPTEDOR (Socio-cultural, Economical, Political, Technical, Ecological, Demographic, Organizatorial and Risk) analysis, major future trends have been identified. The implications of these trends for water supply have been studied and the following categories of hazards have been identified:

- Sabotage and terrorist attacks.
- Conflicts.
- New chemicals and changed chemical pathways.
- Emerging pathogens.
- Public concern.
- Climate changes.
- Aging distribution systems.

For a more detailed description of the future hazards, see "Report on trends regarding future risks" (Rosén and Lindhe, 2007).

The water supply system is subdivided into 12 sub-systems, of which 10 are physical sub-systems representing the installations, one is a non-physical sub-system representing organizational aspects, and one is representing future hazards. A flow-diagram presenting these sub-systems is given in Figure 2. All hazards identified in these 12 sub-systems are listed in separate hazard sub-databases. The breakdown, as given in this flow-diagram, is related to the 'Generic system flow diagram' as presented in the Water Safety Plans, which is described in the third edition of the Drinking Water Quality Guidelines of the WHO (Davison *et al.*, 2005).

The hazard database is presenting the identified hazards in a table at a sub-system level, as in Figure 2, but is also further sub-divided into component and element levels. At the component level the most important elements are given and at the element level the most relevant hazards are given in combination with a description of the cause of the hazard, the hazard type and the consequences. Each hazard has a unique reference number related to the sub-system and component. The format of the hazard database is derived from the Water Safety Plan approach (Davison, *et al.*, 2005). A preview of the TECHNEAU hazard database is given in Figure 3. The hazard database makes use of the definitions given in Table 1.

Testing of the TECHNEAU hazard database

To improve and update the THDB it has been tested and used in six different case studies carried out within the risk assessment and risk management work

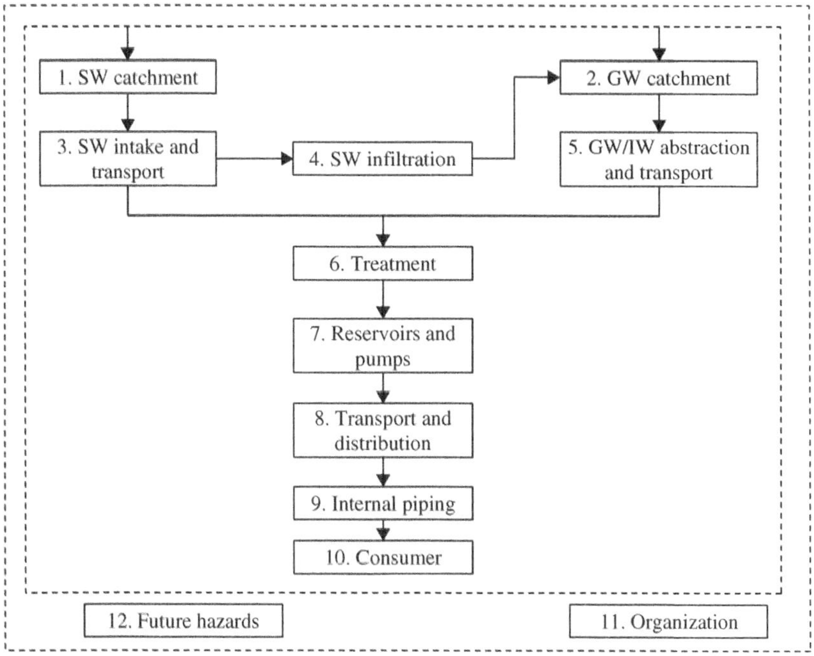

Figure 2. Water supply system subdivided into 12 sub-systems in the THDB (SW = surface water, GW = ground water, IW = Infiltration water).

area (WA4) in the TECHNEAU project. The purpose of each case study was to carry out a risk assessment, in concordance with the WHO Water Safety Plan and to test the different tools and methods developed within the project. The THDB, version 2007, was used in the risk assessment work in combination with brainstorming sessions by the water utility personnel (the local system experts). The database version 2008 contains hazards identified during the hazard identification process and not included in THDB and improved descriptions of the hazards.

In addition to case-studies carried out in WA 4, the THDB was used similarly in three other larger case studies in WA 7 (Integrate, Validate and Demonstrate). In Lisbon, a two-days training seminar on risk analysis was organized where the THDB was used as course material. A hazard identification was performed on the water supply system using brainstorming and the THDB. During brainstorming sessions, thirteen hazardous events have been identified. Applying the THDB resulted in two additional hazardous events. In subsequent risk

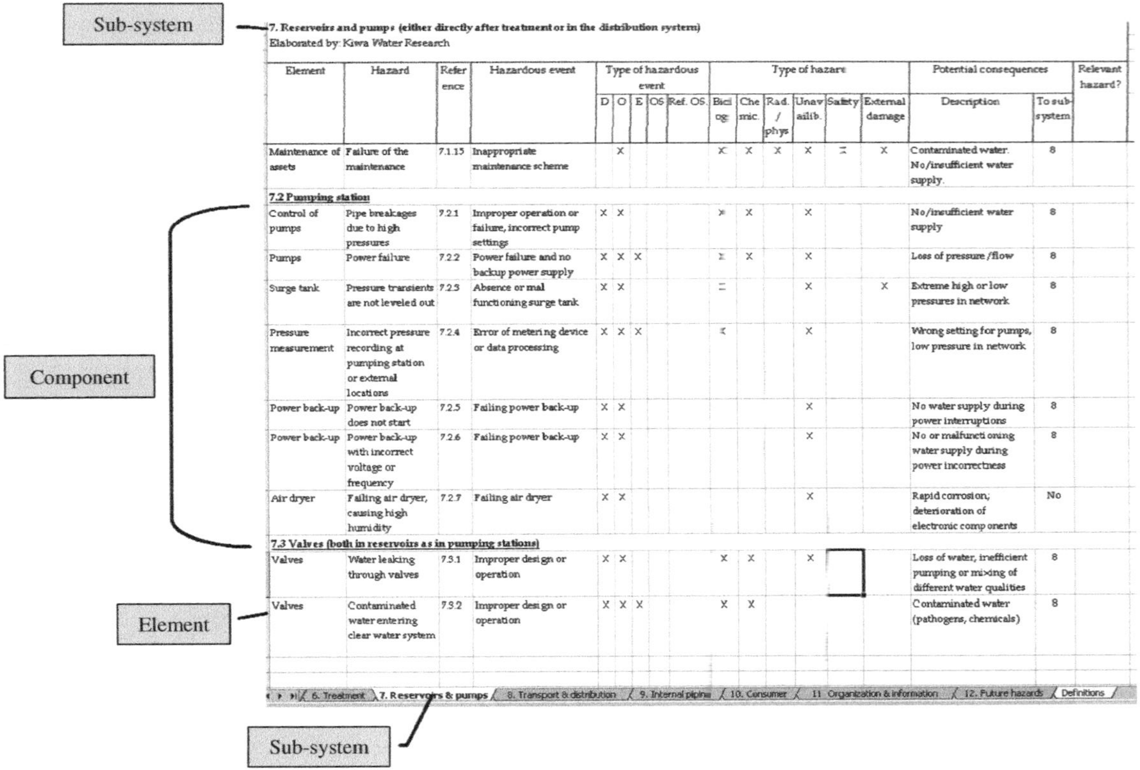

7. Reservoirs and pumps (either directly after treatment or in the distribution system)
Elaborated by: Kiwa Water Research

Element	Hazard	Reference	Hazardous event	Type of hazardous event					Type of hazard						Potential consequences		Relevant hazard?
				D	O	E	OS	Ref. OS.	Biolog.	Chemic.	Rad. / phys	Unavailib.	Safety	External damage	Description	To sub-system	
Maintenance of assets	Failure of the maintenance	7.1.15	Inappropriate maintenance scheme		X				X	X	X	X	X	X	Contaminated water. No/insufficient water supply.	8	
7.2 Pumping station																	
Control of pumps	Pipe breakages due to high pressures	7.2.1	Improper operation or failure, incorrect pump settings	X	X				X	X		X			No/insufficient water supply	8	
Pumps	Power failure	7.2.2	Power failure and no backup power supply	X	X	X			X	X		X			Loss of pressure/flow	8	
Surge tank	Pressure transients are not leveled out	7.2.3	Absence or mal functioning surge tank	X	X				X			X		X	Extreme high or low pressures in network	8	
Pressure measurement	Incorrect pressure recording at pumping station or external locations	7.2.4	Error of metering device or data processing	X	X	X			X			X			Wrong setting for pumps, low pressure in network	8	
Power back-up	Power back-up does not start	7.2.5	Failing power back-up	X	X							X			No water supply during power interruptions	8	
Power back-up	Power back-up with incorrect voltage or frequency	7.2.6	Failing power back-up	X	X							X			No or malfunctioning water supply during power incorrectness	8	
Air dryer	Failing air dryer, causing high humidity	7.2.7	Failing air dryer	X	X							X			Rapid corrosion; deterioration of electronic components	No	
7.3 Valves (both in reservoirs as in pumping stations)																	
Valves	Water leaking through valves	7.3.1	Improper design or operation	X	X				X	X		X			Loss of water, inefficient pumping or mixing of different water qualities	8	
Valves	Contaminated water entering clear water system	7.3.2	Improper design or operation	X	X	X			X	X					Contaminated water (pathogens, chemicals)	8	

Figure 3. Schematic of the TECHNEAU Hazard Database, with the example for number 7 Reservoirs and pumps sub-system.

Table 1. Definitions applied in the TECHNEAU hazard database.

Element:	Lowest level of the system at which hazards are described.
Hazard:	A source of potential harm or a situation with a potential of harm (e.g. a biological, chemical, physical or radiological agent or circumstances that has the potential to have a negative effect on the supply of safe and sufficient water).
Reference:	Reference number of the hazard.
Hazardous event:	An event which can cause harm (e.g. an incident or situation that can lead to the presence of a hazard, what can happen and how).
Type of hazardous event:	Indication of the origin of the hazardous event. D: design-related O: operation-related OS: consequence of a hazard in other sub-system OS ref: reference of other sub-system
Type of hazard:	Indication of the type of hazard. Biolog.: biological Chemic.: chemical Rad./ phys.: radiological or physical (including turbidity) Unavail.: insufficient availability of water supplied to customers Safety: safety to personnel External damage: external damage to third parties, including liability
Consequence description:	Description of potential consequences of the hazard to other sub-systems at firstly and the consumer secondly.
Consequence to sub-system:	Reference of the sub-system affected by the hazard.
Rel. Hazard	Column to be used by the end-user for marking the identified hazards.

ranking, these two hazardous events have both been classified in the group of four most severe risks. Also resulted the Lisbon case-study in an update of the THDB. Similar seminars have been held in Oslo and Pretoria where the THDB also was used.

Water utilities are free to use the THDB and proposals for further improvement are welcomed and have been facilitated through built in feedback function the TKI.

DESCRIPTION OF THE TECHNEAU RISK REDUCTION OPTIONS DATABASE

The objective of the TRRDB is to help end-users in water supply companies to identify relevant risk reduction measures for their system. In the risk management process (Figure 1) the subsequent step after identifying and estimating the hazards within the system is to analyse existing or identify new risk reduction options to mitigate the probability and consequences of the critical hazards. All options are divided into the same 12 sub-systems as the THDB, see Figure 2, and are provided in a catalogue with three different types of risk reducing options, see Table 2. The different risk reduction options have a unique reference number which relates to the reference number for the corresponding

Table 2. Descriptions of the three categories of risk reduction options in the TRRDB (including some all-embracing examples).

Risk reduction option category	Description	Example
Control (C)	Regulations: Measuring systems:	Company-specific, national or international standards Early warning systems (on-line probes and sampling procedures) in the raw water source, treatment plant or distribution system
Education and information (E)	Training programme: Information material or system:	Local or national operator training courses Local company or national industry information material
Physical barriers (B)	Physical barrier installations for removing or reducing hazardous agents:	Treatment process steps, flow regulators in wastewater systems to avoid overflow of sewage into raw water

8. Transport and distribution (from trunk main to the water meter) *Worked by: LNEC, Kiwa, SZU/BUT, Chalmers, (Swartz/WRC)*												
Sub-system	Hazardous event (ref. from THDB)		Risk reduction options									
			Control			Education and Information			Barrier			
	(no.)	(description)	Ref.	Options	Type	Ref.	Options	Type	Ref.	Options	Type	
Pipes	8.1.6	Poor hygiene during pipes installation/repair	C8.1.6a	Implement hygienic guidelines for repairs, including flushing and disinfection best practices	P	E8.1.6a	Adequate continuing education and training of professionals on pipe installation and repair best-practice	P	B8.1.6a	Checking of the microbiological quality of the installed/repaired zone water before putting pipes in service	C	
			C8.1.6b	Prevent waterborne transmissible-disease carriers to be members of the installation/repair crew	P	E8.1.6b	Adequate continuing education and training of professionals on pipe installation and repair best-practice	P	B8.1.6b	Medical control/licensing of the installation/repair personnel	C	
Pipes	8.1.7	Intrusion of contaminated water due to low (negative) pressure in the network, in combination with cracks or leaking joints	C8.1.7a	Maintain appropriate water-pressure throughout the entire network. Coordination with fire fighting services on hydrant use and awareness of which areas may be at risk of loss of pressure	C	E8.1.7a	Inform designers and network operation personnel about the hazards from negative pressures in the network	P	B8.1.7a	Continuous monitoring and automatic control of the pressure throughout the entire network by using SCADA or equivalent systems. Wherever necessary, use pressure boosting pumps.	P	
			C8.1.7b	Minimize pressure surges frequency and intensity	P	E8.1.7b	Inform designers and network operation personnel about the hazards from negative pressures in the network	P	B8.1.7b	Have air vessels installed close to pumps and major valves, control the rate of pumps switching and operate valves and hydrants slowly	C	
Pipes	8.1.8	Too long residence times of water in the network	C8.1.8a	Lower the capacity or increase flow of the system zones with long detention times	C	E8.1.8a	Inform designers and network operation personnel about the hazards from water aging	P	B8.1.8a	Keep effective concentrations of disinfectant residual. Rigorous monitoring of the microbiological quality of the water	P	

Figure 4. Schematic of the TRRDB catalogue with an example taken from number 8 Transport and distribution sub-system.

hazard in the THDB. Each option in the catalogue has one capital letter, depending on risk reduction category (as defined in Table 2) before the reference number and a serial lower-case letter, to distinguish between several options to the same hazard, behind the reference number (a, b, c ...). One example of reference number for a risk reduction option is C8.1.6b (**C** – is a control option and **b** the second option in this category), see Figure 4.

The TRRDB is at this moment in preparation. The first version is expected to be published mid 2009 and will be tested with water companies.

CONCLUSION

One of the first steps to be addressed in risk management is the identification and description of potential hazards for water supply systems. After these hazards have been estimated and evaluated, risk reduction options are to be defined for hazards that are defined as critical. Within WA 4 tools are developed for the identification of hazards (the TECHNEAU Hazard Database – THDB) and for selection of most appropriate risk reduction options (the TECHNEAU Risk Reduction Database – TRRDB).

A preliminary version of the TECHNEAU Hazard databases has been made in the beginning of 2007 (Beuken, 2007). This version has been tested at various water companies, what lead to an updated version in 2008 (Beuken, 2008). As the database is continuously improved and updated, the most recent version can be found on the TECHNEAU Knowledge Integrator (TKI) which is available on the TECHNEAU website (www.techneau.eu). Water utilities are free to use the THDB and proposals for further improvement are welcomed and facilitated in a feedback function within the TKI.

The TECHNEAU Risk Reduction Database has the same structure as the TECHNEAU Hazard Database, with risk reduction options relating to corresponding hazards in the THDB. The TECHNEAU Risk Reduction Database (TRRDB) still is in preparation. The first version is expected to be published mid 2009 and will be tested with various water companies.

REFERENCES

Beuken, R. (2007). *Identification and description of hazards for water supply systems,* TECHNEAU. http://www.techneau.org/index.php?id=124 (assessed 6 April 2009).

Beuken, R. (2008). *Identification and description of hazards for water supply systems, updated version,* TECHNEAU. http://www.techneau.org/index.php?id=124 (assessed 6 April 2009).

Davison, A., Howard, G., Stevens, M., Callan, P., Fewtrell, L., Deere, D. and Bartram, J. (2005). Water Safety Plans, Managing drinking-water quality from catchment to consumer. World Health Organization, WHO/SDE/WSH/05.06.

Rosén, L., Hokstad, P., Lindhe, A., Sklet, S. and Røstum, J. (2007). *Generic Framework and Methods for Integrated Risk Management in Water Safety Plans*, TECHNEAU. http://www.techneau.org/index.php?id=124 (assessed 6 April 2009).

Rosèn, L. and Lindhe, A. (2007). Trend report – Report on trends regarding future risks, TECHNEAU.

Swartz, CD, Törnqvis, M., Öfverström, B. and Pettersson, T.J.R. (2009). Risk Assessment Case Study 2: Upper Mnyameni, South Africa. TECHNEAU Conference 17–19 June, Maastricht, the Netherlands.

Risk assessment case study 1: Březnice, Czech Republic

F. Kožíšek[a], J. Paul[b], D. Weyessa Gari[c], L. Tuhovčák[d] and P. Pumann[c]

[a]National Institute of Public Health, Šrobárova 48, CZ-10042 Prague, water@szu.cz; and the Department of General Hygiene, Third Faculty of Medicine, Charles University in Prague
[b]VAK Beroun, Mostníkovská 255, CZ-26641 Beroun; jiri.paul@vakberoun.cz
[c]National Institute of Public Health, Srobarova 48, CZ-10042 Prague, gari@szu.cz; ppumann@szu.cz
[d]Brno University of Technology, Institute of Municipal Water Management, Žižkova 17, CZ-60200 Brno; tuhovcak.l@fce.vutbr.cz

Abstract This paper presents the results of a risk assessment in a case study, within the TECHNEAU project, that was performed on the drinking water supply system of Březnice town, located in central Bohemia, the Czech Republic. The system uses three groundwater sources and it supplies population about 3500. The risk analysis in this case study covers the system from source to service connection and was focused on identification of all hazardous events which may influence the quality of distributed water, either in terms of non-compliance with drinking water quality standards or risks to compromise consumer's health or satisfaction. Four different approaches were combined to identify the hazards and related undesired events that may occur in the system. For each event, the likelihood and consequence have been assessed using a scale of 5 categories – a risk matrix suggested by the WHO. There have been 47 different hazardous events identified and evaluated, 25 for the sources, 11 for the treatment process (and building), and 11 for the distribution system. 44 hazardous events have been further classified and expressed as 'extreme risk' (1), 'high risk' (15), moderate risk' (16) and 'low risk' (12). The Coarse Risk Analysis method used in this case study seems to be a suitable tool for risk identification and estimation in small water supplies.

Keywords case study, drinking water supply, risk assessment methods, TECHNEAU, water pollution prevention and control, water safety plan

INTRODUCTION

The main objective of the Work Area 4 (Risk assessment and Risk management) of the TECHNEAU project is to integrate risk assessments of the separate parts in drinking water supplies into a comprehensive decision support framework for cost-efficient risk management in safe and sustainable drinking water supply – and thus to support the effort of the World Health Organization (WHO, 2004) and other organizations which emphasize a risk based management of water supplies, including the entire system from source to tap. This is considered necessary in order to guarantee a safe drinking water to the consumers, but additional guidance and examples on how risk assessments can be carried out are needed to help water utilities in this way.

To illustrate and explain the different steps in risk assessment and risk management the TECHNEAU framework has been developed (Rosén *et al.*, 2007). There have been six risk assessment case studies carried out at different drinking water systems in 2007–2008 within this Work Area:

(a) Göteborg, Sweden
(b) Bergen, Norway
(c) Amsterdam, the Netherlands
(d) Freiburg-Ebnet, Germany
(e) Březnice, Czech Republic
(f) Upper Mnyameni, Eastern Cape, South Africa

The purpose of these case studies was to implement and evaluate the applicability of different methods analysing and evaluating risks to different drinking water systems. Some of the risk assessments were carried out using qualitative risk ranking methods and in some case studies quantitative methods were used. A comprehensive presentation of each case study is provided by the six case study reports available at the web site of the TECHNEAU project (www.techneau.org).

This paper includes an overview of the applied methods, the results and experiences of the Březnice case study, where a qualitative risk assessment method was applied as the part of the development of water safety plan for this supply.

METHODS

The basic idea of a qualitative risk assessment is to present the risk without using quantitative measures. This is a simple tool available to all drinking water producers, even those not previously familiar with risk assessment procedures. The most common qualitative risk assessment method is risk ranking using risk matrices. The World Health Organization (WHO, 2004), for example, suggests the use of a risk matrix to prioritise identified hazards. When ranking risks using this method the probability and consequence of each identified hazard are assessed using discretized scales. Such risk ranking using risk matrices is sometimes referred to as semi-quantitative, if the probability and consequence classes are defined using quantitative measures.

Risk ranking by means of risk matrices is easy to perform and the results are also easy to understand and communicate. Hence, this type of analysis is useful in many cases, especially as an initial risk analysis used to identify where further and more detailed studies are needed. Limitations of qualitative risk ranking methods are, for example, the fact that they assume a discrete nature of hazards, it may be hard to consider complex interactions between components of the system and procedures for sensitivity and uncertainty analysis are often lacking.

The water supply system in Březnice was analysed using a coarse risk analysis (CRA) method. Coarse risk analysis, often also referred to as a preliminary risk analysis is a common method for establishing a crude picture of the risk, with relatively modest effort (Aven, 2008).

The basis for the CRA is a description of the water supply system and an identification of hazards and related undesired events that might occur in the system. For each event, the likelihood (probability) and consequence (impact) have been assessed using a scale of 5 categories. The consequent risk for each hazard (hazardous event) is expressed in a semi-quantitative way as extreme, high, moderate, or low through a combination of the probability and the consequence of each hazard (event) by use of a risk matrix as suggested by the WHO in the monographs on the Water Safety Plans (Davison *et al.*, 2005; Bartram *et al.*, 2009).

The risk analysis in these case studies covers the system from source to service connection and was focused on identification of all hazardous events which may influence the quality of distributed water, either in terms of non-compliance with national drinking water quality standards or risks to compromise consumer's health, satisfaction or confidence.

To identify hazards for water quality several different tools have been used: the checklist in the form of the TECHNEAU Hazard Database (Beuken *et al.*, 2007), past experience of the personnel of the water utility and the local public health authority, brainstorming of the working team, and the checklist of main hazards for small water supplies developed by the Swiss Gas and Water Association (SVGW, 2003).

As the water utility operating the supply system in Březnice (VAK Beroun) has not developed any own specific risk tolerability criteriabefore this study has started, the team analysing the risks came with a suggestion. It was agreed to use the classification suggested by Davison *et al.* (2005) and both extreme and high risks should be considered as unacceptable. Consequently, risk reduction options will be developed for all identified extreme and high risks, while all low risks would be considered as acceptable. Regarding the moderate risks, it was agreed to apply the ALARP (As Low As Reasonably Practicable) principle, i.e. the necessity of risk reduction options for these risks would be discussed on a case by case basis. Some of these risks may be accepted if it is economically and/or technically unreasonable to reduce them, but corrective measures and/or monitoring options for such risks should be suggested.

To assess supply reliability (in a selected part of the system – the reservoir) the Preliminary Hazard Analysis (PHA) and consequently the Failure Mode and Effects Analysis (FMEA) and Failure Mode, Effects and Criticality Analysis (FMECA) methods have been used to estimate the risk but data are not presented in this paper.

The risk assessment team (water safety plan team) consisted of representatives of the National Institute of Public Health, the Brno University of Technology, the operator (VAK Beroun), the owner of the infrastructure (municipality), and the local public health authority.

System description

The town Březnice is situated in central Bohemia about 80 km SW of Prague, with a population of 3500. This small town is known and popular among tourists and visitors due to a nice chateau, the old brewery, unique small Jewish quarter and wonderful nature scenery nearby. Neighbouring village (200 people) is also connected to this supply system, but it is supplied exclusively from one source, while Březnice town is supplied from mixture of three sources.

The public water system was built in 1930–1931, originally with only one source; two other sources were explored after the WWII (1946 and 1990). The owner of all utility infrastructures is the municipality (Březnice), but the supply system is operated under license by the professional water company Vodovody a kanalizace Beroun a.s. (VAK Beroun) since 2005.

The basic data on the system and summary of the applied method are provided in Table 1.

RESULTS AND DISCUSSION

Hazard identification and risk assessment

Three detailed field visits covering all sources, treatment plant and reservoirs were carried out. In total, 47 different hazards/hazardous events were finally identified: 25 for sources, 11 for treatment and treatment plant building and 11 for distribution network including reservoirs. However, for three hazards no impact on water quality has been assumed and therefore no risk rating was performed. During the estimation process, several probable hazardous events were investigated in more detail to confirm or exclude suspicion (or assess the extent of the problem).

Table 1. Summary of the applied method and the analysed system in the Březnice case study.

<table>
<tr><th>Method</th><th>Drinking water system</th></tr>
<tr><td>Name: Coarse risk analysis (CRA)</td><td>Consumers: 3 700</td></tr>
<tr><td>Required level of expertise needed: medium</td><td>Source water: groundwater (3 different sources)</td></tr>
<tr><td>Time required for analysis: medium (weeks to month)</td><td rowspan="2">Treatment train of two sources (removal of iron and manganese + disinfection):
Oxidation ($KMnO_4$)
Filtration (open sand fiter)
Chlorination (chlorine gas)
The third source is only disinfected by sodium hypochlorite, without other treatment.</td></tr>
<tr><td>Required level of data details needed: medium</td></tr>
<tr><td rowspan="2">Type of results:
List of all hazards identified for the sources, treatment and distribution parts of the supply system.
Ranking and prioritizing the hazards due to its probability and consequence as extreme, high, medium, or low risks.</td><td>Distribution network: about 33 km</td></tr>
<tr><td>Production: 350–520 m^3 per day</td></tr>
</table>

The hazards (44) have been classified and expressed as “extreme risk” (1), “high risk” (15), “moderate risk” (16) and “low risk” (12). An overview of the risks is provided in Table 2 and some examples are given in Table 3.

Table 2. The ratings and number of risks in each part of the Březnice water supply.

Part of the system	**Risk rating**				
	Extreme (E)	High (H)	Moderate (M)	Low (L)	**TOTAL**
Sources	1	9	7	7	**24**
Treatment	0	2	4	4	**10**
Network	0	4	5	1	**10**
TOTAL	**1**	**15**	**16**	**12**	**44**

For the risks rated as extreme or high, reduction options have been suggested. The risks rated as moderate are discussed with stakeholders whether any risk reduction options are needed or not.

Table 3. Examples of identified hazards in the Březnice water supply and their assessment due to probability and consequence.

Hazardous event/cause	Hazard/ potential consequences to water quality/quantity	Probability	Consequence	Risk
Source 1: Pasture of farm animals (cows) in meadows around lower spring area	Microbially contaminated source water after heavy rainfalls.	Unlikely	Major	High
Source 2: Construction integrity of some wells is in bad conditions – breaks in the ceilings of well-head chamber around the cover; ingress of surface water in the chamber, standing water in the chamber, access of small animals.	Microbiological contamination of water in sources.	Likely	Major	Extreme
Source 3: Arable fields around the well field.	Contamination of source water with pesticides. (We do not know if it is due to hydrogeological conditions possible.)	Unlikely	Minor	Low
Treatment: Unsuitable design of the building (filter beds are situated in the upper floor of the building, beds are leaking).	Compromised statics of the building. Risk of emergency shut down of the building and treatment process.	Rare	Major	High
Distribution network: Unsuitable material of part of the network from water quality point of view (steel with bitumen lining, cast iron without any inner lining, lead service connections).	Contamination of water with chemicals (PAH, lead, iron), corrosion and high turbidity, lower efficiency of chlorine residual.	Moderate	Minor	Moder-ate

Sensitivity analysis

Usually some inherent uncertainty is associated with any hazard identification and risk assessment. Uncertainty can be caused by a lack of knowledge (e.g. missing information on exact hydrogeology of the sources) or by variability in parameters.

The main uncertainty relates to the assessment of the probability and consequence of the identified hazards. We are sure that some of the hazards really exist. In another cases, we know that hazards may be present, but we have no evidence about the occurrence and it is not easy to verify all of them. Finally, the third group of hazards cannot be excluded – they may happen, but they have never happened so far – however, it is very difficult to assess the probability that they will occur.

Method evaluation

Table 4 summarizes some general criteria of method evaluation and provides an overview of the applied CRA method. An *x* indicates low, medium or high for each criterion. Note that the scale (low, medium and high) is only qualitative and no quantification is done.

Lessons learned

- A thorough understanding of the whole water system is vital for a reliable hazard identification as well as risk estimation. The best option is cooperation between operators working daily with the system and experts with good knowledge background in water quality and water contamination sources.
- In case of small and relatively simple systems, the risk analysis can be done quite easily (with the help of guidelines) even by persons with little experience. Even when the reliability may not be very high it is a good start to get the first insight at the risks and have the basis for further risk analysis development.
- If any hazard checklist or database is used for hazard identification, it is recommended to users to read it (proper parts) before visiting the system (or part of the system) and again after the visit. During the field visit there is not usually time enough to study the checklist or database.
- Somebody, who has little experience in risk analysis, can identify many possible (or theoretical) hazards according to the hazard database, but it is necessary to verify or exclude them, e.g. through focused and more detailed

analysis, to obtain a reasonable list of real hazards for further evaluation and work.

- Most hazards might be identified (by means of hazard checklist or database) by somebody with a general knowledge on water supply, but some hazards (e.g. many hazards on special treatment steps like coagulation/flocculation, filtration or filtration) can be identified only by a very good specialist in water treatment and chemistry or in hydraulics.

Table 4. Summary of the method evaluation by means of general criteria.

Criteria	Low	Medium	High
Resouces needed			
Required level of expertise needed		x	
Time required for analysis		x	
Required level of data details needed		x	
Method properties			
Ability to cosider a source-to-tap approach		x	
Ability to include water quantity aspects		x*[1]	
Ability to consider water quality aspects			x
Ability to consider interactions between events, i.e., chains of events		x	
Ability to acknowledge system structure/design		x*[1]	
Ability to consider uncertainities of e.g. probabilities	x		
Ability to consider/model risk reduction options	*[2]	*[2]	*[2]
Ability to be intigrated in the water company management/maintenance routines			x
Updating posibilities, i.e., update when new information becomes availabe		x	
Results			
Ability to provide understandable results to the specific end-user			x
Ability to provide input data to be used in further studies, e.g. more detailed risk analysis	x		

*[1] ability would be high if FMEA/FMECA is used
*[2] the risk-reduction options are not presented and discussed in this paper.

CONCLUSION

Within our assessments, which covered the whole supply system from the sources to service connection, we identified quite comprehensive list of hazards. There were 44 hazards which have been further classified and expressed as "extreme risk" (1), "high risk" (15), "moderate risk" (16) and "low risk" (12). Besides assessing the risks in the Březnice drinking water supply, a main objective of this study was to evaluate the applicability of the methods used in the case study.

The CRA (Coarse Risk Analysis) method used in this case study seems to be a suitable tool for risk identification and estimation in small water supplies. It may be completed or replaced by a more sophisticated method at a later stage, but for the initial risk assessment process it represents a reasonable approach due to expertise, technical and financial requirements on the one side, but with sufficient results and sensitivity on the other.

The national importance of the study may be seen in following aspects:

(a) Březnice case study was the first attempt to prepare a water safety plan in the Czech Republic and therefore this work served as important pilot or model example for further applications in the country.
(b) Březnice case study provided objective information on priority risks and investments needed, and hereby support negotiation between operator (VaK Beroun) and owner of the infrastructure (municipality of Březnice). The operator was and is aware of a number of problems in the supply system, but according to the licence, the operator cannot invest anything into the system, but have to ask the municipality to do so. However, the municipality has other priorities and has invested only inadequate amounts in the system during the last 13 years, because it considered water supply as trouble-free.

ACKNOWLEDGEMENT

The work on Březnice case study was supported by the TECHNEAU project (EU contract No. 018320 GOCE) and partially also by the national project WATER-RISK (Czech Ministry of Education, Youth and Sports; Reg. No. 2B06039).

REFERENCES

Aven, T. (2008). Risk analysis: assessing uncertainties beyond expected values and probabilities. John Wiley & Sons, Ltd.

Bartram, J., Corrales, L., Davison, A., Deere, D., Drury, D., Gordon, B., Howard, G., Rinehold, A. and Stevens, M. (2009). Water safety plan manual: step-by-step risk management for drinking-water suppliers. World Health Organization, Geneva.

Beuken, R., Sturm, S., Kiefer, J., Bondelind, M., Åström, J., Lindhe, A., Machenbach, I., Melin, E., Thorsen, T., Eikebrokk, B., Niewersch, C., Kirchner, D., Kozisek, F., Gari, D.W. and Swartz, C. (2007). Identification and description of hazards for water supply systems. A catalogue of today's hazards and possible future hazards. TECHNEAU report; http://www.techneau.org/fileadmin/files/Publications/Publications/Deliverables/D4.1.1_and_D4.1.2.pdf

Davison, A., Howard, G., Stevens, M., Allan, P., Fewtrell, L., Deere, D. and Bartram J. (2005). Water Safety Plans. Managing drinking-water quality from catchment to consumer. (WHO/SDE/WSH/05.06). World Health Organization, Geneva.

Rosén, L., Hokstad, P., Lindhe, A., Sklet, S. and Røstum, J. (2007). Generic framework and methods for integrated risk management in water safety plans. TECHNEAU report; http://www.techneau.org/fileadmin/files/Publications/Publications/Deliverables/D4.1.3.pdf

SVGW (2003). Recommendations for a simple quality assurance system for water supplies (WQS). Regulation W 1002e. Schweizerischer Verein des Gas- und Wasserfaches (SVGW), Zürich.

WHO (2004). Guidelines for drinking-water quality. 3rd ed., World Health Organization, Geneva.

Risk assessment case study 2: Upper Mnyameni, South Africa

C.D. Swartz[a], M. Törnqvist[b], B. Öfverström[b] and T. Pettersson[b]

[a]Chris Swartz Water Utilization Engineers, PO Box 745, Mossel Bay 6500 South Africa
(E-mail: cswartz@mweb.co.za)
[b]Chalmers University of Technology, SE-412 96, Göteborg, Sweden
(E-mail: Thomas.Pettersson@chalmers.se)

Abstract This paper presents the results of a risk assessment in a case study, within the TECHNEAU project, that was performed on the drinking water supply system of Upper Mnyameni village, Eastern Cape, South Africa. The objectives of the study were to identify hazards in the drinking water supply system (from "source-to-tap"), estimate and evaluate the risks to humans and the development of the society, and evaluate the risk assessment methods that were used. Two types of risk analysis were performed. The first risk analysis was performed by risk ranking of likelihood and consequences and presentation of risks with risk matrices. The second risk analysis was performed by using South African Risk Evaluation Guidelines. The TECHNEAU Hazard Data Base (THDB) was used to facilitate hazard identification for both methods. The major risks were found at the communal water taps (most households do not have taps), unsafe practices for storage of collected water, inadequate hygiene due to lack of easily accessible taps, and unreliable power supply. Suggested risk reduction options were found to reduce the risks significantly. Of the two methods the risk ranking using risk matrices were found to be most useful for the drinking water supply system of Upper Mnyameni. The South African Evaluation Guidelines were found to be less useful, mainly due to the lack of detail when ranking the risk for the raw water source.

Keywords risk assessment, risk matrix, rural water supply, South Africa, TECHNEAU, Upper Mnyameni

INTRODUCTION

Within Work Area 4 (WA4) *Risk Assessment and Risk Management*, in the TECHNEAU project, six risk assessment case studies were carried out at different drinking water systems during 2007–2008. The aim of the case studies was to apply and evaluate the applicability of different methods for risk analysis (i.e. hazard identification and risk estimation) and to some extent risk evaluation of drinking water supplies (Rosén *et al.*, 2007). The case studies provide a number of different examples on how risks in drinking water systems can be analysed and evaluated. Following six drinking water supplies constitute the case study sites where risk assessments were performed in

WA4: Göteborg (Sweden), Bergen (Norway), Amsterdam (the Netherlands), Freiburg-Ebnet (Germany), Březnice (Czech Republic) and Upper Mnyameni, Eastern Cape (South Africa).

The paper presents the results of a risk assessment of the drinking water supply system of the Upper Mnyameni village, Eastern Cape, South Africa. This case study was conducted in collaboration with Amatola Water, the local water utility.

Aims of the study

The objectives of the risk assessment study at Upper Mnyameni were:

- Identifying hazards, estimating and evaluating risks in drinking water systems.
- Investigation of risks for water supply failures and risks for insufficient water quality.
- Consequences for humans and the development of the society.
- Evaluation of the methods that were used.

Risk methods used

The objective of qualitative risk assessment is to present the risks without using quantitative measures, so as to provide a simple tool that can also be used at small water treatment plants. The most common qualitative risk assessment method is risk ranking using risk matrices. In this method, the probability and consequence of each identified hazard are assessed using scales which are quantified. This risk matrix method is therefore also referred to as a semi-quantitative risk assessment method, if the probability and consequence classes are defined using quantitative measure.

The water supply systems in both case studies, in Březnice and Upper Mnyameni, were analysed using the coarse risk analysis (CRA) method. Coarse risk analysis, often also referred to as a preliminary risk analysis is a common method for establishing a crude risk picture, with relatively modest effort (Aven, 2008).

Risk estimation has also been performed using the South African Evaluation Guidelines. The evaluation guidelines divide the drinking water supply into risk evaluation into five different categories. Each category is ranked into three different risk fields and from these a weighting is performed to get the total risk of the system. The method is described in more detail in the section on *Risk Analysis*.

System description of Upper mnyameni drinking water supply

Upper Mnyameni water treatment plant is situated in a rural area in the Amatola Mountains in the province Eastern Cape (Figure 1). The small villages Upper and Lower Mnyameni are low income communities with a very high unemployment rate (approximately 80 percent) (Dyer, 2007). The major industry in the villages is forestry. There is also some non-formalized low scale farming activities such as roaming cattle in the villages.

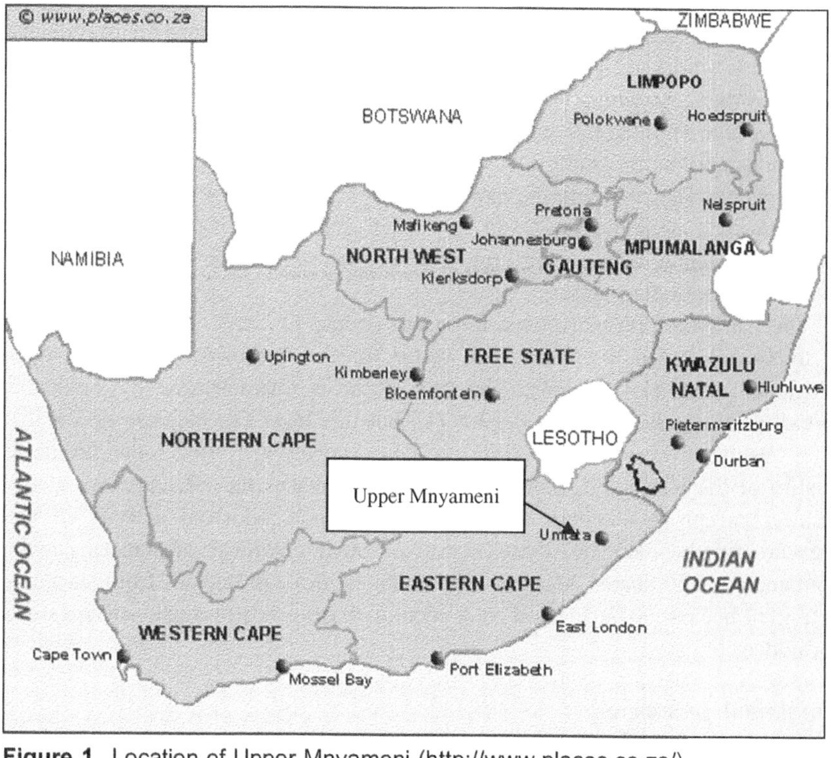

Figure 1. Location of Upper Mnyameni (http://www.places.co.za/).

Amatola Water Board is the local drinking water supplier for Amahlati Municipality, to which Upper and Lower Mnyameni belongs. It is one of totally 20 water boards and utilities in South Africa. Amatola Water Board belongs to South African Association of Water Utilities and has the mandate from the South African government to supply municipalities and other consumers with water.

Upper and Lower Mnyameni are two rural villages in the Eastern Cape province in South Africa, about 80 kilometers from the south east coast. The villages are supplied with drinking water by a water treatment plant that takes its water from the Mnyameni dam. Altogether the water treatment plant supplies approximately 2500 people with water. These communities are very rural and no major industries or other commercial activities are supplied with water from the Upper Mnyameni water treatment plant.

Raw water source

The raw water source is the Mnyameni Dam which is situated approximately 1 kilometer from the Upper Mnyameni village. The outflow from this dam is the source of the Keiskamma River which has its outflow to the Indian Ocean close to the village Hamburg, South Africa. The dam is surrounded by precipitous mountains and the incoming water comes from rain and snow melting, but also from groundwater flow into the dam. The fact that the dam is partly supplied by groundwater makes the supply of raw water stable for long periods with dry weather. The flow of water into the dam is much higher than the demand from two water treatment plants.

The dam was originally used for water supply for large scale farming, but is today only used to supply raw water for the treatment plants in Upper Mnyameni and Masincedane. The raw water is characterized by moderate turbidity levels, usually around 10 NTU, but this level can increase drastically as a result of heavy rains. High turbidity negatively affects the aesthetic quality of the water and, more importantly, decreases the effectiveness of the disinfection process. There are neither roads nor industrial activity in the area around the dam and therefore oil or other chemical substances cannot contaminate the water source. The dam is also protected from farming activities because it is located in a natural reserve where such activities are forbidden.

Treatment processes

The water gravitates from the dam to the water treatment plant (WTP). Figure 2 shows a schematic flow diagram of the water supply system in Upper and Lower Mnyameni. The water flows by gravity to a balancing chamber located just outside the treatment plant. In the balancing tank pre-chlorination takes place, but the main purpose of the tank is to have a flow into the WTP with static head. Chlorination is not effective at this stage because the turbidity is too high for successful disinfection to be achieved. From the balancing tank the water gravitates down to the filter house. Before the water reaches the filter house, a coagulant is pumped into the pipe. Inside the filter house there are two parallel flocculation tanks and six parallel pressure sand filters. The function of the flocculation tanks is to increase the time for the flocculation process to prevent flocs forming after the water has passed through the filters. The flocs are then filtered out in the lower sand layer of the filter. A booster pump then carries the water to the Clean Water Reservoir (CWR) at the plant. Inside the CWR chlorination takes place through chlorine tablets.

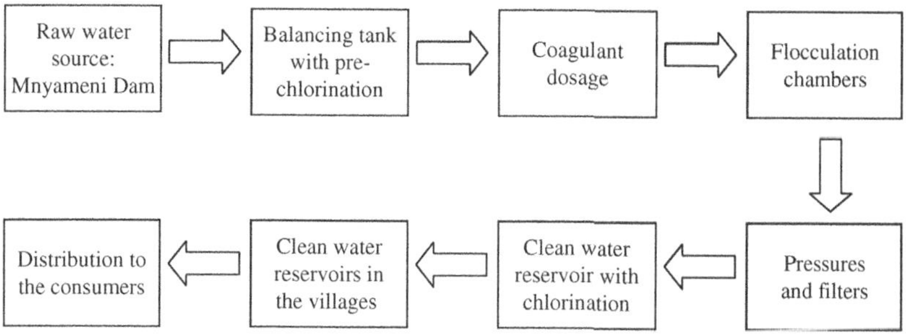

Figure 2. Flow diagram of the water supply in Upper and Lower Mnyameni.

Distribution

Chlorinated water is pumped from the WTP to a high altitude reservoir in the upper part of Upper Mnyameni. From this reservoir parts of the water gravitates to a smaller reservoir in the middle of the village and parts of the water is directly distributed to consumers. Lower Mnyameni has a similar distribution system but there are three reservoirs. There are two electrical pumps at the treatment plant, one that pumps to each village. Both pumps have the possibility to pump to the other village if necessary. There is no back-up power supply for the pumps in case of power failures. Most taps are outside and there is approximately one tap per 20 houses. Villagers usually have some kind of container that they fill up with water. Some of the households have water inside the houses, of which some are illegal (Generally there are 5–20 households per tap in the rural villages of the Amatola Mountains). When this study took place there was no test made for retention time in reservoirs and pipes. Experts at Amatola Water estimated the retention time to not more than 3 hours during the day-time, but longer during the night-time as demands drops. The total drinking water production is approximately 240 m^3/day.

Operation

The plant works 24 hours a day all year around, but it is only manned during office hours. During this time samples of water are taken and analyzed every hour. The analysis includes turbidity and pH-value in the raw water and for the treated water analysis includes turbidity, pH-value and amount of residual chlorine. The raw water samples are taken from the pipe between the balancing tank and the flocculation tanks and the clear water sample is taken in the CWR. The sand filters are backwashed in the morning and in the afternoon. At weekends, the plant is manned for shorter periods of time and at least one backwashing of the sand filters is done. The levels in two of the reservoirs on the distribution system, one in Upper Mnyameni and one in Lower Mnyameni are monitored in the operator's office at the plant. The water level in the CWR at the plant is also measured. This information is also available in the control room at the water treatment plant in Sandile, which is manned 24 hours a day. If there is a pump failure at the Upper Mnyameni treatment plant this can also be noticed in Sandile. The pumps cannot be controlled from Sandile, only monitored. In case of a breakdown of one of the two pumps the other pump can alternate between the villages, but it can only be done manually at the WTP in Upper Mnyameni.

RISK ANALYSIS

To cover the whole water supply system a "source-to-tap" approach has been used. A system description was made after discussion with Amatola Water, gathering of data and by site visits. Hazards were identified by both using the TECHNEAU Hazard Database (Beuken *et al.*, 2007) and by brainstorming from discussions with Amatola Water. Two different methods have been applied to analyse the risk. The first method was risk ranking, where the likelihood and consequence was assessed for each identified hazard and the risk presenting using risk matrices with ALARP (As Low As Reasonable Practicable) zones. Rating of likelihood and consequences for the risk matrix were made in collaboration with experts at Amatola Water. The second risk assessment method was performed by using South African Risk Evaluation Guidelines (Water Research Commission, 2007).

Hazard identification

The following eleven hazardous events were identified from the brainstorming session and by using the TECHNEAU Hazard Database:

1. High turbidity causing ineffective chlorination.
2. Contaminated taps due to animals leaning/scratching against them.
3. Inadequate hygiene due to low water availability at homes.

4. Contaminated groundwater leaking into pipes.
5. Poor storage of water.
6. Lack of treated water leading to use untreated water.
7. High turbidity when the WTP is unmanned causing high bacterial amount.
8. Ineffective mixing of chlorine leading to high bacterial amount.
9. Sabotage at any part of the system.
10. Incorrect actions due to lack of enough operational skills.
11. Pump failure when the plant is unmanned.

The eleven possible hazards that can affect the drinking water were rated by experts at Amatola Water. The hazards were rated by likelihood and consequence of occurrence. There are two consequence ratings, one focuses on human health and one on number of people affected.

Risk estimation and presentation of risks with risk matrices

Both the probability (likelihood), that the different hazards will occur, and the consequence, of the events, were ranked on a scale from one to five. The consequences of the events were ranked with respect to health consequences (*Risk Matrix 1*) and number of people affected (*Risk Matrix 2*). The two risk matrices were weighed and merged to a total risk matrix (*Risk Matrix 3,* Figure 3).

	Little or no consequence	Minor consequence	Considerable consequence	Major consequence	Catastrophic consequence
Frequent			2,5		
Occasional		6			
Possible					
Rare				3	
Has not occured		1,4,9	7,8,10,11		

Figure 3. Risk matrix 3 (total risk) estimates risks for water supply in Upper and Lower Mnyameni.

The matrices are divided into three zones. The green field shows risks that are considered to be acceptable and the red field indicates that the risks are unacceptable and could not be tolerated, i.e. must be reduced immediately. The yellow field indicates the ALARP (As Low As Reasonably Practicable) region. That means that the risk can be accepted if it is economically and technically unreasonable to reduce it.

Risk Matrix 1 – health effects

This matrix shows that the most serious problem for water quality is at the taps and the handling of the water. The risks that concern the water treatment processes are considered fairly low. The reason for this is that the raw water is fairly clean and could be used untreated. There are no known major problems with the treatment process, but the risk of problems with for example poor working filters or dosage of chemicals must be calculated for periods when the plant is unmanned for long time. In terms of quantity, the major risks have to do with the pumps. While both pumps are electrical and no back-up power supply is available, longer power failures always lead to lack of water for the consumers.

Risk Matrix 2 – number of people affected

The possible hazardous events have also been arranged in a matrix regarding the number of people affected by a certain hazardous event. That matrix shows the possibility of the event occurring compared to how many people that would be affected. This matrix is also divided in to three zones in the same way as for the quality risk matrix but the zones look a bit different. For example, the red zone only includes three fields. The reason for this is that the population is too small to make the risks unacceptably high just by quantity matter. Generally the hazardous event will affect the whole population in the villages, approximately 2500 people. The decision to enlarge the green field was made by reasoning that an event that never has occurred is more severe if it is classified as major health risk than an event of low severity that affects a large number of people. How the fields of the risk matrix are chosen may vary.

Risk Matrix 3 – total risk matrix

In order to combine the two factors *health* and *number of people affected* of the two different matrices the two factors have been weighted in order to create a third "total" risk matrix (Figure 3). Decision was taken to make the health aspect more important than the number of people affected. The health consequences are weighted as three times as important as the number of people affected by an incident. It is reasonable to think that severe health consequences for a few people result in larger consequences than a minor discomfort for many people.

Sensitivity analysis

There are several uncertainties for the input information. Upper Mnyameni is a small village in an underdeveloped country and the information about the system is limited. The lack of information, e.g. pipe information, number of taps, and number of people in the villages leads to more reliability on the experts. For a system that is not very well documented it also requires a more thorough field study, though a lot of information will still be missing. Likelihood, health effects and number of people affected by a certain hazardous event will always be difficult to estimate. The experts at Amatola Water had long experience in the field of water engineering and thorough knowledge about small drinking water systems in rural areas, which make the reliability of the experts' estimations greater. Communication with local people can also lead to uncertainties due to lack of language skills or misunderstanding in other ways.

RISK EVALUATION

Risk tolerability

There is no single way that the zones of the risk matrix must be put and it is important to consider and motivate how the fields are chosen. The fields of the risk matrices were chosen differently depending on what the different consequences were. Health effects for one individual, presented in *Risk Matrix 1*, were considered most severe and therefore the red field is largest in this matrix. *Risk Matrix 2* considered the possible number of people being affected by an event and was given larger green zone and smaller red zone. The total risk matrix, *Risk Matrix 3*, was given the same zones as the risk matrix that considers heath effects.

Risk reduction options

This sections describes what measures can be taken to decrease the risk of the different scenarios. The risk matrix in Figure 4 shows how the risks move in the matrix when measures are taken. The scenarios that falls under the green field does not need any measures and will thus not be described in this section.

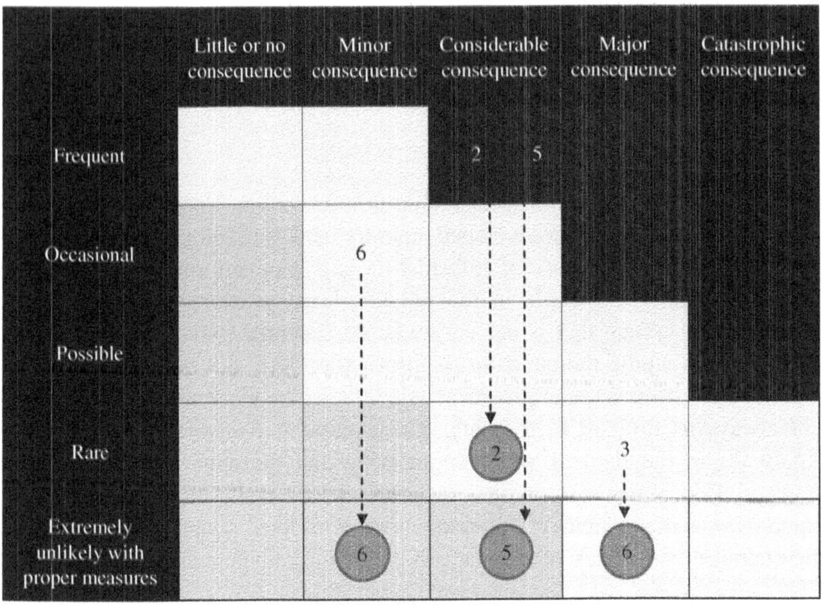

Figure 4. Risk matrix that shows how the hazardous events move after risk reduction measures have taken place.

Scenario 2: Contaminated taps caused by animals leaning against the taps. Contamination may spread when people, especially children drink from taps.

This hazardous event is in the red field in Risk Matrix 3 with ALARP zones. That means the risk is not acceptable and should be reduced as soon as possible. The best thing would be to make household connections for all the households in the villages. This is a long term ambition and will be an extensive project that will require large investments. It must also be taken into consideration that household connections probably will increase the water demand. A more simple measure would be to add some protection for the taps, e.g. fences, etc. to prohibit that animals can come in contact with the taps.

Scenario 3: Few taps make water accessibility low (one tap per 5 up to 20 houses and distances up to 200 meter to the nearest tap), which might lead to inadequate hygiene. For example: improper washing after using latrines which might cause bacterial spreading.

This risk scenario falls into the yellow zone in Risk Matrix 3 and should then be prevented if it is reasonable. This risk will be minimized if all the villagers get household connections, which is the long term ambition for the water supply in South Africa.

Scenario 5: Poor storage of water. For example storage in open buckets or dirty bottles.

This risk scenario falls into the red zone in Risk Matrix 3 and needs to be prevented. This risk will be minimized if all the villagers get household connections, which is the long term ambition in South Africa. It can also be reduced if people get information on how to handle and store their water in a hygienic way (Jagals, 2004).

Scenario 6: Long-time power failure or other incident leading to lack of water. People might use the untreated water at Mnyameni River instead (quality related problem).

This risk scenario falls into the yellow zone in Risk Matrix 3 and should therefore be prevented if measures are reasonable. As it would need major upgrades at the power supply system to prevent the occurrence of long time power failures it would take major large scale actions and it would fall outside feasible measures for this study. It is also difficult to reduce the time it takes to restore power back as these failures are often a result of storms which also affect access to the villages. A power generator at the WTP is the only reasonable short-term option in order to ensure enough drinking water.

Risk estimation with South African Risk Evaluation Guidelines

Risk estimation has also been performed using the South African Evaluation Guidelines. The evaluation guidelines divide the drinking water supply into risk evaluation into five different categories. Each category is ranked into three different risk fields and from these a weighting is performed to get the total risk of the system.

In the guidelines it is recommended that the first a rapid assessment based on the material available takes place. After the rapid assessment a more detailed assessment shall take place to determine the potential risk of the system, which results in a risk level. Whenever the system changes or there is additional information the guideline should be consulted again and the risk level be revised.

When performing the risk ranking the system is divided into five subsystems, or *categories,* each ranked from 1–10 individually. The categories are:

(a) Water source
(b) Design

(c) Operation
(d) Reporting
(e) Operator

After ranking of each risk all categories are weighted according to the assessment of its importance to give the overall risk:

(a) Water source 10%
(b) Design 30%
(c) Operation 30%
(d) Reporting 10%
(e) Operator 20%

Overall risk = (0.1 × Water source risk) + (0.3 × Design risk) + (0.3 × Operation risk) + (0.1 × Reporting risk) + (0.2 × Operator risk)

The low risk level (1–4) is systems with minor deficiencies and the water quality mainly meet the requirements of SA quality drinking water guidelines.

The Medium Risk level (5–7) does not require immediate action, but measures should to avoid future problems.

The high risk level (8–10) needs immediate attention and corrective actions should be taken. In this risk level region it is a combination of major deficiencies that threatens safe drinking water.

RESULTS AND DISCUSSION

Method evaluation

Two types of risk analysis have been performed. The first risk analysis was performed by risk ranking of probability (likelihood) and consequences and presentation of risks with risk matrices. The second risk analysis was performed by using South African Risk Evaluation Guidelines. TECHNEAU Hazard Data Base was used as hazard identification for both methods.

The hazardous events that are analysed in the case study are developed from the examples in the TECHNEAU Hazard Data Base. Even though THDB is developed primarily for urban communities in developed areas, it is also applicable in rural communities such as Upper and Lower Mnyameni. THDB was found to be a useful and extensive tool for identifying hazards.

Ranking consequences can be difficult due to that a consequence can be hard to define. It can e.g. include the number of people affected and the severity of a certain event for one individual. The risk matrices were found to be a useful tool for presentation of risks.

When using South African Evaluation Guidelines the outcome might be misleading. An example is the assessment gave a high risk of the water source. If some adjustments are made in the South African Risk Evaluation Guidelines it might become a useful and efficient tool for risk assessment.

CONCLUSIONS

Risk estimation with risk matrices is a useful and efficient tool. It is easy to understand and present data. When choosing what consequences are of importance it is vital to think it through thoroughly. For this case study it was decided to use *health* and *number of people affected* by a certain hazardous event as consequence factors.

The major risks were found at the water taps (most households do not have taps), and from insufficient storage of water, inadequate hygiene due to lack of easily accessible taps and for the lack of power-supply. Suggested risk reduction options were found to reduce the risks significantly.

Of the two estimation methods the risk matrices were found to be most useful for the drinking water supply system of Upper Mnyameni. The South African Evaluation Guidelines were found to be less useful in this case study, mainly due to the lack of detail when ranking the risk for the raw water source.

REFERENCES

Aven, T. (2008). Risk analysis: assessing uncertainties beyond expected values and probabilities. John Wiley & Sons, Ltd.

Beuken, R., Sturm, S., Kiefer, J., Bondelind, M., Åström, J., Lindhe, A., Machenbach, I., Melin, E., Thorsen, T., Eikebrokk, B., Niewersch, C., Kirchner, D., Kozisek, F., Gari, D.W. and Swartz, C.D. (2007). Identification and description of hazards for water supply systems – A catalogue of today's hazards and possible future hazards, TECHNEAU, Deliverable no. D 4.1.1, D 4.1.2.

Davison, A., Howard, G., Stevens, M., Allan, P., Fewtrell, L., Deere, D. and Bartram, J. (2005). Water Safety Plans. Managing drinking-water quality from catchment to consumer. (WHO/SDE/WSH/05.06). WHO, Geneva. http://www.who.int/water_sanitation_health/en/

Dyer, A. (2007). Personal communication. http://www.places.co.za/

Jagals, P. (2004). The effect of engineering and educational interventions on the microbiological quality of container-stored water in developing-world households. Paper presented at the Water Institute of Southern Africa Biennial Conference, Cape Town, May.

Rosén, L., Hokstad, P., Lindhe, A., Sklet, S. and Røstum, J. (2007). Generic framework and methods for integrated risk management in water safety plans, TECHNEAU, Deliverable no. D 4.1.3, D 4.2.1, D 4.2.2, D 4.2.3.

SANS 241:2005, *South African National Standards for Drinking Water.*

SVGW (2003). Recommendations for a simple quality assurance system for water supplies (WQS). Regulation W 1002e. Schweizerischer Verein des Gas- und Wasserfaches SVGW, Zürich.

Water Research Commission (2007). *Risk Evaluation Guidelines*. Draft Report.

Distribution

Conceptual model for discolouration in drinking water systems: Who's to blame and what to do?

J.H.G. Vreeburg[a], J. Menaia[b], L. Branco[c], M. Benoliel[d], P. Aprisco[d], N. Rebola[c] and B. Cordeiro[c]

[a]KWR Watercycle Research Institute, PO Box 1072, 3430 BB Nieuwegein, The Netherlands (E-mail: jan.vreeburg@kwrwater.nl)
[b]National Laboratory for Civil Engineering – LNEC, Av. Brasil 101, 1700-066 Lisbon, Portugal (E-mail: jmenaia@lnec.pt)
[c]EPAL Operation Department, Av. Berlim, 15, 1800-031 Lisbon, Portugal (E-mail: luisbran@epal.pt)
[d]EPAL Central Laboratory, Rua do Alviela, 12, 1170-012 Lisbon, Portugal (E-mail: maria.benoliel@epal.pt)

Abstract Discolouration is the most common reason for customers to complain at the water company about water quality. The discolouration is mostly caused by resuspended solid matter that may originate from several sources as suspended particles in the treated water, which include upstream treatment, pipe's corrosion or detachment of organic matter from the pipes' biofilm. Systematic research on the origin of the accumulated particles revealed that in many cases the suspended particles in the treated water leaving the water production plant are the major source for the discolouration risk. The hydraulic circumstances in the network, however, allow the particles to settle in the pipe, causing discolouration risks. A conceptual model for the particle sources and behaviour clarified the possibilities for managing the accumulation rate of the particles: improve the operation of treatment to minimise the particle load to the system and manage the corrosion rates. Managing discolouration risks can also be attained by the implementation of a judicious pipe cleaning program. In order to methodologically answer these questions an extended conceptual model was conceived. The model offers a good framework to quantitatively approach the relation between velocity profiles and particles behaviour in the network. Such relation is the basis of the Resuspension Potential Method (RPM), which was implemented and tested on the EPAL distribution system in Lisbon as a tool to assess discolouration risks at a series of the network zones. In addition to being easily tailored, this new measuring methodology proved to be an effective tool to locate and rank sediment hotspots which need to be cleaned in order to prevent discolouration risks.

Keywords discolouration, drinking water network sediments, particle load and behaviour

INTRODUCTION

Discoloured water is one of the main causes for customers to complain to the water companies (Vreeburg and Boxall, 2007). Nonetheless, what the customer perceives as being discoloured water is suspended matter. In fact, if a sample of "discoloured" water is left to stand for a long period (>24 h) the water clears and the suspended particulate matter will deposit. The deposition of this matter also occurs in the drinking water transport and distribution networks. The resuspension of this sediment is probably the most important cause of discoloration incidents. Resuspension occurs as velocities in the network change drastically inducing shear stress rises on sediment layers. Typically such "hydraulic incidents" arise from pipe burst, hydrant use and valve exercise (Vreeburg *et al.*, 2004b).

The particles responsible for discolorations events in the network have different sizes, densities and may have its origin outside the network, or in processes occurring within it. The particle related processes explaining the discolouration risk causes are summarised in a conceptual model that is presented in Figure 1. The complexity of the model makes clear that there is no unique cause for the build up of sediment layers in the network. The root cause of a discolouration complaint is the resuspension of the particle layer in combination with a customer noticing the increased turbidity and motivated enough to actually notify the water company. This shows the limited value of customer complaints as a tool to representatively evaluate the discolouration risk of a network.

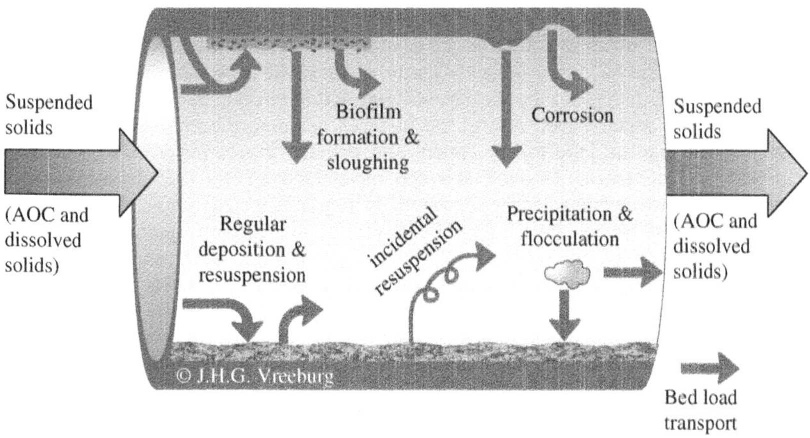

Figure 1. Conceptual model for particle related processes in a drinking water network (Vreeburg, 2007).

In what concerns discolouration, the TECHNEAU program focuses on customer perception and complains, as well as on appropriate technology for the water companies to influence this positively. The goal of the research is to develop tools to manage the discolouration risk. The first step is to understand and quantify the processes as pictured in Figure 1. Eventually the knowledge on the processes and their interaction will lead to a more quantitative modelling which allows for the prediction of vulnerable locations. Concurrently more practical research is done into possibilities for managing the layer of particles in the network. Three different approaches are followed to develop operational tools and concepts:

- Minimizing the particle originating processes
  This involves the reduction of suspended solids and biological agents in the treated water. It is part of the TECHNEAU program, which focuses partly on new technologies, but also on optimising conventional treatment processes.
- Managing the particle accumulation processes
  The accumulation is facilitated by very low velocities in looped distribution networks. Self cleaning networks with sufficiently high velocities are not considered within the TECHNEAU program, but are a promising development that can help to manage the discolouration problem in a sustainable way.
- Managing the sediment layer itself
  Reducing the amount of particles in the network can be done with efficient and timely cleaning.

In this paper the focus will be on the practical and operational aspects of the discolouration of drinking water, other than controlling the corrosion process. First the connection between the treatment performance and the discolouration problems is described in an extended conceptual model. A case study in the network of EPAL, the Lisbon Water Company, is introduced in which the first step to come to practical solutions is made. The discolouration risk in the network is analysed with the adjusted RPM (Resuspension Potential Method), showing the practical value of the tool. Next steps in quantifying the actual sediment load will be analysed in the near future following the extended conceptual model application.

EXTENDED CONCEPTUAL MODEL: THE HYPOTHESIS ON WHAT TO DO

The model presented in Figure 1 shows the particle associated processes and consequently the corresponding theoretical possibilities to manage discolouration. The goal of TECHNEAU is to fulfill these theoretical possibilities quantitatively and to materialize them into practical tools and strategies. The field for improvement are the treatment and the operation of the network in the form of cleaning. Figure 2 shows the schematic presentation and cohesion of the measurements. In this figure the corrosion of unlined cast iron pipes is excluded.

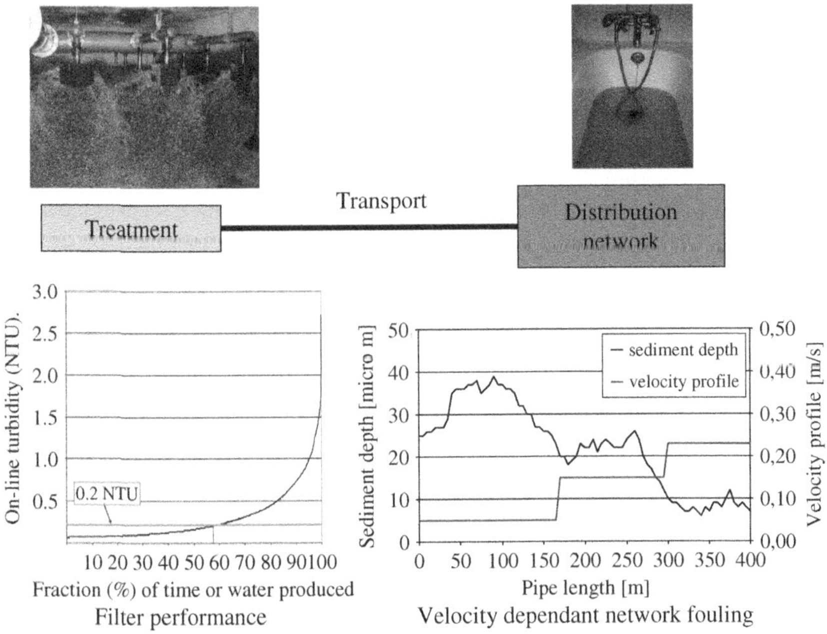

Figure 2. Schematic representation of the coherence of the involved parameters.

The core of the figure comprehends the processes of treatment-transport-distribution. If the treated water still contains particles they can settle in the distribution network. The characterisation of the performance of the treatment towards to continuity of the turbidity level is being developed in TECHNEAU. A sample of such a characterisation is given in the left lower corner of Figure 2. This is a cumulative frequency distribution of the turbidity over a longer period. The median of this distribution is around 0.2 NTU and the average performance meets every present standard. Nevertheless, peaks in turbidity need to be particularly considered, as they depict noteworthy inputs to the particle load of the system. The technology that is developed within the TECHNEAU research will focus on the reduction of the peaks and the continuity of the performance of the treatment.

The particles will be transported through the transport system and because of the relative high velocity in these systems the accumulation of particles will probably be limited. In the distribution network, however, the velocity is generally very low, with some variation during peak demands. In the right lower graph of Figure 2 the effect of the variation in velocity on the sediment layer is visualised. The horizontal axis is the stretched length of a pipe in a distribution system. The red, more or less straight line gives a theoretical velocity characterisation of the flow pattern in this pipe. Every part of the pipe has it own flow pattern over the day depending on the demand pattern of the consumers connected to that particular piece of pipe. One of the challenges to meet is to analyse the velocity profile and characterise this into one figure. Interesting developments were published recently in this field (Blokker and Vreeburg, 2005; Blokker *et al.*, 2006) and this will add to the overall result.

The hypothesis on the distribution of sediment over the pipe length and the relation with the velocity profile is given by the blue, more fickle, line in the right lower graph of Figure 2. This comes down to the suggestion that the amount of sediment found in a pipe has a reversed relation with the velocity profile. The connection between the performance of the treatment and the sediment level is that the absolute amount of sediment is treatment type specific and that the downstream distribution of the sediment is velocity dependant. In other words: the absolute level of the blue, fickle line that represents the level of sediment is determined by the type of treatment. In the case studies that are part of the TECHNEAU project these hypotheses are challenged.

METHODS

To asses the discolouration risk a modified version of the Resuspension Potential Method (Vreeburg *et al.*, 2004a) is implemented. This method evaluates the turbidity effect of an increase of the velocity by 0.35 m/s in an isolated pipe. The modification towards the original method is that the duration of the disturbance is limited to 5 minutes instead of 15 minutes. The turbidity is measured with high frequency (a few measurements per minute) and the trace of the turbidity is rated following the categories given in Table 1. An alternative can be to average the turbidity over the total measuring time of 25 minutes. For measurements the water company of EPAL has developed a convenient trailer with the equipment and flow control valve mounted on it (see Figure 3).

Table 1. Rating categories for the adjusted RPM.

Rate	0	1	2	3	4
Maximum during disturbance [NTU]	0–2	2–4	4–8	8–16	>16
Average during disturbance [NTU]	0–2	2–4	4–8	8–16	>16
Average during resettling time [NTU]	0–2	2–4	4–8	8–16	>16

Figure 3. Measuring equipment as developed by EPAL mounted on a trailer.

In a number of test cases, both within and outside of the TECHNEAU scope, the methodology has been applied and results are available. In this paper the case study of the water company of Lisbon EPAL is used. There, experiments were designed to assess the discolouration risks in the network as a support tool for decisions on the prioritisation of network zones cleaning. The ultimate corporate goal is to decrease the number of customer complaints on discolouration.

In 9 zones of the network 5 to 6 RPM measurements were performed yielding a data base of over 35 measurements. All the experimental zones have the same drinking water source and do not include unlined cast iron pipes, so could be excluded corrosion . None of the zones was recently cleaned. This experimental approach was used to substantiate the conceptual model and to get a wider range of data on sediment layers in the network. Following the measurements the zones with the highest discolouration risk are selected for cleaning.

RESULTS AND DISCUSSION

Recently, the effect of particle free water versus normal drinking water in a common drinking water distribution network has been published in a partly TECHNEAU sponsored publication (Vreeburg *et al.*, 2008). The experiment concerned a comparative trial in two similar distribution networks that were both cleaned prior to the actual experiment. One of them was fed by normal drinking water that met all the Dutch legal requirements, meaning that is was fit to be distributed without a residual disinfectant. The other network was fed with the same drinking water, but now in situ post-treated with an ultra filtration membrane system. The experiment showed that in the network with particle free water hardly any discolouration risk could be measured after one-and-a-half year supply, while the other network developed a significant discolouration risk in the same period.

This experiment showed convincingly that there is a relation between the particle load originating from the treatment and the development of a discolouration risk in the distribution network. It also showed that the discolouration risk is determined by resuspendable particles, irrespective of their origin. This supports the likelihood of the model presented in Figure 1. Accordingly and referring to the title of this paper it is easy to blame the treated water for the discolouration problems for the conditions of the tested system.

The extended conceptual model of Figure 2 shows the effect of the hydraulic situation in the network. The particles are able to settle due to the low velocities that occur within the network. So, also the hydraulics in the network can be blamed. Consequently the improvement of that would lead to self cleaning networks. In the Netherlands this has led to so-called self cleaning networks that cater primarily for drinking water and less for fire fighting water. The success of this formula shows that it is possible to follow that path, though it demands for a lot of flexibility both at the fire fighting departments and at the water companies (Slaats *et al.*, 2002; Vreeburg 2007).

The results of the RPM analyses in Lisbon are summarised in Table 2. The different ways to interpret the RPM data show the same outcome, though the turbidity data gives a better insight in the extent of the discolouration. For instance the first two zones, 3070 and 3490 gave in average RPM ratings nearly the same results: 9.4 and 9.8 respectively. The standard deviation shows some difference, indicating that zone 3490 has some more variance in the measurements. This is ascribable to the maximum turbidity during disturbance, which was significantly higher in zone 3070. However, in this zone sediments settled more easily, as it is shown by the lower turbidity values during the resettling time. Moreover, the two zones display a distinct discolouration risk, as the visibility threshold for the turbidity is from 5 to 10 NTU (Slaats *et al.*, 2002).

The relation between the total score and the total average is very good, as is shown in Figure 4. (R^2 = 0.77) This means that both the criteria may be used for making decisions about the discolouration risk in the different zones.

The experiments showed the suitability of the RPM methodology to effectively assess the discolouration risk in identifying and rating critical zones for discolouration. Cleaning those zones will reduce such risk considerably. Analysis of the flushed water will further give an insight into the amount and characteristics of the sediment accounting for the discolouration risk. Because all the trial zones receive water from the same treatment and have no unlined cast iron pipes, the velocity profile will explain the differences in the results of the RPM.

Table 2. Results of RPM analyses of 9 zones in EPAL network.

					RPM rating					
	Disturbance		Reset	Total	Disturbance					
	Max	Avg	Avg	Avg	Max	Avg	Reset	Tot	St	n
Zone	[NTU]	[NTU	[NTU	[NTU]					dev	
3070	115.93	38.95	8.11	14.28	3.8	3.4	2.2	9.4	2.07	5
3490	96.7	37.42	20.62	23.98	3.4	3.4	3	9.8	3.83	5
1220	48.69	9.72	2.56	3.99	3.4	2.6	0.6	6.6	2.07	5
3410	138.52	28.83	8.34	12.44	3.75	3.5	2.25	9.5	2.38	4
3440	86.31	13.01	5.95	7.36	3.4	2.6	1.4	7.4	4.04	5
3530	113.33	24.77	7.39	10.87	3.2	2.6	1.8	7.6	3.78	5
4100	50.34	15.68	6.78	8.56	3.33	2.67	1.67	7.67	3.93	6
5140	48.76	12.23	3.06	4.9	3.75	2.25	1	7	2.94	4
5170	161.99	44.09	7.68	14.96	3.67	3	2	8.67	4.04	3
1239	67.75	13.42	10.11	10.77	3.8	2.6	1.4	7.8	2.77	5

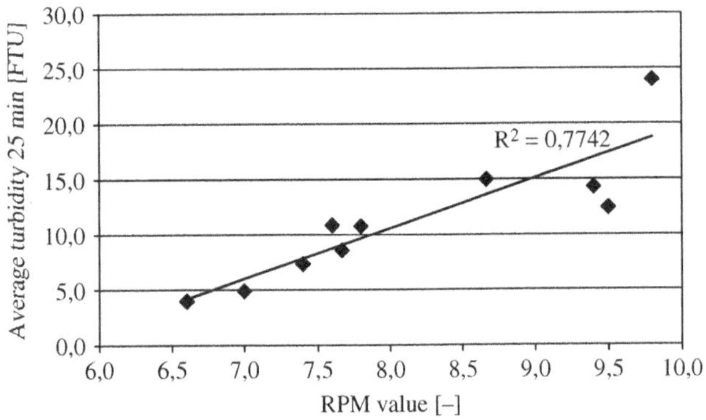

Figure 4. Relation between the average turbidity during the total RPM test and the RPM ranking.

CONCLUSIONS

The TECHNEAU research on discolouration in drinking water distribution systems is focussed on practical measures to effectively support the management of discolouration risks, then diminishing customer complaints. Practical analysis tools like the RPM are available and are tailored to the needs of the individual companies. Results from the first test case proved that the methodology can readily be implemented and gives reliable results.

The trials developed on the EPAL distribution network showed that the interpretation of the RPM results can be done through the ranking system accrediting points to a number of categories or by using the overall averaged turbidity. There is a good relation between the two. Probably the averaged turbidity is easier to implement as it represents an actual physical parameter.

The conceptual model for particle behaviour in the network is a good working frame to asses the discolouration problems. The extended conceptual model, incorporating the possible measures also proves to be a suitable frame for action and research.

Further research will give the relation with the treatment type, but possibly also with other factors as pipe material and pipe level. In short time this data will come available and the hypothesis will be challenged.

ACKNOWLEDGEMENTS

Acknowledgements are due to Joaquim Sereno and Nuno Dias, members of the EPAL Operation Division staff, for their valuable assistance in RPM performing.

REFERENCES

Blokker, E.J.M. and Vreeburg, J.H.G. (2005). Monte Carlo Simulation of Residential Water Demand: A Stochastic End-Use Model. World Water Congress, Reston, VA, USA, American Society of Civil Engineers.

Blokker, E.J.M., Vreeburg, J.H.G. *et al.* (2006). Combining the probabilistic demand model SIMDEUM with a network model. 8th annual water distribution system analysis symposium, Cincinnati, Ohio, USA.

Slaats, N., Rosenthal, L.P.M. *et al.* (2002). Processes involved in generation of discolored water, American Water Works Association Research Foundation/Kiwa, The Netherlands.

Vreeburg, J.H.G. (2007). *Discolouration in drinking water systems: a particular approach*. PhD Thesis, TU Delft, ISBN: 978-90-74741-91-0.

Vreeburg, J.H.G. and Boxall, J.B. (2007). Discolouration in potable water distribution systems: A review. *Water Research*, **41**, 519–529.

Vreeburg, J.H.G., Schaap, P. and van Dijk, J.C. (2004a). Measuring discolouration risk: Resuspension potential method. Proceedings IWA Leading Edge Conference, May 2004, Prague.

Vreeburg, J.H.G., Schaap, P.G. and van Dijk, J.C. (2004b). Particles in the drinking water system: from source to discolouration. *Water Science and Technology*, **4**(5–6), 431–438.

Vreeburg, J.H.G., Schippers, D., Verberk, J.Q.J.C. and van Dijk, J.C. (2008). Impact of particles on sediment accumulation in a drinking water distribution system. *Water Research*, **42**(16), 4233–4242.

Water treatment: optimization with respect to what?

B. Eikebrokk

SINTEF Building and Infrastructure, N-7465 Trondheim, Norway (E-mail: Bjornar.Eikebrokk@sintef.no)

Abstract Most existing water supply systems will be in operation also for decades to come. Thus, optimization of operation performance with respect to safety, sustainability and cost-efficiency should be a priority effort. However, the remaining optimization challenges and unexploited optimization benefits still seem significant. Optimization efforts are often restricted to water treatment plants, and are often motivated from water quality and cost issues. Useful optimization criteria may include: 1) Maximum compliance with the current water quality standards and national regulations, 2) Maximum treatment and disinfection barrier efficiency, stability and safety, 3) Minimum use of resources like chemicals and energy, and 4) Maximum cost-efficiency. Optimum water treatment performance can also be a solution to many problems in distribution, e.g. corrosion, biofilm formation, post precipitation and particle deposition and re-suspension. In this paper a roadmap to optimization is suggested, and important water supply optimization issues and criteria for the overall water supply system are discussed. Specific attention is given to the different sources for particle discharge from water treatment facilities, along with possible means of reducing direct and indirect particle discharges. A major challenge in such a holistic optimization effort is to find the right balance between different and sometimes conflicting and system specific optimization criteria.

Keywords coagulation, distribution, filtration, operation, optimization, particle deposition, water treatment

INTRODUCTION

Coagulation is a vital part of most water treatment processes. A failure in coagulation can have severe consequences for the water treatment and for the entire water supply system, e.g. poor removal of particles, pathogens and natural organic matter (NOM). This may decrease water safety and result in distribution problems as well, e.g. solids deposition, corrosion, regrowth and biofilm formation. Coagulation has at least seven objectives: 1) to maximize removal of turbidity and particles, 2) to maximize removal of pathogens, i.e. to maximize treatment barrier efficiency and stability, 3) to maximize removal of NOM (TOC) and disinfection by-product (DBP) precursors, i.e. "enhanced coagulation", 4) to minimize coagulant residuals, 5) to minimize production of sludge and waste products, 6) to minimize use of resources like chemicals and energy, and 7) to minimize operating costs.

 TECHNEAU: Safe Drinking Water from Source to Tap. Edited by Theo van den Hoven and Christian Kazner. ISBN: 9781843392750. Published by IWA Publishing, London, UK.

Optimum coagulation can be defined as the coagulation conditions of dosage and pH that achieve maximum compliance with the given objectives (Edzwald and Tobiason, 1999). Hence, sub-optimum coagulation can lead to poor operation performance, e.g. reduced quality of treated water and reduced treatment barrier efficiency. When evaluating operation performance, a comprehensive knowledge of the characteristics of the treatment process in question is required. Figure 1 shows the water quality characteristics from a filter during a typical filter run, with filter ripening, stable operation and breakthrough. The detrimental effects of sub-optimal coagulation, hydraulic steps (e.g. as a result of backwash of parallel filter units) on filtered water quality is also illustrated.

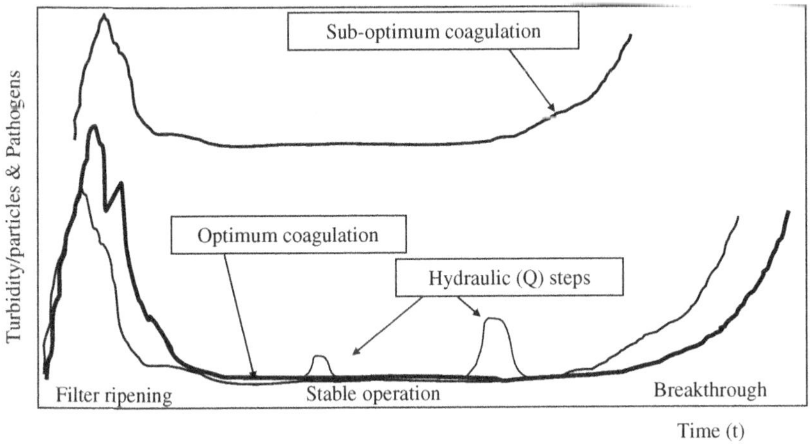

Figure 1. Effects of sub-optimal coagulation, hydraulic steps, filter ripening and breakthrough in a typical filter run.

According to Vreeburg *et al.* (2008), the particles originating from the treatment plant play an important and potentially dominant role in the generation of deposits in distribution systems. Therefore, optimization and improvement of treatment processes with respect to the short peaks in turbidity and particle content during filter ripening, hydraulic steps and particle breakthrough is imperative.

Vreeburg and Boxall (2007) stated that particles in water distribution networks, either as loose deposits or trapped in cohesive layers, are the main factors related to discoloured water which is the reason for a large proportion of customer complaints. Figure 2 illustrates well the range of physical, chemical and biological processes that can take place during distribution.

The relative importance of the different processes is site and system dependent, i.e. depending on water quality, hydraulic conditions, pipe materials, etc.

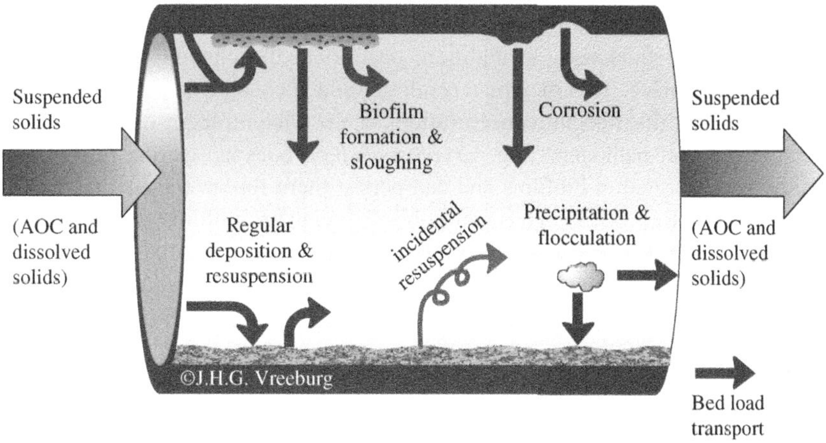

Figure 2. Particle related processes in drinking water distribution systems (Vreeburg, 2007).

Even more serious than discolouration, particle deposits ("soft deposits") are considered the key site for microbial growth in drinking water distribution networks, and Zacheus *et al.* (2001) found high numbers of heterotrophic bacteria, actinomycetes and fungi in deposit samples from 16 drinking water distribution networks located at eight towns in different parts of Finland. Also coliform bacteria were often isolated from deposit samples. After one year, the viable microbial numbers in new deposits were almost as high as in the old deposits before the first mechanical cleaning.

In terms of response to hydraulic steps, there are great similarities between Figures 1 and 2, i.e. between filter pores and distribution pipelines. A hydraulic step, i.e. an increase in velocity will disturb the balance between attachment and detachment of particles/deposits in both cases. As a result of this, turbidity peaks will appear in filter effluents, and re-suspension will occur in distribution systems.

The availability of nutrients is a key driver for the formation of biofilms in distribution systems. Biofilm bacteria predominantly use organic carbon as their main nutrient source, and biodegradable organic carbon is often considered as the limiting nutrient for bacterial growth in drinking water. Phosphorus has also been shown to limit bacterial growth in drinking waters containing large quantities of organic carbon or extremely low levels of phosphorus. Therefore, control of biodegradable organic carbon (BDOC) through adequate water treatment is important to control biofilm formation in the distribution system. This is of specific importance in systems where chlorination is not used for this purpose. Biofilms are normally eroded continuously by shear forces thereby limiting biofilm thickness. In addition, occasional detachment (sloughing) of biofilms takes place. This may be a result of sudden changes in shear conditions (flow velocity), disinfectant concentrations or the bacteria themselves. It is also documented that pathogens (e.g. *Cryptosporidium* oocysts) can be readily and rapidly incorporated in biofilms and can persist there for several months before detachment (CRC, 2005). Thus, microbes growing in biofilms deteriorate the quality of water when released into the water, and some microbes occurring in biofilms can also cause taste and odor problems (Nagy and Olsson, 1985; Astier *et al.*, 1995; CRC, 2005) or be pathogenic (Armon *et al.*, 1997; Quignon *et al.*, 1997; Warnecke, 2006).

On this background, selection of overall operation optimization criteria and objectives should be based on specific water treatment as well as distribution systems characteristics. A holistic optimization strategy for operation of the entire water supply system needs to address and minimize also the risk for harmful and/or unwanted conditions in the distribution system, e.g. corrosion, re-growth, biofilm formation, post-precipitation of coagulant residuals, particle deposition and re-suspension, etc.

Sustainable solutions for controlling problems related to discoloured water and particle deposits may be found through preventing the causes of particles, such as improved (optimized) treatment and control of post-precipitation, regrowth and corrosion in the network. However, according to Vreeburg and Boxall (2007) the quantification and evaluation of such strategies in different situations is beyond current methods and understanding, thus representing main challenges to meet in the future.

On this background, a major purpose of this paper is to establish links between optimization of water treatment processes, i.e. coagulation and filtration, and the formation and control of particle deposits in distribution systems.

OPTIMIZATION OF OPERATION PERFORMANCE OF WATER SUPPLY SYSTEMS

Given the facts that most water treatment plants (WTPs) and distribution systems (DS) are already in place, and a significant proportion of the documented waterborne disease outbreaks are caused by sub-optimal operation of existing WTPs and DS, surprisingly little R&D efforts are devoted to the optimization of operation performance of existing water supply systems.

Within the integrated EU-project Techneau, however, one out of a total of seven work areas is devoted to optimization and optimum operation of WTPs and DS in a holistic source-to-tap approach. Enhanced coagulation, ozonation-biofiltration, nanofiltration and ultrafiltration with pre-coagulation are the selected treatment processes to be studied with respect to identification of best operation practices, optimization strategies and routines, as well as identification and exploitation of optimization potentials and optimization benefits.

Optimization of coagulation and filtration operation performance

This paper will focus on optimization of the operation performance of coagulation and filtration processes with emphasis on the typical removal characteristics of NOM and particles/turbidity that is considered important with respect to biofilm formation and particle deposition and re-suspension in distribution systems. Possible sources for particles will be presented and discussed as well.

Given the multiple objective coagulation process presented above, it is not obvious that all the objectives can be met. Selection of the best optimization criteria is normally site- and system specific, and compromises have to be made with respect to selection of the overall optimum operating conditions.

From a public health view pathogen removal during water treatment is generally more important than NOM removal. On the other hand, NOM should be removed because NOM controls and/or influences on most water treatment processes and produce undesirable and potentially harmful DBPs. NOM can also significantly contribute to corrosion and biofilm formation processes as well as taste and odour compounds in the distribution system. NOM can also negatively affect downstream disinfection processes by increasing chlorine and ozone disinfectant demand and/or increase UV-absorbance. In addition, both metal coagulant residuals (Fe) and NOM can contribute to fouling of UV-lamp sleeves and UV-sensor windows. Coagulant dose levels and sludge production rates should not be minimized if pathogen and NOM removal is sacrificed. If the distribution system is over-designed due to reduced water consumption and/or reduced leakages, thus allowing for extensive particle deposition/re-suspension and lack of self-cleaning, particle removal can be an optimization criteria of high priority.

OPTIMIZATION EFFORTS: RESULTS AND DISCUSSION

Some results from pilot and full-scale optimization investigations are presented below, with presentation and discussion of important optimization aspects, optimization criteria and effects of failures in coagulation. In addition, some areas of possible conflicts between the optimization candidates are discussed as well. The relations between water treatment optimization and particle deposition/re-suspension are emphasised.

Pathogen removal and treatment barriers

Although there is no direct correlation between water turbidity and pathogen concentrations, efficient removal of pathogens during coagulation and filtration processes is normally associated with low filter effluent turbidities. Thus, efficient treatment barriers are normally associated with consistent low turbidity levels. As an illustrating example, the Norwegian drinking water quality guidelines provide specific coagulation barrier efficiency indicators: 1) Turbidity <0.2 NTU, or <500/mL of 2–400 μm sized particles (on-line turbidity or particle monitoring from all filter units), 2) Colour <5 mg Pt/L, 3) TOC <3 mg/L, and 4) Coagulant residual <0.15 mg /L (Al or Fe).

Thus, durability plots of filter effluent turbidity (Figure 3) can be used to illustrate treatment barrier efficiency and stability in coagulation and filtration facilities. Figure 3 represents an extreme case from a full-scale facility due to specific circumstances, and is not at all representative for a normal situation. However, filter effluent turbidity durability plots can be used not only for evaluation of water treatment performance, but also for assessments of the contribution from treated water to particle deposition and re-suspension in distribution networks.

NOM fractionation and removal of NOM fractions

According to Edzwald and Tobiason (1999) coagulation is controlled by NOM (and not by turbidity or some other parameters) when the specific UV-absorption (SUVA) is 4 or greater. In this SUVA-range NOM is dominated by high-molecular weight aquatic humic substances with high hydrophobicity. DOC removals are normally 50% or greater in this range. For SUVA-values in the range of 2–4, coagulation is also influenced by NOM. Only for SUVA values below 2, the corresponding NOM dominated by low molecular weight, non-humic and hydrophilic substances have little influence on coagulation. In this SUVA range, the obtainable DOC removals are normally poor, 25% or lower.

As indicated above, NOM removal is related to SUVA levels, i.e. to NOM composition. A rapid fractionation technique (Chow *et al.*, 2004) may give valuable information about NOM composition and treatability by coagulation. Here, four fractions of DOC are identified: 1) Very Hydrophobic Acid (VHA),

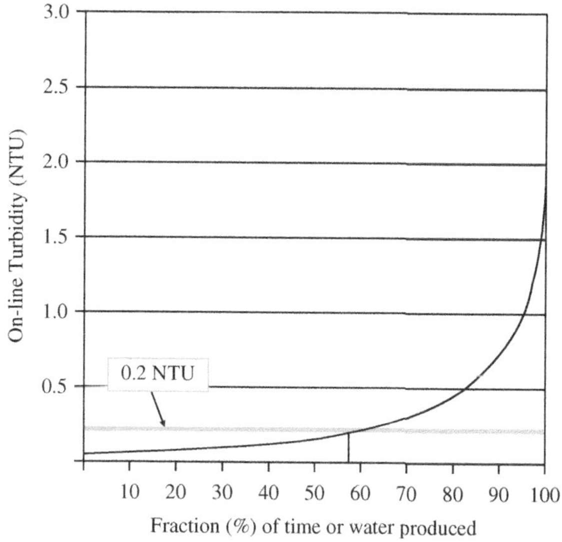

Figure 3. Durability plots of filter effluent turbidity as a treatment barrier indicator.

Slightly Hydrophobic Acid (SHA), Charged Hydrophilic Acid (CHA), and Neutral Hydrophilic Acids (NEU). Typically, the VHA and SHA fractions of DOC include the humic and fulvic substances, the CHA fractions include the proteins, and the NEU factions include the polysaccharides. The hydrophobic fractions (VHA and SHA) are often associated with good treatability by coagulation and high reactivity (high DBPFP). The hydrophilic fractions (CHA and NEU) on the other hand, may be associated with biofilm formation and membrane fouling potentials. Thus, predictions of removable and recalcitrant NOM (DOC) fractions by coagulation can be made on the basis of seasonal NOM fractionation analyses. In addition, the hydrophilic fraction concentrations (CHA, NEU) can be related to growth and biofilm formation potentials.

Figure 4 illustrates an example of the seasonal variability in NOM-fraction concentration and treatability. Raw water DOC levels are within the range of 1.4–2.8 mg DOC/L, while treated water DOC levels are reduced down to 0.5–0.9 mg DOC/L as a result of coagulation and contact filtration. As shown, there is a great seasonal variability in NOM concentration and NOM composition, which is indicative of the challenges with respect to adaptation of the operating conditions to the variations in raw water (NOM) quantity and quality, i.e. the process control challenges.

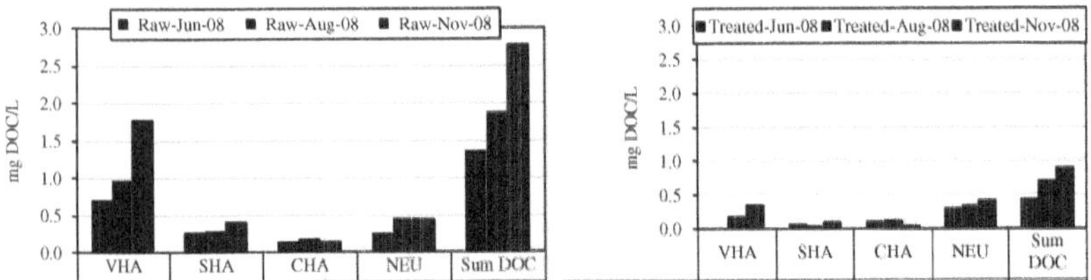

	VHA	SHA	CHA	NEU	Sum DOC
Raw-Jun-08	0.6995	0.2650	0.1454	0.2370	1.3469
Raw-Aug-08	0.9678	0.2826	0.1746	0.4426	1.8676
Raw-Nov-08	1.7791	0.4015	0.1453	0.4454	2.7713

	VHA	SHA	CHA	NEU	Sum DOC
Treated-Jun-08	−0.0286	0.0625	0.0924	0.2974	0.4237
Treated-Aug-08	0.1823	0.0430	0.1256	0.3385	0.6894
Treated-Nov-08	0.3508	0.0945	0.0363	0.4057	0.8873

Figure 4. NOM fraction concentrations in raw and treated water samples taken in June, August and November 2008. Results from Bergen WTP, Norway.

Figure 5 shows an example on the effects of coagulant dose levels on NOM-fractions removal by coagulation. Great proportions of the VHA, SHA and CHA fractions can be removed provided that proper coagulant doses and coagulation pH values are used. The NEU-fraction however, is not removed by coagulation. In fact, the NEU fraction concentration increases in many cases, as a result of transformation from other fractions and poor treatability.

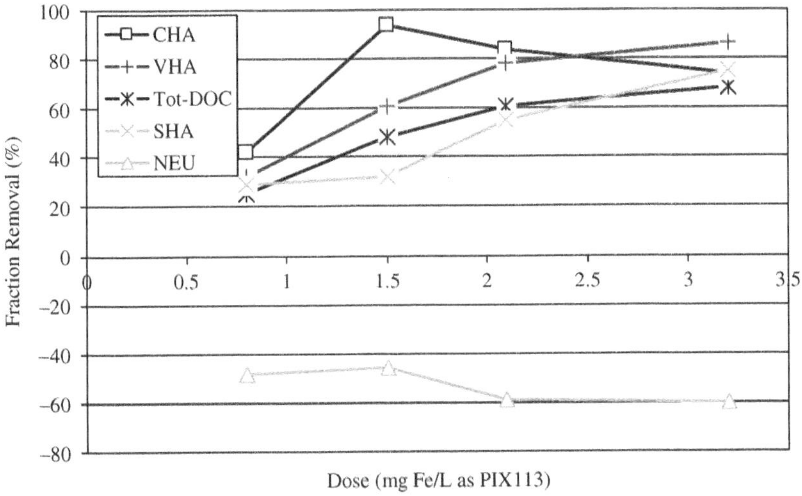

Figure 5. NOM-fraction removal by coagulation-contact filtration. Results from VIVA pilot, Trondheim.

NOM fractions and biodegradability

Figure 6 shows the results of NOM-fractionation and BDOC-analysis on a number of raw and treated water samples from water supply systems in different parts of Norway, i.e. raw water, ozonated water, ozonated and biofiltered water, as well as coagulated water. More information on the applied BDOC analysis tool is presented elsewhere (Eikebrokk *et al.*, 2007). However, a modified BDOC procedure was applied here, with six glass bead containing and pre-adapted BDOC columns were operated in series. The pump capacity was adjusted to give an EBCT of about 1.5 hour in every column, thus yielding a total EBCT of 9 hours. Sampling and DOC analysis was performed on inlet water and effluent water from all columns. The BDOC were calculated as the differences in DOC between inlet water and the different effluent samples. The resulting curve showing BDOC versus EBCT gives valuable information not only on terminal BDOC levels but also on degradation kinetics.

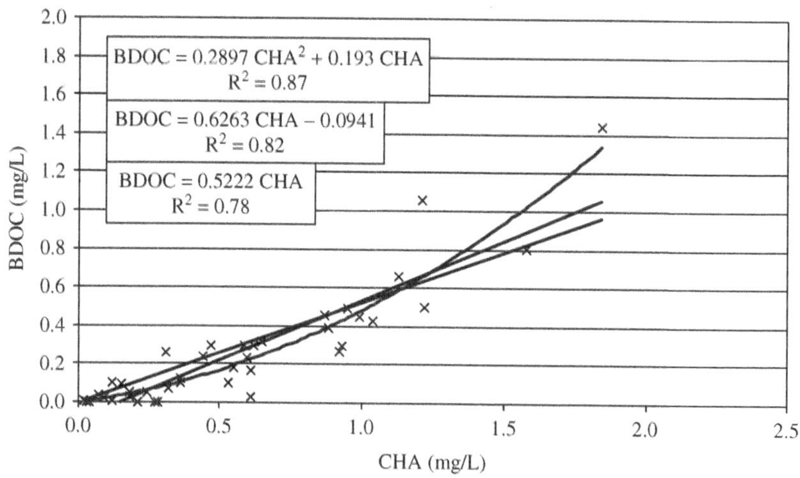

Figure 6. Correlation between BDOC and CHA-fraction concentrations. Results from raw and treated surface waters (coagulated; ozonated; biofiltered) in Norway.

As presented in Figure 6, a good correlation between BDOC and the CHA fraction of NOM was found in these experiments. The ozonated samples dominate the higher values, while coagulated water showed low values of both parameters. Thus, control of hydrophilic DOC fractions, i.e. NEU and CHA in specific can be a good strategy for controlling biofilm formation in the distribution system.

Coagulant residuals and particle/turbidity removal

An adequate coagulation process control system is required in order to adapt the process conditions to the variations in raw water quality. Sub-optimum coagulant dosage and coagulation pH can not only lead to reduced treatment barrier efficiency, but also to post-precipitation of metal coagulant residuals and particle formation in later treatment steps and/or in the distribution system.

A typical example on the influence of coagulant dose and coagulation pH on NOM-removal efficiency (as TOC and colour), residual turbidity and residual iron coagulant concentration is shown in Figure 7. High residual metal coagulant concentrations may not influence on the on-line turbidity readings until post precipitation occurs in later treatment steps (e.g. alkaline filter layers, GAC-filters) or in the distribution system. When employing iron coagulation at low pH-levels (pH 4–5) in specific, the difference may be quite significant between on-line turbidity readings and laboratory results from water samples. This is due to precipitation of colloidal iron hydroxide precipitates as a result of pH-increase and time/storage during sampling and analysis. Stripping of CO_2 is likely relevant in this process. Thus, transformation of metal coagulant residuals from water treatment may be a significant source of particles in the distribution system.

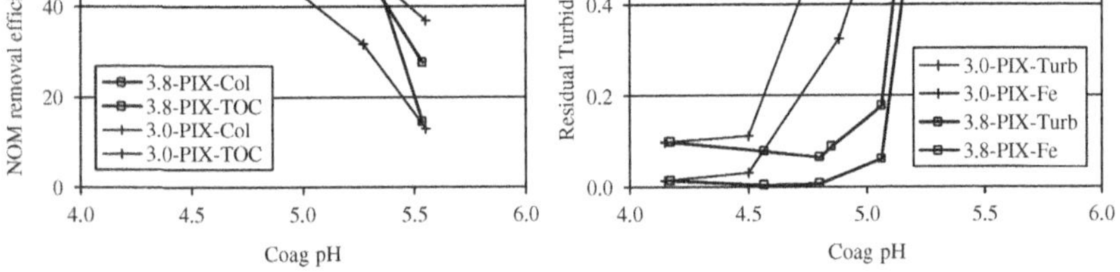

Figure 7. Influence of applied Fe-coagulant dose levels and coagulation pH on NOM removal efficiency, residual turbidity and residual coagulant concentrations.

The results presented in Figure 7 also demonstrate that the optimum coagulation pH-window is widened when the coagulant dose level is increased. Operation at absolute minimum or sub-optimum coagulant dosage levels can therefore increase the risk for non-overlapping pH-windows for removal of NOM and turbidity/residual metal, respectively. This is most relevant when operating in the upper part of the optimum pH-window for the given coagulant. From Figure 7 it can be seen that a coagulant dose of 3 mg Fe/L at pH 5.0 results in significant colour and TOC removals in excess of 80% and 40%, respectively. However, residual turbidity and Fe concentrations are high under these operating conditions (close to 1 NTU and 0.6 mg Fe/L, respectively). When the coagulant dose is increased from 3 to 3.8 mg Fe/L, the non-overlapping optimum pH-ranges are practically removed, and good NOM removal can be obtained while maintaining the turbidity and residual coagulant concentrations at low levels.

Thus conflicts between optimum treatment conditions for the different optimization objectives/parameters will exist in practice, e.g.: 1) the struggle for the lowest possible chemicals consumption, sludge production rates, and operating costs but also the corresponding risk for non-overlapping optimum pH-ranges for NOM and turbidity/residual coagulant residuals on one side, and 2) process stability and robustness, as well as ease of operation and operator satisfaction on the other side.

From this it should be concluded that close to minimum coagulant doses can result in narrow optimum pH-windows and elevated turbidity as well as residual coagulant concentrations, even though NOM removal may not be significantly affected. Therefore, specific care should be taken to avoid sub-optimum coagulation conditions in distribution networks where particle deposition and re-suspension problems prevail. The risk for elevated turbidity and potential particle-forming residual metal coagulants seems highest in the upper part of the optimum coagulation pH range, as illustrated in Figure 7.

Alkaline post-filtration can significantly reduce metal coagulant residuals, especially iron. Alkaline post-filters or alkaline filter layers as part of 3-media filters are widely used for integrated enhanced coagulation-filtration and corrosion control in Norway. When using iron coagulants, the low coagulation pH (4–5) is ideal for a rapid dissolution of the calcium carbonate, thereby increasing pH, alkalinity and calcium content to requested levels even at low design EBCT. In addition, residual metals coagulants are effectively precipitated and removed within the wide range of pH-values encountered in the alkaline filter layer.

An analysis of eight-year data on total and dissolved aluminium, turbidity, DOC and applied alum coagulant dosage at a full-scale WTP in Canada, showed that when alum/DOC ratio is less than 7 (i.e. Al/DOC <0.6), insufficient alum addition led to incomplete coagulation resulting in discharge of colloidal material mostly consisting of organic aluminium in particulate form (Srinivasan *et al.*, 1999).

Eikebrokk (1999) found that the residual coagulant concentration determined the required coagulant dosage when treating low turbidity, soft and humic raw waters by enhanced coagulation and contact filtration. He defined absolute and practical minimum coagulant dosages that are used to assess coagulation conditions and operation performance at several WTPs in Norway. The example presented in Figure 8 shows the model predictions of the absolute and practical minimum coagulant doses, as well as the ranges of raw water colour and the almost fixed applied coagulant dose levels for five WTPs in Bergen, Norway in the year 2007. It is indicated in the data that some plants may be under-dosing in limited periods of the year, thus operating with reduced water quality, reduced treatment barrier efficiency and increased residual metal concentrations. On the other hand, over-dosing of coagulant seems to prevail when the raw water colour is low, thus consuming more chemicals and more energy (increased filter backwash frequencies) than required. In addition, the WTPs will produce excess sludge during this period. Figure 8 illustrates well also the need for an adequate on-line process control system with better adaptation of coagulation conditions to variations in raw water quality.

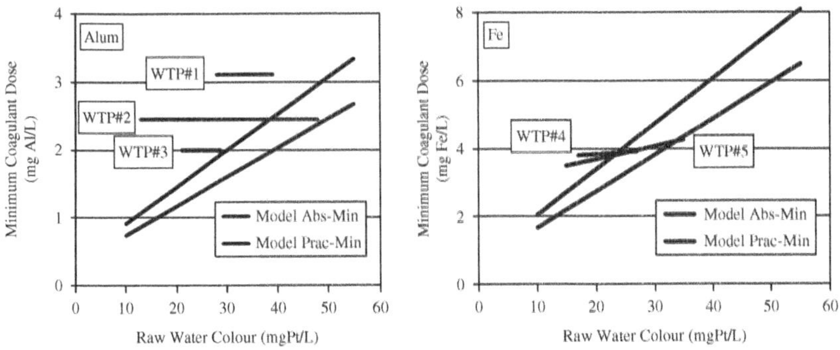

Figure 8. Variability in raw water colour and applied coagulant dose levels compared to minimum dose model predictions for five WTPs in Bergen, Norway in 2007.

Additional optimization criteria

Additional optimization criteria for optimum operation performance of WTPs like sludge and waste production, chemicals and energy used, carbon dioxide footprint, as well as cost-efficiency are not further addressed in this paper. It should however be pointed out that significant optimization benefits seem to exist out there, even at very well operated facilities. As an example, full-scale optimization trials at a coagulation-direct filtration plant in Norway revealed great optimization potentials, e.g. a 40% reduction in the use of coagulation chemicals, a 40% reduction in sludge production, and 5–10% reduction in energy use, mainly due to prolonged filter run times, less frequent backwash and less water wasted (Eikebrokk *et al.*, 2007).

A ROADMAP TO OPTIMIZATION

Optimization of water treatment processes is no longer only a question of compliance with water quality standards and regulations. Although it may be challenging to meet regulatory requirements, especially for small systems, supplementary optimization criteria need to be included and some vital questions should be asked:

1. Is the WTP operated in such a way that treatment barrier efficiency/water safety is maximized?
2. Is the WTP operated with a minimum use of resources, e.g. energy and chemicals?
3. Is the WTP operated with a minimum production of sludge/waste?
4. Is the WTP operated at maximum cost-efficiency?
5. How can optimization of the treatment solution reduce problems in distribution?

Responsible WTP operators need to know the answers to these questions. If not, they need to know how to find out. Optimization efforts should start with a mapping of today's situation as a reference basis, i.e. the current operation performance status. Comprehensive analyses of raw and treated water quality and quality variations are also important. For full-scale operation optimization trials, a systematic approach is needed and great care should be taken in order not to compromise produced water quality. Small steps should be taken in optimization trials, and adequate water sampling and analysis is important. Furthermore, the optimization potentials and possible benefits should be identified, and the implementation of new and revised treatment conditions should be done on the basis of careful assessments of the major implications for treatment as well as distribution.

For enhanced coagulation-filtration facilities, a five-step optimization procedure is suggested, including evaluations of possible effects and consequences for the distribution system as well:

1. Mapping of operation performance status. This includes Initial operation performance evaluation and comparison to model predictions

2. Water quality characterization (Diagnostic tools). This includes water quality and variability characterisation, e.g. rapid NOM fractionation, size exclusion chromatography, biodegradability testing (AOC, BDOC), zeta potential measurements in coagulated water, etc.
3. Optimization trials (Curative tools). This includes planning and performance of experimental trials including water sampling and analysis, with systematic variation of coagulant dose and pH – without compromising produced water quality
4. Identification of optimization potentials and possible benefits. This is done on the basis of the results from steps 1–3
5. Implementation of selected operation conditions. This is done on the basis of an overall assessment of potential advantages and disadvantages – for the treatment plant as well as the distribution system.

CONCLUSIONS

Existing water treatment and distribution systems will undoubtedly be in operation for decades to come. However, operation optimization efforts seem yet to be scarce, with no dominating position at the R&D area. This is rather peculiar since water safety relies on adequate operation performance and because pilot and full-scale trials clearly indicate that significant optimization potentials and benefits seem to exist.

Optimization of water treatment plants and distribution systems has to rely on selected optimization criteria that are site and system specific. Comprehensive knowledge of the specific characteristics, limitations and challenges of a water supply system is required in order to design and perform full-scale optimization studies and trials. An optimization procedure (a road map to optimization) for enhanced coagulation-filtration facilities is suggested.

Compliance with national regulations and water quality standards is of course a driving for optimization efforts for water treatment facilities. This may however not be enough to maximize safety of a water supply system. Even though the optimization efforts of treatment and distribution system have been more integrated in holistic source-to-tap safety mode in the last few years, even more integration is needed when it comes to operations. Water treatment and distribution should be evaluated together in a holistic mode, since specific water quality aspects related to water treatment may lead to user complaints and impose risks in the distribution system. In this paper a few "holistic" optimization criteria are presented and discussed, in specific those related to particle discharge/deposition and biofilm formation/sloughing in distribution systems.

In addition to corrosion control efforts, e.g. control of pH, alkalinity and calcium contents in treated water, specific coagulation-filtration optimization efforts that can significantly reduce problems related to particle deposition and re-suspension during distribution, include the following:

1. Application of filter-to-waste procedures, initial coagulant overdosing or other means of reducing the length and/or negative effects of filter ripening
2. Avoidance of hydraulic steps and corresponding particle breakthrough by reprogramming process control systems, inlet pumps, etc. to maintain constant filtration rates also during filter backwash
3. Strict control of single filter effluent turbidity to avoid breakthrough, i.e. secure backwash at the right time
4. Application of filter backwash routines that keep the filter bed clean and prevent accumulation of filter deposits over time
5. Application of coagulant dose levels that are well adapted to variations in raw water quality (NOM) and sufficiently high to avoid narrow optimum pH-windows and conflicting pH ranges for optimum removal of various parameters/substances
6. Control of hydrophilic NOM fractions (NEU and CHA in specific) to reduce the potential for re-growth and biofilm formation in the distribution network.

ACKNOWLEDGEMENT

This paper was prepared as part of the EU funded project Techneau. The author is grateful for valuable contributions from the co-workers within Techneau: T. Juhna, Riga Technical University, and E. Melin, S.W. Osterhus and G. Thorvaldsen, SINTEF.

REFERENCES

Armon, R., Starosvetzky, J. Arbel, T. and Green, M. (1997). Survival of *Legionella pneumophila* and *Salmonella typhimurium* in biofilm systems. *Wat. Sci. Tech.*, **35**(11–12), 293–300.

Astier, F., Paquin, J.L., Mathieu, L., Morlot, M. and Hartemann, P. (1995). Study of development of the musty taste in water according to its ageing process in a pilot plant. *Environ. Tech.*, **16**, 955–965 (In French, English abstract).

Chow, C., Fabris, R. and Drikas, M. (2004). A rapid fractionation technique to characterise natural organic matter for the optimisation of water treatment processes. *J. Wat. Supp.: Res. Tech. – AQUA*, **53**(2), 85–92.

CRC for Water Quality and Treatment (2005). Biofilms: Understanding the impact on water quality and water treatment processes. ISBN 1876616458, Australia.

Edzwald, J.K. and Tobiason, J.E. (1999). Enhanced coagulation: US requirements and a broader view. *Wat. Sci. Tech.*, **40**(9), 63–70.

Eikebrokk, B. (1999). Coagulation-direct filtration of soft, low alkalinity humic waters. *Wat. Sci. Tech.*, **40**(9), 55–62.

Eikebrokk, B., Juhna, T., Melin, E. and Osterhus, S.W. (2007). Water treatment by enhanced coagulation and ozonation-biofiltration. Techneau-Report D5.3.2A, Dec. http://www.techneau.eu

Nagy, L. and Olson, B. (1985). Occurrence and significance of bacteria, fungi and yeasts associated with distribution pipe surfaces. *Proc. AWWA Technology Conf. on Advances in Water Analysis and Treatment*, Houston, TX, 213–237.

Quignon, F., Kiene, L., Levi, Y., Sardin, M. and Schwartzbrod, L. (1997). Virus behaviour within a distribution system. *Wat. Sci. Tech.*, **35**(11–12), 243–253.

Srinivasan, P.T., Viraraghavan, T. and Bergman, J. (1999). Factors influencing residual aluminium levels at the Buffalo Pound Water Treatment Plant, Saskatchewan, Canada. *J. Wat. SRT – AQUA*, **48**(4), 167–175.

Vreeburg, J.H.G. (2007). Discoloration in drinking water systems: A particular approach. Delft 978-90-74741-91-0.

Vreeburg, J.H.G. and Boxall, J.B. (2007). Discoloration in potable water distribution systems. A review. *Wat. Res.*, **41**, 519–529.

Vreeburg, J.H.G., Schippers, D., Verberk, J.Q.J.C. and van Dijk, J.C. (2008). Impact of particles on sediment accumulation in a drinking water distribution system. *Wat. Res.*, **42**, 4233–4242.

Warnecke, M. (2006). Cryptosporidium oocyst interactions with drinking water pipe biofilms. Research report No. 5, Sydney water, Oct. CRC for Water quality and Treatment, ISBN 1876616067.

Zacheus, O.M., Lehtola, M.J. Korhonen, L.K. and Martikainen, P.J. (2001). Soft deposits, the key site for microbial growth in drinking water distribution networks. *Wat. Res.*, **35**, 1757–1765.

Corrosion and corrosion modeling

S.W. Osterhus

SINTEF Water and Environment, Klabuveien 153, N-7465 Trondheim, Norway (E-mail: stein.w.osterhus@sintef.no)

Abstract In distribution networks containing a significant part of unprotected iron pipes, iron corrosion may be the major cause of particles and discoloration incidents in the network. Many factors are affecting iron corrosion, and it is an extremely complex process. The objective of this paper is to address the importance of iron corrosion as a source of particles in the distribution system, and outline how iron corrosion is affected by physical and chemical parameters, and how this can be incorporated into the overall quality network model as a source of particles formation. Predicting or modeling iron corrosion is very difficult, and at the present time it is not possible to quantitatively predict or model iron corrosion rate for a universal distribution system. However, it is shown here that by using corrosion test rigs, site specific models can be developed on a form that can be used for estimation of particles generation by iron corrosion, which in turn can be applied to a hydraulic model for particles in the network. However, such corrosion models need extensive calibration and validation, and testing of the models in case studies are needed to assess their usefulness and to further develop the models.

Keywords corrosion, distribution system, drinking water, iron pipes, modeling, water quality

INTRODUCTION

Re-suspension of sediments may lead to discolored water at the consumer's tap, which is a major cause of customer complaints. However, not only the aesthetics of discoloration are of concern, as sediment dispersal is accompanied with a release of microorganisms into the water along with undesirable components, then triggering possible threats to the human health. Water companies urgently need a practicable understanding of the processes and mechanisms leading to discoloration incidents occurring in the water distribution network, and to develop management tools and techniques. Vreeburg and Bauxall (2007), did a comprehensive review of techniques to assess the discoloration risk and the strategies available to control this.

Discoloration is associated with the mobilization of accumulated particles from within distribution networks. Such particles have different sizes and densities and hence probably have different origins, often characterized as either external sources or from processes occurring within the system. Particles can enter the distribution network as background concentrations of organic and

 TECHNEAU: Safe Drinking Water from Source to Tap. Edited by Theo van den Hoven and Christian Kazner. ISBN: 9781843392750. Published by IWA Publishing, London, UK.

inorganic material from the source water (Lin and Coller, 1997; Kirmeyer *et al.*, 2000; Slaats *et al.*, 2002; Ellison, 2003), due to incomplete removal of suspended solids at the treatment plant (Gauthier *et al.*, 2001; Vreeburg *et al.*, 2004) or be added to the water by the treatment plant itself, such as carbon and sand particles, alum or iron flocs and bioparticles originating from biofilters. The distribution system itself can also produce particles such as from pipe and fitting corrosion and lining erosion (Stephenson, 1989; Ruta, 1999; Gauthier *et al.*, 2001; Clement *et al.*, 2002; Slaats *et al.*, 2002; Boxall *et al.*, 2003), biological growth (LeChevallier *et al.*, 1987; Stephenson, 1989; Clark *et al.*, 1993; Meches, 2001) and chemical reactions (Stephenson, 1989; Sly *et al.*, 1990; Walski, 1991; Lin and Coller, 1997; Kirmeyer *et al.*, 2000); or external contamination that may occur during operations such as pipe repairs (Gauthier *et al.*, 1996; Slaats *et al.*, 2002), intrusion (Gauthier *et al.*, 1999; Kirmeyer *et al.*, 2000; Prince *et al.*, 2001) and backflow.

The formations as well as the resuspension of the deposits are not understood in detail. Therefore, a general approach describing all of the important processes is not available. Within the EU-project Techneau, WP 5.5, the aim is, based on extensive literature review and investigations in networks and pilot plants, to develop a mathematical approach which describes the relevant processes of deposits formation and behavior in the network, and thereafter link the mathematical approach to a hydraulic model (Wricke *et al.*, 2007). This model can be used to enable a better understanding of the most important processes and, hence, to open the way for efficient operation and maintenance measures.

In distribution networks containing a significant part of unprotected iron pipes, iron corrosion may be the major cause of particles and discoloration incidents in the network. Many factors are affecting iron corrosion, and it is an extremely complex process. Because of the large variability in distribution system conditions, a particular factor may be critical in one system but relatively unimportant in another system. Moreover, corrosion itself has several different manifestations and can be evaluated in many ways. Previous studies on iron pipe corrosion have focused on different aspects of iron corrosion: pipe degradation (measured by weight loss, oxygen consumption or corrosion current), scale formation (measured by head loss or scale deposition), and by-product release (measured by iron concentration, color, staining, turbidity, or number of customer complaints). Thus, it can be difficult to compare conclusions or theories from different studies (McNeill, 2000).

Numerous investigations have been made of factors and water quality parameters affecting iron corrosion, which in turn has been summarized and reviewed by several others (AWWA, 1985; McNeill, 2000; Böhler *et al.*, 2005; Broo *et al.*, 2006; Teasdale *et al.*, 2007). They all conclude that build-up of corrosion scale is important to the following iron corrosion and that many water quality parameters are affecting the corrosion and scale build-up, with the most important parameters being pH, alkalinity, buffer capacity, dissolved oxygen, aggressive anions (Cl^- and SO_4^{2-}), and calcium, in addition to temperature and flow velocity. Some have also reported important effects of Natural Organic Matter, NOM (Böhler *et al.*, 2005) and of the type of NOM (Broo *et al.*, 2006), residual chlorine (Teasdale *et al.*, 2007), silicate (McNeill, 2000; Broo *et al.*, 2006), and of microbial growth and the operational conditions such as flow stagnation, sludge accumulation, etc. (Böhler *et al.*, 2005). Further more, the mechanism described by the siderite model is well documented and accepted (AWWA, 1985), as well as the phenomena of inter-stationary corrosion (AWWA, 1985; Böhler *et al.*, 2005). However, there are considerable discrepancies in the reported results both when it comes to the effect of the different parameters and their magnitude. The discrepancies are partly caused by the large variability in the distribution system conditions, and to the fact that the corrosion may be evaluated differently as discussed above (McNeill, 2000). For example, the effect of a specific parameter may be entirely different when evaluating by-product release than when evaluating weight loss. Moreover, the effect of many parameters would largely depend on the original water quality. For example, the effect of changes in important corrosion parameters like pH, alkalinity and calcium would be very different in a soft, low pH, low alkalinity water containing NOM than in mineral rich hard water. Another reason for discrepancy could be the time used for the evaluation and how the data is interpreted, as long term effects of a parameter could be very different from short term effects observed on relatively freshly exposed material (Broo *et al.*, 2006).

The objective of this paper is to address the importance of iron corrosion as a source of particles in the distribution system. This will include an outline on how iron corrosion is affected by physical and chemical parameters, and how this can be incorporated into the overall quality network model described by Wricke *et al.* (2007), as a source of particles formation.

METHODS

In order to obtain data to determine the effect of various factors on iron corrosion, corrosion test rigs are used. Such test rigs can be used for general iron corrosion investigation, or to determine the corrosion effects of certain parameters under specific conditions. However, the test rigs can also be used at a specific water work site to determine expected iron corrosion rates in the following distribution network. Two different test rigs have been used: 1) test rig for oxygen consumption and iron release; 2) test rig for weight loss coupons.

Test rig for oxygen consumption and iron release

When using the test rig for oxygen consumption and iron release pipes (1 m long DN 20 steel pipes) where pre-corroded in a separate rig under the desired condition (AWWA, 1985) before installed in the test rig. Prior to the pre-corrosion, the test pipes are treated with hydrochloric acid to remove any residual layers to create starting conditions as consistent as possible. Before the pipes are placed in the test rig the iron discharge is determined using a flush step procedure where iron and manganese concentration, as well as the turbidity, are measured in each flush step (Böhler *et al.*, 2005). A schematic sketch of the test rig is shown in Figure 1. When the pipe is placed in the test rig it is flushed with fresh water for 3 minutes at 0.89 m/s (1000 L/h).

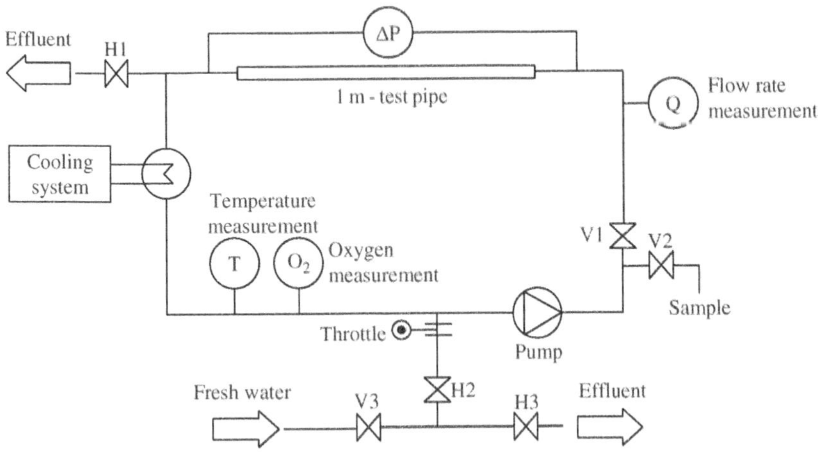

Figure 1. Schematic sketch of the test rig used for oxygen consumption and iron release (Wricke *et al.*, 2006).

The corrosion rate can now be calculated from the oxygen consumption which is determined from the slope of the straight line obtained by plotting the continuous measured oxygen concentration for a period of 15 minutes. Simultaneously to recording the corrosion rate during circulation, parallel measurements of the iron release rates are carried out. During the 15-minute monitoring period, samples are taken at intervals of 2 minutes in the first 10 minutes, followed by intervals of 5 minutes.

Test rig for multiple weight loss coupons

The test rig for multiple weight loss coupons consists of four parallel lines each with different flow conditions as follows: 1) constant high flow rate (0.2 m/s),

2) constant low flow rate (0.03 m/s), 3) alternating high flow rate (0.2 m/s for 13 hrs) and complete stagnation (for 11 hrs), and 4) alternating high flow rate (0.2 m/s for 13 hrs) and low flow rate (0.03 m/s for 11 hrs). The flow rate and alternating flows are controlled by adjustment valves, solenoid valves and timers. A schematic sketch of the set-up is shown in Figure 2. In each line 12 steel coupons (St 37) are mounted and separated by a material which is an electrical insulator (Bakelite). The size of the coupons is 15 mm times 20 mm with a thickness of 1 mm. Both sides of the coupons are exposed to the water.

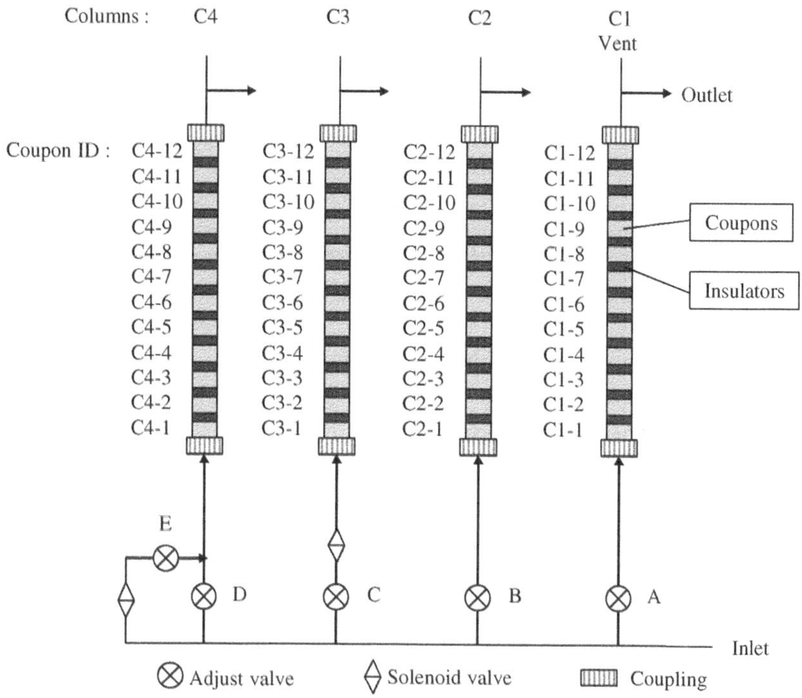

Figure 2. Schematic sketch of the test rig for multiple weight loss coupons.

Before the coupons are inserted to the rig they are degreased (using methanol or ethanol), rinsed (in distilled water), dried (at 105°C for 2 hrs) and weighted (the weight is called W1). Coupons are extracted from the coupon rack at various exposure times. Normally three coupons are extracted from each line at each time, i.e. three coupons from each line every third month for a total of 12 months. The extracted coupons are dried (at 105°C for 2 hrs) and weighted (called W2). Thereafter, the corrosion products are removed by using a stirred concentrated HCl-bath for 2–5 minutes, followed by brushing of the coupon, and another 2–5 minutes in the stirred concentrated HCl-bath. After the HCl-bath the coupon is thoroughly flushed and rinsed (in distilled water) before it is dried (at 105°C for 2 hrs) and weighted (called W3). The exposure time is recorded and the weight loss of metal (W1–W3) and the weight of accumulated corrosion products on the coupon (W2–W3) are determined.

RESULTS AND DISCUSSIONS

Predicting or modeling iron corrosion is very complex, and at the present time it is only possible to indicate general effects of various parameters on iron corrosion. Examples of such general effects are summarized and shown in Table 1. However, the magnitude of the effects and the significance of the various parameters are difficult to predict as several combined effects may occur, and in some waters not all the parameters listed may have an effect, and in other waters parameters not accounted for may play a role. Consequently, it is currently not possible to quantitatively predict or model iron corrosion rate for a universal distribution system. In order to investigate the effect of different parameters, either to improve the general corrosion knowledge or to obtain data for site specific corrosion effects, the test rigs described above can be and have been used.

Data from corrosion test rigs

The data obtained from the rig for oxygen consumption and iron release, and the rig for weight loss coupons measures slightly different aspects of corrosion. The former measures oxidation rate and iron release at the time of measurements, while the latter measures the average loss of metal and accumulation of corrosion products during a longer time period. Data from both methods may be converted to corrosion rates. Critical for both methods are the pre-corrosion time or the duration of the test as it is the long term corrosion occurring after a well developed scale has formed which is of interest.

After development of corrosion product layers, it is assumed that some kind of long term steady state is established. This is not entirely true as some systems will continue to grow corrosion layer having a net long term accumulation. However, this net long term accumulation would normally be very small and negligible. Therefore, it is assumed that corroded iron (at this long term steady state) is either released into bulk as soluble Fe(II), which may be transported to tap, oxidized and deposited, re-suspended, etc. Or the iron may be directly oxidized and incorporated into the scale, transported as particles to

Table 1. General effects of some parameters on iron corrosion (AWWA, 1985; McNeill, 2000; Böhler *et al.*, 2005; Broo *et al.*, 2006; Teasdale *et al.*, 2007).

Parameter	Effect[1]	Comments
pH	+ −	the effect depends on water quality
Alkalinity	+ +	
Calcium	+	relevant for soft waters
Temperature	−	
$Cl^- + SO_4^{2-}$	− −	
NOM	+	the effect depends on the type of NOM and the water quality
Silicate	+	relevant for pristine water qualities
Chlorine	−	
Oxygen	− +	mixed effect: increase iron oxidation, but promotes formation of protective scale
Flow rate	+ −	the effect depends on water quality
Flow stagnation	−	
Calcite Indices	+	relevant for soft waters
$(Cl^- + 2SO_4^{2-})/HCO_3^-$	− −	
Inhibitors	+	examples : silicate, phosphate, polyphosphate, zinc phosphate, etc. The effect depends on water quality
Biological activity	−	
Buffer intensity	+	

[1]+ means reduced corrosion and – means increased corrosion as the parameter increases.

tap, deposited, re-suspended, reduced to soluble Fe(II), etc. These different mechanisms are illustrated in Figure 3. Consequently, it is assumed that the long term iron corrosion rate also is the potential for particles generation and some soluble Fe(II).

Evaluation of data from multiple weight loss coupons

The challenge can be to obtain long term corrosion rates from the test rig data. This can be illustrated by dividing the corrosion progress into three phases: 1) the initial phase, 2) the build-up phase, and 3) the steady state phase. During the initial phase the metal surface is relatively unprotected and consequently the

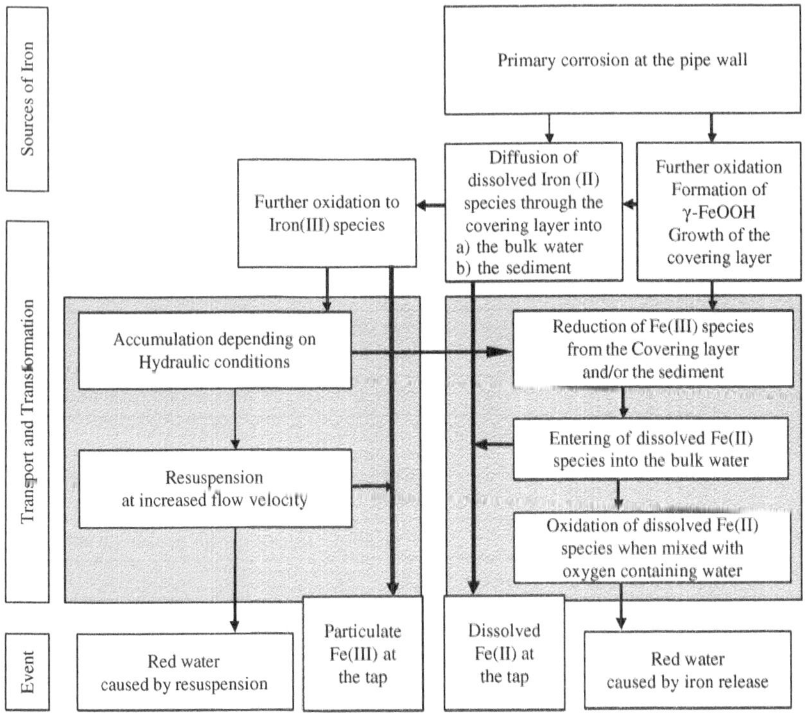

Figure 3. Scheme of the processes leading to discoloration and iron release caused by corrosion (modified after Wricke *et al.*, 2007).

corrosion rate is high. The duration of the initial phase is normally short. In the build-up phase, the corrosion rate declines with time as the corrosion product layer develop. The duration of this period will depend on the corrosion conditions (water quality and flow conditions), but it can be relatively long. Often corrosion rig experiments are conducted entirely within this phase. The steady state phase represents the relative constant long term corrosion rate occurring in the network.

Normally the corrosion rate is determined as the tangent of the curve for weight loss versus exposure time for the last 6–12 months or 9–12 months of exposure. This provides sufficient accurate corrosion rates in some cases. However, if the build-up phase is long, the corrosion rate will be overestimated, or if there is scatter in the 9 or 12 months weight loss data, the corrosion rate determination may become very inaccurate.

In order to improve the evaluation of weight loss data, it is suggested to describe the build-up phase as follows (Broo *et al.*, 2006): It is assumed that the corrosion occurs at a rate, r, at the available non-blocked surface area, a:

$$r = k_1 \cdot a \tag{1}$$

where k_1 is the rate constant. As the corrosion proceeds, and corrosion products develops, the remaining surface area available for corrosion is reduced with time. The time dependant rate of corrosion can then be written as:

$$r = k_1 \cdot a_0 \cdot (1 - k_2)^t \tag{2}$$

where a_0 is the total surface area, t is the exposure time, and k_2 is a constant describing the surface coverage. When this is integrated from time 0 to t, the weight loss, Δm, during the exposure time can be written as:

$$\Delta m = k_1 \cdot a_0 \cdot \left(\frac{(1 - k_2)^t - 1}{\ln(1 - k_2)} \right) \tag{3}$$

k_1 (rate of corrosion of available non-blocked surface) and k_2 (coverage of blocked surface) can then be determined from the experimental data by curve fitting. A low k_1 and a high k_2 would give a low long term corrosion rate. In order to obtain a corrosion rate, Δm is calculated over a short time in the end of the experimental period. This gives a more accurate determination of the corrosion rate, and has been used to evaluate the weight loss data from the test rig in this work.

This model for data interpretation could probably be improved by also taking into account factors like pitting/non-uniform corrosion, the degree of blockage, etc.

Corrosion modeling example

Below is an example of an empirical corrosion model developed based on data obtained in experiments with the test rig for multiple weight loss coupons, for Lake Jonsvatnet, the water supply for the city of Trondheim. The model results depend on the flow rate.

Low flow rate ($\leq$0.1 m/s):

$$\mathrm{CR} = 2.156 \cdot \mathrm{T}/(\mathrm{pH} \cdot \mathrm{Ca}^{0.3} \cdot \mathrm{Alk}^{0.4}). \tag{4}$$

High flow rate (>0.1 m/s):

$$CR = 32.34 \cdot T/(pH^{2.0} \cdot Ca^{0.6} \cdot Alk^{0.8}) \quad (5)$$

where CR = Corrosion rate [g/m^2 day], T = temperature [°C], Ca = calcium concentration [mg Ca/L], and Alk = alkalinity [mmole/L]. CR is the total average iron corrosion products formed (as Fe) and represented by the first upper box in Figure 3. However, as can be seen from Figure 3, the corrosion products may accumulate at the corrosion site as solids, or may enter the bulk solution as soluble Fe(II) or as particulate Fe(III). Soluble Fe(II) may later be oxidized and form particulate Fe(III). In addition, particulate Fe(III) may sediment and re-suspend, or it may be reduced and re-oxidized. For this modeling purpose, where the focus is on particles generation by corrosion, we assume a release factor (RF) which is the fraction (between 0 and 1) of corroded iron that is transported away from the corrosion site. RF consists of a soluble part (RF_{sol}) and a particulate part (RF_{part}). RF_{sol} and RF_{part} represent the state at tap or at the point of measurement. Likewise, the fraction of corroded iron accumulated at the corrosion site would be (1-RF). If MW_F is the mole weight fraction of corrosion product to iron, the amount of corrosion products formed is the sum of released and accumulated corrosion products, and can be written as:

Release of corrosion products:

$$CR \cdot RF = CR \cdot RF_{sol} + CR \cdot RF_{part} \cdot MW_F. \quad (6)$$

Accumulation of corrosion products:

$$CR \cdot (1 - RF) \cdot MW_F = CR \cdot (1 - RF_{sol} - RF_{part}) \cdot MW_F. \quad (7)$$

For corrosion products consisting of FeOOH the factor $MW_F = 1.59$. By combining Equations 6 and 7 the total particulate corrosion products (both including particles causing red water incidents and particles causing slightly increased turbidity levels) is given by:

$$\text{Particulate corrosion products} = CR \cdot (1 - RF_{sol}) \cdot MW_F. \quad (8)$$

Equation like 4 to 8 can then be incorporated into a hydraulic network model with a water quality module (Wricke *et al.*, 2007), like the MSX extension to the standard EPANET 2 hydraulic model (Bruaset, 2008). The detachment and transport of corrosion product particles will largely depend on the hydraulics and the nature of the corrosion products. Consequently, extensive calibration and validation is needed. This would include measurements of soluble and particulate iron, in addition to assessments based on the re-suspension potential method, RPM (Vreeburg and Bauxall, 2007; Wricke *et al.*, 2007).

CONCLUSIONS

Predicting or modeling iron corrosion is very complex, and at the present time it is not possible to quantitatively predict or model iron corrosion rate for a universal distribution system. Only indication of qualitative effects of various parameters can be done. However, it is shown that by using corrosion test rigs, site specific models can be developed on a form that can be used for estimation of particles generation by iron corrosion, which in turn can be applied to a hydraulic model for particles in the network.

However, such corrosion models need extensive calibration and validation, and testing of the models in case studies are needed to assess their usefulness and to further develop the models.

REFERENCES

AWWA Research foundation (1985). Internal corrosion of water distribution systems, Cooperative research report.

Böhler, E., Hofmann, D. und Tränckner, J.(2005). Entwicklung von Methoden zur Selektion effizienter Spülregime für unterbelastete Sektoren in bestehenden Wasserversorgungsnetzen zur Vermeidung der Rostwasserbildung. (A methodology for the selection of efficient flushing programs in stagnating sectors of distribution systems to avoid red water formation). Veröffentlichungen aus dem Technologiezentrum Wasser Karlsruhe, p. 27.

Boxall, J.B., Skipworth, P.J. and Saul, A.J. (2003). Aggressive flushing for discolouration event mitigation in water distribution networks. *Water Sci. Technol. — Water Supply*, **3**(1/2), 179–186.

Broo, A.E., Berghult, B., Osterhus, S.W. and Hem, L.J. (2006). Iron Corrosion in Drinking Water Pipes – the Effect of Water Flow Rate and the Method for NOM Removal, Research report no 2006-05, VA-Forsk, Sweden.

Bruaset, S. (2008). Optimization of water network operation and maintenance. Master thesis, Norwegian University of Science and Technology, Trondheim, Norway.

Clark, R.M., Grayman, W.M., Males, R.M., Hess, A.F. (1993). Modeling contamination propagation in drinking water distribution systems. *J. Environ. Eng. ASCE*, **119**(2), 349–354.

Clement, J., Hayes, M., Sarin, P., Kriven, W.M., Bebee, J., Jim, K., Beckett, M., Snoeyink, V.L., Kirmeyer, G.J. and Pierson, G. (2002). Development of red water control strategies. Awwa Research Foundation and American Water Works Association, USA.

Ellison, D. (2003). Investigation of pipe cleaning methods. Report No. 90938, Awwa Research Foundation and American Water Works Association, Denver, USA.

Gauthier, V., Rosin, C., Mathieu, L., Portal, J.M., Block, J.C., Chaix, P. and Gatel, D. (1996). Characterization of the deposits in drinking water distribution systems. In: Proceedings of the Water Quality Technology Conference of American Water Works Association.

Gauthier, V., Gérard, B., Portal, J.M., Block, J.C. and Gatel, D. (1999). Organic matter as loose deposits in a drinking water distribution system. *Wat. Res.*, **33**(4), 1014–1026.

Gauthier, V., Barbeau, B., Millette, R. and Prévost, M. (2001). Suspended particles in the drinking water of two distribution systems. *Wat. Sci. Tech.: Water Supply*, **1**(4), 237–245.

Kirmeyer, G.J., Friedman, M., Clement, J., Sandvig, A., Noran, P.F., Martel, K.D., Smith, D., LeChevallier, M., Volk, C., Antoun, E., Hiltebrand, D., Dyksen, J. and Cushing, R. (2000). Guidance manual for maintaining distribution system water quality. Report No. 90798, AWWA Research Foundation and American Water Works Association, Denver, USA.

LeChevallier, M.W., Babcock, T.M. and Lee, R.J. (1987). Examination and characterization of distribution system biofilms. *Appl. Environ. Microbiol.*, **53**(12), 2714–2724.

Lin, J. and Coller, B.A. (1997). Aluminum in a water supply, Part 3: domestic tap waters. *Water. J. Aust. Water Assoc.*, 11–13.

McNeill, L.S. (2000). Water quality factors affecting iron and lead corrosion in drinking water. Ph.D. dissertation at Virginia Polytechnic Institute and State University, USA.

Meches, M. (2001). Chapter 13—Biofilms in drinking water distribution systems. Controlling disinfection by-products and microbial contamination in drinking water, R.M. Clark and B.K. Boutin (eds.), National Risk Management Research Laboratory, Office of Research and Development, US Environmental Protection Agency, Cincinnati, Ohio, USA, pp. 13.1–13.10.

Prince, R., Goulter, I. and Ryan, G. (2001). Relationship Between Velocity Profiles and Turbidity Problems in Distribution Systems. World Water and Environmental Resources Congress, American Water Works Association, Orlando, Florida, p. 9.

Richardt, S., Korth, A. and Wricke, B. (2009). Supplementary studies about the corrosion process in drinking water pipes. Techneau report, deliverable D5.5.11.

Ruta, G. (1999). Assessing Water Quality at Customers Internal Taps. In: Australian Water and Wastewater Association, 18th Federal Convention Proceedings.

Slaats, P.G.G., Rosenthal, L.P.M., Sieger,W.G., van den Boomen, M., Beuken, R.H.S. and Vreeburg, J.H.G. (2002). Processes involved in generation of discoloured water. Report No. KOA 02.058, American Water Works Association Research Foundation/Kiwa, The Netherlands.

Sly, L.I., Hodgkinson, M.C. and Arunpairojana, V. (1990). Deposition of manganese in drinking water distribution system. *Appl. Environ. Microbiol.*, **56**(3), 628–639.

Stephenson, G. (1989). Removing loose deposits from water mains: Operational guidelines. Water Research Centre, Swindon, UK.

Teasdale, P., O'Halloran, K., Doolan, C. and Hamilton, L. (2007). Literature Review on Discoloured Water Formation and Desktop Study of Industry Practices. Research report no 51, CRC for Water Quality and Treatment, Australia.

Vreeburg, J.H.G., Schaap, P.G. and van Dijk, J.C. (2004). Particles in the drinking water system: from source to discolouration. *Wat. Sci. Technol.*, **4**(5–6), 431–438.

Vreeburg, J.H.G. and Boxall, J.B. (2007). Discolouration in potable water distribution systems: A review. *Wat. Res.*, **41**, 519–529.

Walski, T.M. (1991). Understanding solids transport in water distribution systems. In: Proceedings Water Quality Modelling in Distribution Systems, AWWA Research Foundation, USA, pp. 305–309.

Wricke, B., Henning, L., Korth, A., Vreeburg, J., Schaap, P., Osterhus, S.W., Juhna, T., Hammes, F. and Coelho, S. (2007). Particles in relation to water quality deterioration and problems in the network – State of the art review. Techneau report, deliverable D5.5.1 and D5.5.2.

Model for the calculation of optimised flushing concepts

S. Richardt, A. Korth and B. Wricke

DVGW TZW Dresden, Wasserwerkstraße 2, D-01326 Dresden (E-mail: richardt@tzw-dresden.de)

Abstract To avoid discoloration events in water distribution networks a model for the calculation of optimized flushing intervals has been developed. The model is based on the network-specific deposit situation as well as the hydraulic situation in the investigated network area and simulates the deposit growth in each pipe segment with consideration the mobilization. Together with the algorithm for the determination of the data base for the calculation, the model allows highly efficient network maintenance, which leads to a distinct reduction in customer complaints. The model was transferred in a self-running program that can be linked to any hydraulic simulation software. This enables a wide use of the research results under varying conditions.

Keywords distribution network, flushing, maintenance, water discoloration

INTRODUCTION

One of the most frequent reasons for quality-based customer complaints in the distribution of potable water are water discoloration events caused by the mobilization of accumulated particles (Powell and Brandt 2003; Vreeburg and Boxall 2007; Korth and Richardt 2008). To handle this problem it is common practice to pass off the discolored water by opening a hydrant. Possible problems because of this procedure are:

- the generation of more complaints because of changing hydraulic conditions from opening one or more hydrants without separating pipe segments
- inefficient removal of deposits due to low flow velocities
- high expenditures for urgent measures and standby services
- lack of a systematic management strategy.

The following concept describes a method for calculating optimized flushing intervals for pipe segments in a distribution network. With the optimized flushing intervals it is possible to flush each pipe segment before a mobilization (and with this a possible discoloration event) takes place. So it is part of a sustainable maintenance practice.

Edited by Theo van den Hoven and Christian Kazner. ISBN: 9781843392750. Published by IWA Publishing, London, UK.

DATA BASE

The basis for network maintenance by using optimized flushing concepts is the use of the unidirectional flushing procedure with clear water front (Antoun *et al.*, 1999; Friedman *et al.*, 2002). This practice ensures that there is no displacement of existing deposits from one network area into another during the cleaning process (Ellison *et al.*, 2003).

The model for the calculation of optimized flushing intervals is based on the determination of the existing deposit situation in a separate network area which was built up over a defined time interval. To measure this deposit situation it is necessary to flush the investigated network area two times (see Figure 1).

Figure 1. Flushing with gathering of flow rate and deposit situation.

The first flushing generates the base for the determination of the deposit growth over the defined time interval. If there are exact information available about the last network flushing (with a verified deposit discharge), it is possible to directly start the second flushing. With the second flushing, the deposit situation in the investigated network area will be recorded by measuring the turbidity and the deposit composition over the flushing time for every flushing segment. To extrapolate the turbidity of the segment length parallel to the turbidity measurement, the flow rate needs to be measured and linked to time by the pipe volume (see Figure 2). The deposit composition will be determined by a continuous sampling throughout the flushing time. The flushing samples must be analysed for their iron and manganese content (Korth *et al.*, 2008).

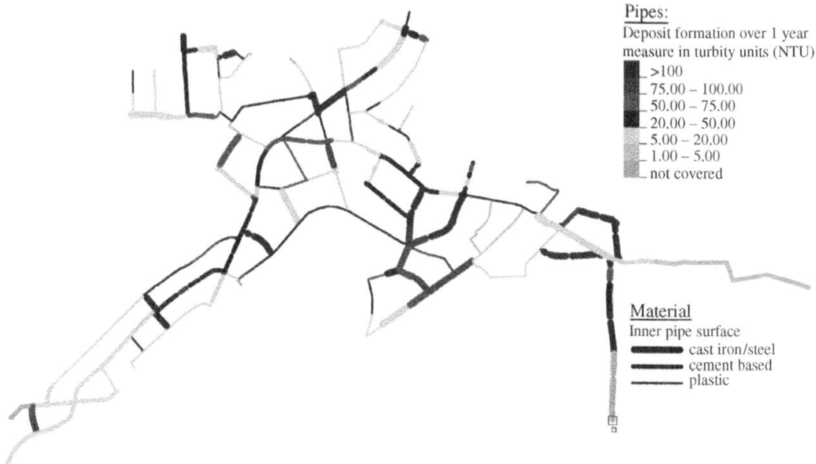

Figure 2. Deposit situation (build-up over 1 year) as turbidity units.

In addition to the deposit information, it is necessary to have detailed knowledge about the hydraulic circumstances in the investigated network area. For this, a well-calibrated model needs to be used.

MODEL APPROACH

The model for the determination of optimized flushing intervals is based on the processes of deposit formation and mobilization, which were cleared up in TECHNEAU as well as in a joint research project by the German government and the DVGW. A special function was developed, describing the dependency of the maximum possible deposit level on the hydraulic conditions (Richardt *et al.*, 2007; Richardt, 2008). The function is based on the results of more than 120 measurements and was determined with an accuracy of 0.84. This function allows the calculation of deposit mobilizations depending on the existing flow velocity for each pipe segment.

The mentioned data and parameters are linked in the model approach which is depicted in a simplified way in the algorithm below (see Figure 3).

The first step of the calculation is the differentiation of the measured deposit in a fraction because of corrosion and the second fraction because of load from the treatment plant respectively from network areas in front of the investigation area. The differentiation will be made by using a network-specific iron to manganese basic ratio. The network-specific iron to manganese basic ratio describes the treatment-specific load. This value needs to be determined, for example, from the deposits of a protected pipe close to the treatment plant.

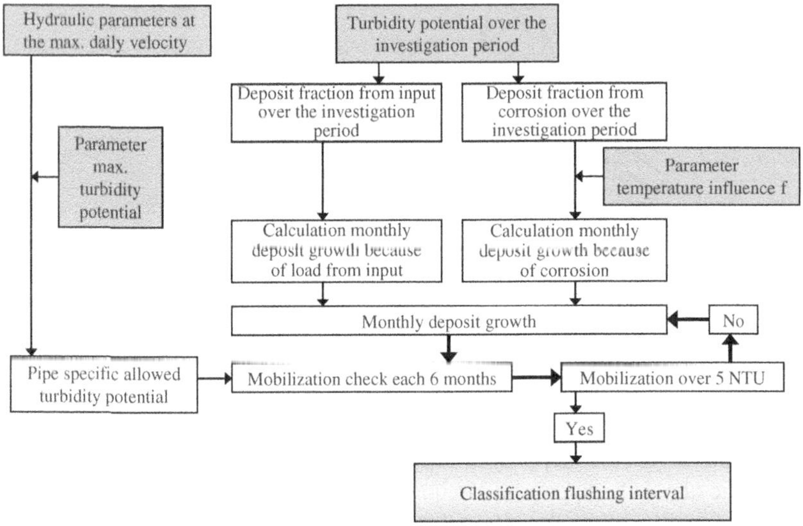

Figure 3. Model algorithm.

Afterwards, the monthly deposit growth will be calculated by implementing a factor for the temperature influence on the deposit formation due to corrosion, which was also determined by the investigations carried out in the research projects mentioned.

Parallel to the calculation of the deposit growth, the maximum possible deposit level will be determined with the data ascertained by the hydraulic simulation. A security factor is used to define the deposit level allowed in each pipe.

After a 6-month-simulation, the first mobilization check is carried out to prove if there is a mobilization over a visual border of 5 NTU (Vreeburg *et al.*, 2004; Böhler *et al.*, 2005). If there is no calculated mobilization over 5 NTU, the deposit growth will be calculated for another 6 months. Afterwards, the next mobilization check will be carried out. This procedure will run as long as a mobilization over 5 NTU for each pipe segment is to be determined or a defined basic flushing interval (normally 8 years) is reached.

If the mobilization over 5 NTU is recorded for the pipe-segment, the optimized flushing interval is defined by the time-period of 6 months before, because the mobilization should be prevented by flushing. Consequently, the optimized flushing interval describes the status of a discoloration event just having been prevented, with consideration of the defined security factor.

Calculation program

The model approach was transferred to a self-running program (see Figure 4) and can be linked to any hydraulic simulation software.

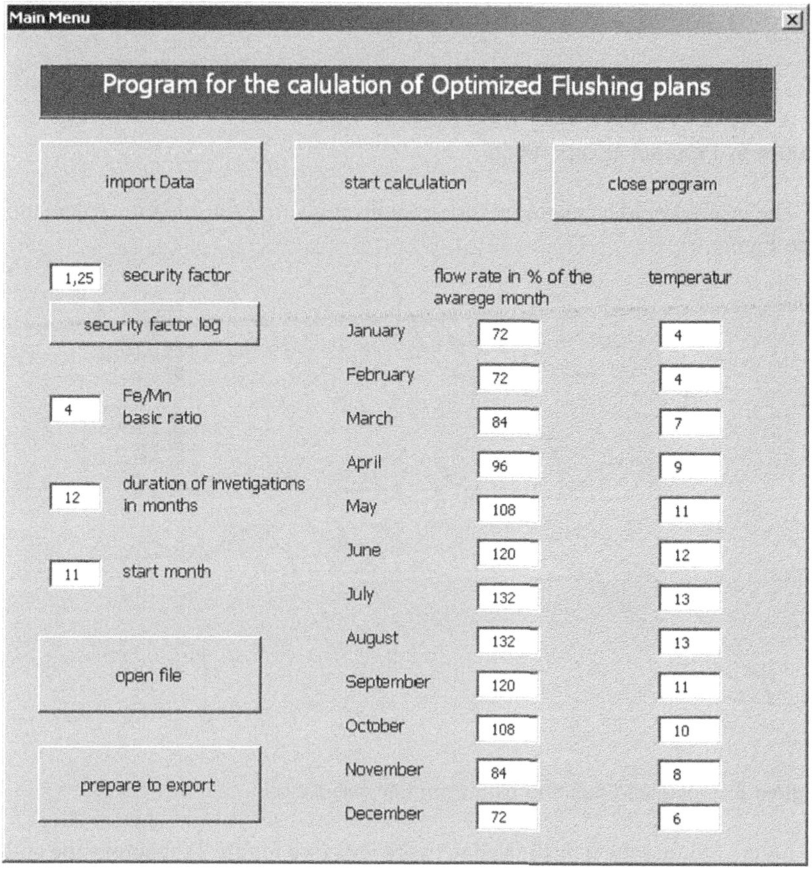

Figure 4. Program for the calculation of optimized flushing intervals.

The results of the hydraulic simulation and the deposit situation will be imported for the calculation. Afterwards, the area-specific data will be implemented, and the calculation itself can start (see Figure 5).

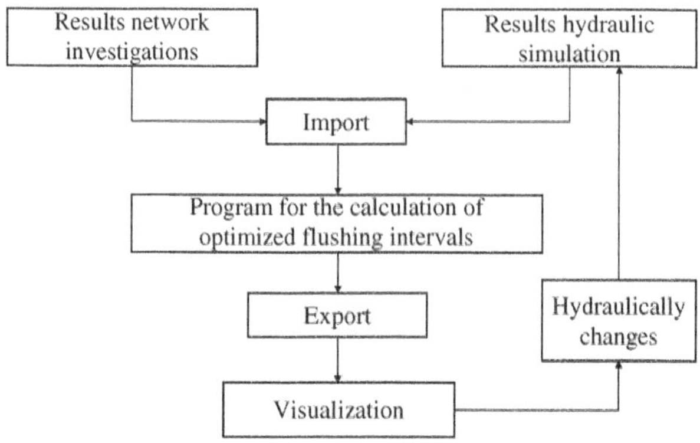

Figure 5. Program incorporation.

The results could be exported for a visualization after finishing the calculation (see Figure 6).

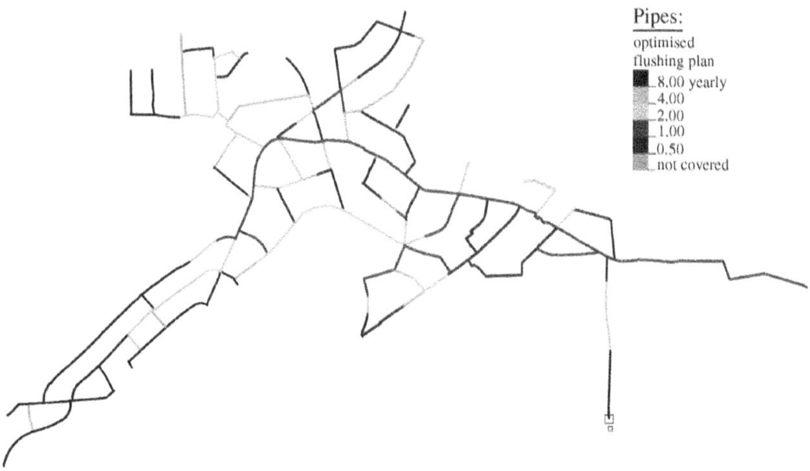

Figure 6. Optimized flushing plan from the calculation.

Figure 6 shows the investigated network area (see Figure 2), whereby the color of the pipe segments describes the flushing interval at one-half to 8 years. It becomes clear that mainly the network entrance pipes and the direct connection pipes to the main distribution pipe need to be flushed at shorter intervals (one or two times a year). In contrast to this it is sufficient to flush major parts of the end areas of the distribution network at larger intervals such as 4 to 8 years.

CONCLUSIONS

Optimized flushing concepts are an efficient tool for preventing water discoloration. A model was developed to calculate optimized flushing intervals for pipes. It is based on the determination of the specific deposit situation in a distribution area as well as the hydraulic situation. The model is designed as a self-running program.

After starting the network maintenance with optimized flushing in four areas, there was a major decrease in customer complaints. In contrast to this approach, dead-end flushing, which still is a frequently used technique (Perez and Huren An, 2003), is an inefficient strategy that generates more problems than it solves.

As a next step the program should be implemented directly into the TECHNEAU modelling software and used for TECHNEAU case studies.

REFERENCES

Antoun, E.N., Dyksen, J.E. and Hiltebrand, D. (1999). Unidirectional flushing: a powerful tool. *Journal of the American Water Works Association (AWWA)*, **91**(7) , 62–71.

Böhler, E., Hofmann, D. and Tränckner, J. (2005). *Spülung von Wasserversorgungsnetzen zur Vermeidung der Rostwasserbildung.* DVGW-Technologiezentrum Wasser. Karlsruhe. (Veröffentlichungen aus dem Technologiezentrum Wasser Karlsruhe, 27).

Ellison, D., Duranceau, S.J., Ancel, S., Deagle, G. and McCoy, R. (2003). *Investigation of Pipe Cleaning Methods.* Denver, USA: AWWA Research Foundation (90938).

Friedman, M.J., Kirmeyer, G.J. and Antoun, E.N. (2002). Developing and implementing a distribution system flushing program. In: *Journal of the American Water Works Association (AWWA)*, **6**, 48–56.

Korth, A. and Richardt, S. (2008). *Spülverfahren und Spülstrategien für Trinkwasserversorgungsnetze.* In: Roscher, Harald (Hg.): Rehabilitation von Wasserversorgungsnetzen. 2. Aufl. Berlin: Vulkan Verlag, pp. 373–378.

Korth, A., Richardt, S. and Wricke, B. (2008). Strategy for the development of optimized flushing plans. *Water Asset Management International*, **4**(3), 9–12.

Perez, A. and Huren An, J. (2003). *Automatic flushing: a solution for enhancement of distribution system water quality.* In: AWWA Research Foundation (Hg.). Proceedings AWWA WQTC Conference, Philadelphia, pp. 1–9.

Powell, J. and Brandt, M. (2003). Identification of discoloration risks. In: *AWWA Research Foundation (Hg.): Proceedings AWWA WQTC Conference*, Philadelphia, pp. 1–18.

Richardt, S. (2008). *Model for Optimized Flushing Concepts.* Presentation at TECHNEAU work area 7 meeting. Lisbon, LNEC Lisbon.

Richardt, S., Korth, A., Donath, O. and Wricke, B. (2007). *Corrosion-based Particle Input and Deposit Formation in Drinking Water Distribution Netzworks Presentation on the* "High Quality Drinking Water" Workshops TU Delft. Herausgegeben von J. Q. J. C. Verberk. Technische Universität Delft.

Vreeburg, J.H.G. and Boxall, J.B. (2007). *Discolouration in potable water distribution systems: A review. doi: 10.1016/j.watres.2006.09.028.* In: Water research (Water res.) **41**(3), 519–529, first published: 14. December 2006.

Vreeburg, J.H.G., Schaap, P. and van Dijk, J.C. (2004). *Measuring Discoloration Risk. Resuspension Potential Method.* In: IWA Leading Edge Conference Prague.

Influence of NOM chlorination on halophenols: appearance and control on a water treatment plant

P. Roche, C. Tondelier and D. Benanou

Centre de Recherche sur l'Eau, VEOLIA WATER, Chemin de la Digue, 78600 Maisons-Laffitte, France (E-mail: pascal.roche@veolia.com, christophe.tondelier@veolia.com, david.benanou@veolia.com)

Abstract This study deals with the appearance and control of halophenols on a water treatment plant. For drinking water production, the ultimate disinfection step using disinfectants as chlorine guarantees the potability of the water from the station to the consumer's tap. However, the chlorination step involving reactions of chlorine with residual DOC could also lead to halophenols, which are involved in tastes and odours. Halogenated phenols formation and speciation (mono- to tri-, chloro, bromo and mixed phenols) are directly linked to the disinfection parameters and physico-chemical properties of the matrix as pH, temperature, bromide concentration, chlorine level, disinfectant nature, contact time. The three NOM type – fulvic or humic acid, lignin – have led to the same DBPs but with a different distribution. Stir Bar Sorptive Extraction (SBSE) followed by Gas Chromatography/Mass Spectrometry (GC/MS) appears to be a suitable technique for the analysis of trace amounts of halophenols in aqueous samples after in situ derivatization, and allows reaching low limit of detection levels, closed to the ng/L. Different treatments can be used on a WWTP to remove halophenol precursors in the raw water, as coagulation, adsorption, ozonation and nanofiltration. In terms of perspective, pre-oxidation with chlorine dioxide followed by NaOCl desinfection has led to interesting results.

Keywords chlorination, halophenol, NOM, odour, SBSE, taste

INTRODUCTION

When designing a drinking water treatment plant a particular focus has to be given to the high content of natural organic matter (NOM) which could be responsible of the deterioration of the quality of the water. One aspect of the problem concerns the organoleptic qualities and is linked to taste and odour (T&O) compounds. Chlorine reacts with dissolved natural organic matter and numerous so-called disinfection by-products (DBPs) can result from this reaction. Among these DBPs, trihalomethanes (THMs) and haloacetic acids (HAAs) were the first chlorine DBPs reported and are currently regulated. Today, about 600 DBPs are identified. Among them some highly toxic compounds are found such as iodo and bromo compounds, halonitromethanes and N-nitrosodimethylamine. These individual

 TECHNEAU: Safe Drinking Water from Source to Tap. Edited by Theo van den Hoven and Christian Kazner. ISBN: 9781843392750. Published by IWA Publishing, London, UK.

DBPs or mixtures of DBPs could represent a potential human health risk, and, on a lower scale, be responsible of Taste & Odor problems (Benanou *et al.*, 2003; Corbi, 2006; Bruchet *et al.*, 2008).

Halophenols

Phenols are chemicals commonly present in the environment of surface waters. They are also generated during lignin degradation (Davidson, 1996) and decomposition of organic matter. Unfortunately, chlorine oxidants used in water treatment induce formation of chlorinated derivatives such as chlorophenols. As reported by Michalowicz *et al.* (2007), these compounds exhibits high toxicity, are mutagenic carcinogenic and could also bring medicinal taste and odour to the water at the ng/L concentration level. One of the most potent taste and odour compound is the 2,4,6-trichloroanisole which may be formed by microbial methylation of 2,4,6-trichlorophenol (Nyström *et al.*, 1992). Other organic compounds with particularly low taste and odor threshold are 2,4-dichloroanisole (80 ng/L), 2-chlorophenol (41 ng/L), 2-bromophenol (12 ng/L) and 2,6 dichlorophenol (5.5 ng/L) (Young *et al.*, 1996) (Piriou *et al.*, 2007). In bromide-containing waters, chlorination leads to hypobromous acid formation. Hypobromous acid is more reactive than hypochlorous acid, especially with phenolic compounds. Under these conditions, bromination can be highly significant and brominated products can be formed. The bromophenols have the same type odour as chlorophenols (iodoform, phenolic, medicinal), and the odour threshold concentration (OTC) of bromophenols follow approximately the same pattern as the chlorophenols. 2-bromophenol and 2,6-bromophenol have the lower OTCs (Piriou *et al.*, 2007) in the ppt range. With a statistical model Al-Samarrai *et al.* (2004) indicated that high levels of chlorine, phenol and temperature were associated with high odour intensities. Recent works (Acero *et al.*, 2005; Piriou *et al.*, 2007) carried out with phenol have shown that chlorination doses and contact time, could be a mean to control intermediates products such as 2,4 dichlorophenol (2,4-DCP) or 2,6-DCP. Acero *et al.* (2005) studied the effect of temperature on the formation of these compounds and explained that the rate of chlorination and bromination decreases by approximately a factor of 2 as the temperature decreases by 10°C. With lower temperatures, intermediates species (as mono- or dibromophenols) could have a longer life and generate T&O problems during longer periods. Piriou *et al.* (2007) observed that under chlorination, higher phenol concentrations lead to higher T&O intensity and that high pH level also had a positive effect on taste intensity ($pH = 8.5 \gg pH = 6.5$). Chlorination of phenol leads to the formation of halophenols and the further halogenation of these compounds could lead to the formation of non-phenolic molecules such as THMs, HAAs and haloacetonitriles (Hirose and Okitsu, 1982; Acero *et al.*, 2005; Lu *et al.*, 2009). However, more limited work has been done to link halophenols appearance with chlorination of NOM present in water. Given that chlorination of NOM also leads to THMs and HAAs, intermediate steps, including a halophenol as haloforms precursors, should exist. It is precisely the aim of this work, the main ideas being to highlight the influence of the matrix and the operating conditions on the generation of different halophenols. To control their formation, a strategy could involve managing the drinking water plant to decrease NOM concentrations in the resource. This strategy leads to evaluate the performances of the different steps of a drinking water treatment towards removal of halophenols precursors.

Organic matter (OM)

A conventional water treatment plant could be composed of coagulation, sedimentation, sand filtration, ozonation, GAC filtration and disinfection. As chlorine reacts with organic matter to generate DBPs, dissolved organic carbon (DOC) is a key parameter in the field of water treatment, since its value and composition will determine the reagent dosage (coagulant, oxidant and adsorbent) and the DBPs level. Its removal thus constitutes a prime objective of treatment. About 50% of the raw water is surface water with varying TOC (total organic carbon) concentrations from 2 to 10 mg/L. Whereas turbidity can be removed up to 99%, the removal of dissolved organic carbon (DOC) can not reach the same level and more than 2 mg/L can be found in the effluent of water treatment plants (WTPs). DOC had been removed through coagulation followed by sedimentation up to 60%. The doses of coagulant are generally around 2 mg Fe/mg TOC for iron salts and 1 mg Al/mg TOC for aluminium salts. The NOM fraction not removed corresponds to the most hydrophilic fraction. Rizzo *et al.* (2005) studied the effectiveness of coagulation measuring TOC, UV_{254} and potential Total THM formation during a series of jar tests. The TTHMFP (total THM formation potential) elimination results were greatly reduced, up to 65%, by use of 60 mg/l $FeCl_3$. Adsorption on PAC highly limits THM occurrence but does not adsorb bromide ions, resulting in a percentage of brominated THM higher in the treated water (Amy *et al.*, 1991). Adding PAC prior to the coagulation stage has been shown to reduce THM precursors by over 70%. Ozone oxidation does not really eliminate TOC but transforms it into new species which are more polar, more hydrophilic, richer in oxygen, and more acidic. On the other hand, ozone facilitates later elimination by coagulation (Philipot and Henriet, 1991) or biological filtration by increasing the biodegradability of the organic matter. With ozonation, Guay *et al.* (2005) show that THM production is significantly less compared to slow filtration on sand without upstream ozonation. In a study of different NOM elimination treatments, GAC trials showed the elimination of 60% of THM precursors after six months of operation (Jacangelo *et al.,* 1995). Membrane filtration, particularly nanofiltration can control THM formation potential, but require pre-treatment to avoid clogging (Taylor *et al.*, 1987). TOC must therefore be lowered for the most part during clarification stages, which must be optimised by adjusting the pH, adsorption and reactant dosage. In a study carried out by Krasner *et al.* (1989) chlorine dose was roughly proportional to TOC concentration. As TOC was removed through the treatment train, the chlorine dose decreased and TTHM formation declined. The removal of DOM and the reduction of DBPs during water treatment are related to specific physico-chemical properties of this material. For example, coagulation treatment has been shown to be more effective in removing the high molecular weight and non-polar fraction of DOM (Collins *et al.*, 1986; Liang and Singer, 2003). The use of membrane for the removal of DOM is influenced by the ability of specific fraction to cause fouling while its speciation (hydrophilic and hydrophobic) will affect the formation of specific type of DBPs (Hua and Reckow, 2007).

As a conclusion, the bibliographic background gathered shows that eliminating DBPs precursors is effective at each stage of the treatment process. By analogy with THMs, it appears that halophenol control could greatly depend on DOC control.

EXPERIMENTAL

Tests have been carried out with Evian water spiked with different fraction of organic matter: humic acids (HA), fulvic acids (FA) and lignin (L). The different parameters studied concern the quality of the water, pH, temperature, organic matter, bromides, and the control parameters such as contact time, quality and dose of the reagent.

Standards and material

Sodium hypochlorite was purchased from Aldrich (high quality chlorine), collected on a water plant or was a commercial product. Stock solutions were prepared and stored at 5°C. Before use, they were diluted with Evian Water and their exact concentration was determined before use. Chlorine dioxide was generated in our laboratory before use. Humic acid (Suwannee River Humic Acid, 2S101H) lignin and fulvic acids (Suwannee River Fulvic Acid, 1S101F) were purchased from IHSS (US). Concentrated solutions were prepared in Evian water and TOC was measured. For experiments, Evian water was spiked by dilution with these stock solutions stored at 5 °C. pH measurements were carried out with a WTW 340i/SET pH-meter which was calibrated with standard buffer solutions (pH 4 and 7). DO was measured with an HACH DR 4000 spectrophotometer at 254 nm with a 5 cm quartz cell. TOC was measured with a Bioritech 1010 OI Analytical TOCmeter with a detection limit of 0.1 mg/L. Bromide used for spiking was KBr from ALFA AESAR 99.8% purity. For the evaluation of treatment steps efficiency, waters were collected on a plant; raw water, settled water and filtered water (sand and GAC). Evian was the reference water for the other trials.

Chlorination experiments

The halophenol formation potential was investigated using 500 ml of water spiked with 1 to 3 mg/L of HA, L or FA. For the evaluation of halophenol precursor removal, water was chlorinated before and after each treatment step. Experiments were performed at room temperature. Samples were adjusted to pH 7.5 and homogenised during the experiment with a stir bar on an agitation plate. The water free chlorine content was determined using the N,N-diethyl-p-phenylenediamine procedure with a photometric detector pocket Hach DR 820. After a contact time varying from 1 to 180 minutes, the excess of chlorine was quenched by addition of sodium thiosulfate.

Extraction of halophenols

A sorptive extraction technique that uses a stir bar coated with PDMS and known as stir bar sorptive extraction (SBSE) was developed (Corbi, 2006; Tondelier *et al.*, 2009). SBSE with in situ derivatisation has been applied for the analysis of phenolic compounds. 1g of potassium carbonate was introduced in 100 ml sample spiked with 40 μl of a mixture of 5 (deuteriated mono to pentachlorophenol) internal standards. After introducing 500 μl of acetic anhydride, as derivatization reagent, a stir bar was added and a Teflon-coated silicone septum cap was placed on the vial without crimping. SBSE was performed at room temperature for 60 min while stirring at 700 rpm.

GC-MS determination

The analytical system consists of an Agilent 6890 gas chromatograph coupled to an Agilent 5973 mass spectrometer detector (Agilent Technologies, Santa Clara, CA, USA). This GC–MS system was equipped with a thermal desorption unit

(TDU) and a PTV (programmable temperature vaporizer) injector (CIS 4, cooled injection system 4) from Gerstel (Mülheim a/d Rhur, Germany). The operating conditions were the following: 20 mm length by 0.5 mm PDMS layer thickness stir bars were used for the extraction step. Stir bars were thermally desorbed in splitless mode at 30°C for 0.8 min and then heated to 250°C at 60°C/min and held for 8 min at this temperature. The desorbed solutes were cryofocused in the CIS 4 at −50°C. After the stir bar desorption, the PTV injector was programmed to 300°C at 12°C/s and held for 2 min. The injection was done in solvent vent mode. The carrier gas was helium at 1.5 mL/min constant flow. The compounds were separated on a 30 m × 0.25 mm I.D. × 0.25 μm film thickness HP5-MS capillary column. The oven was programmed from 50°C (2 min) to 200°C at 10°C/min then to 300°C at 25°C/min (7 min). Detection was achieved with the MS equipped with an Electronic Impact source, in dual SIM/SCAN mode: the SIM mode was used for quantification and the SCAN mode for confirmation.

Treatment experiments

The objective of this study was to monitor the fate of the precursors of odorous compounds such as chlorophenols and especially 2,4,6 trichlorophenol during drinking water treatment, and to evaluate the efficiency of different treatment stages in eliminating them. At each step, the matrix was spiked with HA, FA or lignin. The target treatment stages were tested as follows:

Coagulation/flocculation/decantation

Series of jar-test were carried out with raw water from Oise River on a floculator (type FLH6 Hydrocure). The pH was adjusted to 5, 6, 7 or 8 with 1N sulfuric acid or 1N NaOH. Coagulant was 30 mg/L of WAC HB or $FeCl_3$. 3 steps were operated: agitation at 200 rds/min for 3 minutes, 50 rds/min for 15 minutes and settling during 60 minutes. UV 254 nm and TOC were measured before and at the end of the experiment.

Adsorption onto activated carbon

10 to 20 mg/L of a wood based powdered carbon Picasorb 16–15/35 μm from PICA – were introduced in settled water of Oise River and mixed during 15 and 30 minutes. Then samples were filtered on Nucleopore 1 μm membrane.

Ozonation

Ozone tests were carried out with the OZOTEST method, which is approved by Europe/Africa IOA Group. It is described elsewhere (Roche *et al.*, 1994) and allows to obtain a complete assessment of the oxidant. The matrix was sand filtered water.

Nanofiltration

Tests were carried out with a flat sheet equipped with Dow Chemical membrane: NF 90 (nanofiltration) or BW30 LE (RO membrane). The surface of the membrane was 460 cm^2. Ultra-pure water spiked with HA and 500 mg/L $CaCl_2$, $2H_2O$. pH was adjusted to 7.5 and temperature was 20°C. Time experiment was up to 72 hours.

Chlorine dioxide

Chlorine dioxide was generated by mixing sodium chlorite with hydrochloric acid with a Prominent Bello Zon. A 1 g/L stock solution was prepared for the trials, its concentration was measured and it was stored at 5°C.

The purpose of these experiments was to evaluate the potential for chlorophenol formation by chlorination of inlet and outlet waters of each treatment. Evaluating this THPF (Total Halophenol Potential formation) should allow us to better understand the generation of these compounds and adapt or optimise existing processes.

RESULTS AND DISCUSSION

Halophenol generation

In order to identify the halophenols generated by NOM chlorination, a first set of experiments was carried out with contact times between 2 and 360 minutes in Evian water spiked with 2 mg/L HA. After 180 minutes, the concentration of tri-halophenol was stable. The excess of chlorine was quenched for each contact time using thiosulfate and halophenols were extracted as described in Experimental section. As presented in Table 1 the reactions lead to the formation of mono-, di- and tri-substituted species, with similar evolution pattern until 30 minutes. For our conditions and after this point, a noticeable decrease of the mono- and di-halogenated species can be observed for free chlorine concentrations over 1.5 mg/L. On the contrary, tri-halogenated phenol concentration is increasing as already described by different authors. Initial halogenation of phenol occurs at the aromatic ring preferentially in ortho or para positions to the hydroxyl group. Moreover, successive halogenation could lead to the formation of 2,4 and 2,6-di-halophenol and then to 2,4,6-tri-halophenol. Our results highlight the simultaneous formation of mono-, di- and tri-halogenated species. It also should be noted that the major compound after 15 min of reaction is

2,4,6-trichlorophenol. Karlsson *et al.* (1995) reported the formation of this compound during the oxidation of aromatic compounds derived from humic substances by sodium hypochlorite. The presence of mixed dichlorobromophenols at high level is also remarkable, even with a bromide concentration below the quantification limit (25 μg/L).

Table 1. Halophenols concentration (ng/L) generated in Evian water spiked with HA 2 mg C/L, Temperature 20°C, pH 7.2 – NaOCl 3 mg/L.

Time (min)	0	1	5	15	30	60	180
4-chlorophenol	0.0	0.0	64.4	85.6	88.8	65.0	5.5
2,6-dichlorophenol	0.0	8.4	24.3	44.1	45.6	28.5	0.0
2,4-dichlorophenol	0.0	11.9	37.7	77.6	106.9	94.9	33.9
2,4,6-trichlorophenol	2.2	16.8	52.1	156.5	340.6	622.5	1103.8
2,4-dichloro-6-bromophenol	0.0	0.0	10.6	45.5	108.9	194.5	330.7
2,6-dichloro-4-bromophenol	0.0	0.0	3.0	10.7	24.2	41.1	65.6
2,6-dibromo-4-chlorophenol	0.0	0.1	0.8	2.3	5.1	7.1	9.9
2,4-dibromo-6-chlorophenol	0.0	0.1	0.8	2.0	5.6	7.5	9.6
2,4,6-tribromophenol	0.0	0.1	0.2	0.4	0.6	0.8	0.8

The absence of phenol and 2-chlorophenol should be noted. The presence of 2-bromophenol, and also 2,6-bromophenol which could be critical for drinking water production (see presence of bromides) have been detected during the first 60 minutes at the ppt level. Tetrachlorophenol and pentachlorophenol have been referenced by different authors as possible DBPs (Acero *et al.*, 2005). Tetrachlorophenol has not been detected during this study and pentachlorophenol only at very low concentration in some series. These results were, qualitatively, the same for fulvic acids or lignin.

Influence of the quality water

As reported in literature, the halophenols formation depends on several factors such as OM concentration, temperature, pH and bromide level. The contact time also appears as a major parameter. In our experiments, DBPs concentrations were stable after 180 minutes reaction time with a chlorine residual as high as 1.3 mg/L (Figure 10), allowing us to study the influence of others parameters. For this contact time and in a first approach, it can be considered that all the reactive sites have reacted.

pH and temperature influence

Trials, carried out for pH varying from 5 to 9, show a higher tri-halophenol concentration for pH level around 7 as seen in Figure 1. The degree of sodium hypochlorite dissociation depends on the pH value of the solution. Dissociation is not favourable because OCl-anion has much lower oxidative capacity than undissociated HOCl. The value of pK_a of sodium hypochlorite is 7.5 which means that at pH 7.5 50% of this compound is dissociated into OCl-anion. Increase in pH values leads to further dissociation of the acid thus lowering its oxidative capacity. Decrease in pH has an opposite effect but the different pK_a values of the phenolic compounds have also to be taken into account. Highly halogenated phenols have a lower pK_a value and therefore the available concentration of the reactive phenolate species is higher at neutral pH (Acero *et al.*, 2005).

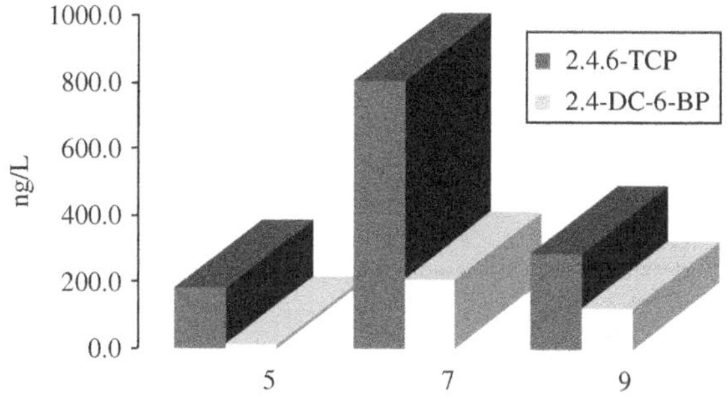

Figure 1. pH effect.
Conditions: 2 mg/L HA, 180 min, 20°C, TT 3 mg/L NaOCl

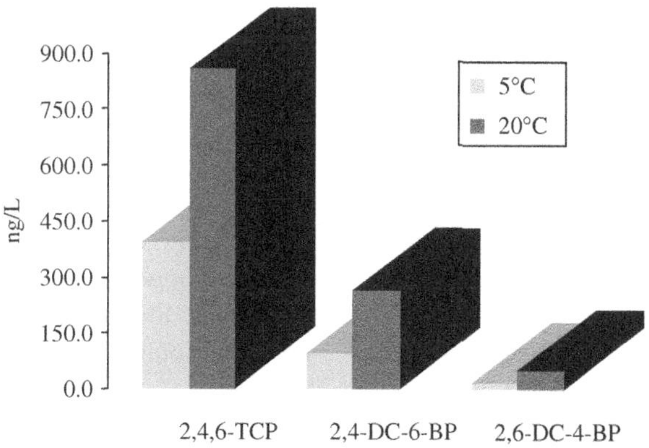

Figure 2. Temperature effect, pH 7.5.
Conditions: 2 mg/L HA, 180 min, 20°C, TT 3 mg/L NaOCl

Temperature also has a significant influence as presented in Figure 2. Tests were carried out at 5, 10, 15 and 20°C. They show an increase of all the di- and mainly tri-halogenated compounds for a higher temperature. The rate of chlorination and bromination decreases by approximately a factor of 2 for 15°C decrease in temperature which is close to the observations of Acero *et al.*, 2005. The reaction is slower at lower temperature thus the maximum concentration of tri-halogenated phenols appears at later reaction times. In the same way, di-chlorophenols concentrations decrease more slowly at lower temperature. This means, that temperature influences the reaction kinetic but does not change the final equilibrium. Temperature will be an important factor in the evolution of species in the network. Differences could then exist in terms of distribution of species between the oultlet of the plant and the tap of the consumer. This spatial variability is very important especially if the compounds are bio-transformed and give tastes and odors. It also means that with low temperature, intermediate species (as mono and dibromophenols) could have a longer life and generate T&O problems during longer periods of time.

Presence of bromide ions

In presence of bromide ions, the distribution of halogenated phenols is quite different because of the reaction of phenol with hypobromous acid (HOBr), produced by oxidation of bromide with chlorine (Piriou *et al.*, 2007; Uyak and Taroz, 2007) and leading to brominated phenols. NOM chlorination in bromide-containing water was investigated in pure water (HPLC) and in Evian spiked with bromides at 50 and 100 μg/L. Concentration of bromides in natural waters rarely exceeds such concentrations. The chlorine dose was 3 mg/L and the residual concentration, after 180 minutes, was about 0.7 mg/L. The experiments were performed at room temperature and samples were quenched after reaction. The tri-halogenated compounds are the major products as highlighted before (Table 1) but the increase in bromides concentration also increases their total concentration (Figure 3). Figure 4 presents the mixed tri-halophenols distribution. For a 180 min contact time and for a 100 μg/L bromide concentration only the six tri-halogenated compounds of the Table 1 have been detected. For these conditions, the main compound is the 2,4-dichloro-6-bromophenol whereas 2,4,6-trichlorophenol is the one in pure water. In fact, as seen in Figure 5 to 8, the kinetic of reactions with bromide is quite higher than with chlorine. In Figure 8 the generation of brominated compounds, when bromide concentration increases, is linked to a decrease of the final concentration of TCP highlighting the higher kinetic of the bromide reaction. The same trend is observed for intermediate species such as 4-chlorophenol (Figure 7) or 2,4-dichlorophenol, the evolution of these compounds depending on the contact time and the level of bromides concentration. The evolution of brominated compounds as 4-bromophenol (Figure 8) or di-bromophenol to more halogenated compounds is also faster as for chlorinated compounds. High bromide levels in the presence of phenol facilitates the rapid bromination pathway which leads to further bromination of 4-bromophenol and the odorous 2,6-dibromophenol to higher brominated compounds.

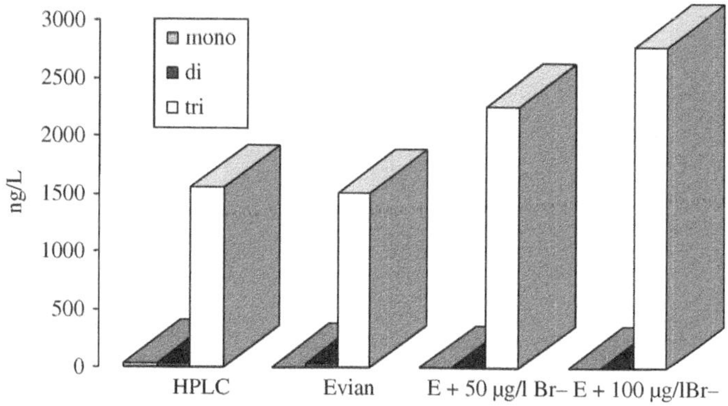

Figure 3. Bromide ions influence on halogenated phenols distribution.

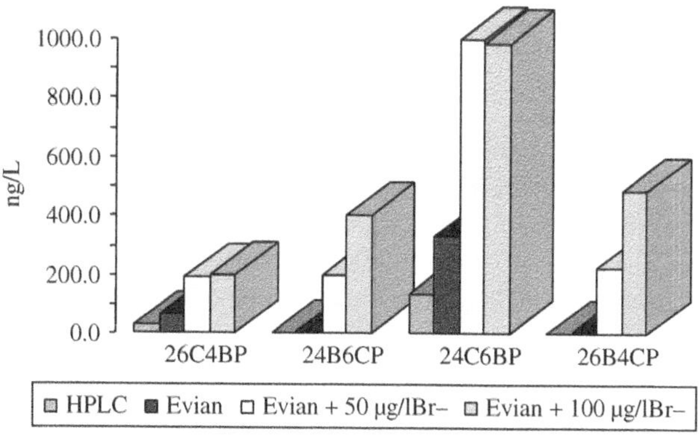

Figure 4. Bromide ions influence on mixed tri-halogenated phenols distribution.

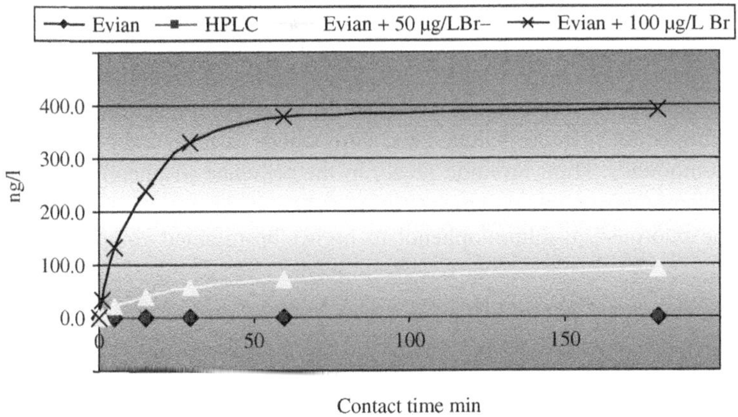

Figure 5. Bromide ions influence on tri-bromophenol formation.

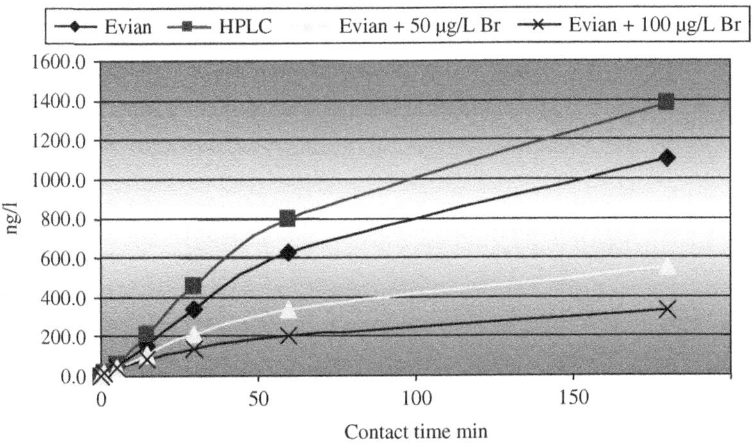

Figure 6. Bromide ions influence on tri-chlorophenol formation.

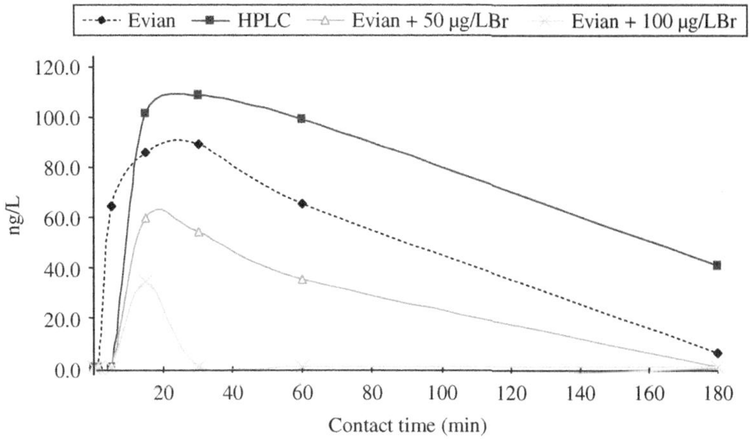

Figure 7. Bromide ions influence on 4-chlorophenol evolution.

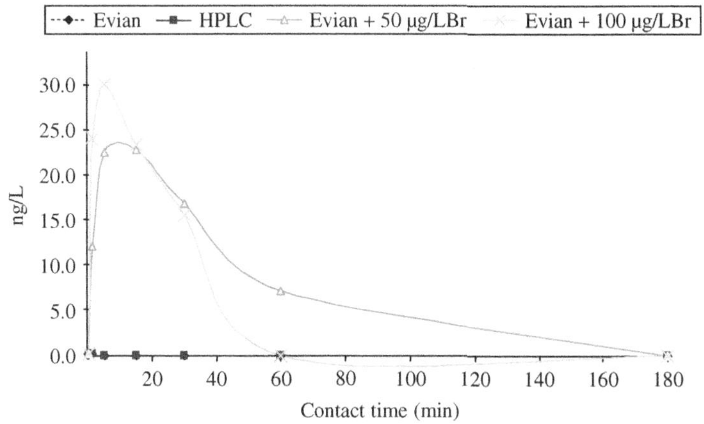

Figure 8. Bromide ions influence on 4-bromophenol evolution.

Note: Figures 3 to 8 present results at 2 mg/L HA, 180 min, 20°C, TT 3 mg/L NaOCl conditions.

From these results it can be concluded that the oxidation of bromide by chlorine is faster than the oxidation of phenols. The rate constants for the reactions of HOBr with phenolate ions have been referenced as about two orders of magnitude higher than those of HOCl (Piriou *et al.*, 2007; Acero *et al.*, 2005). During the chlorination of phenol originated from OM, and bromide-containing natural water, the bromination of phenols is the predominant pathway. Mono- and di-chlorophenols concentrations are negligible. Bromination of mono-bromophenols leads to mixed bromo/chloro derivatives (Figure 4). It also appears that high chlorine doses (residual 0.7 mg/L for 180 min) lead to low concentrations of intermediate odorous chloro- and bromophenols.

Control of halophenols formation

The dose of sodium hypochlorite introduced, the use of chlorine dioxide rather than sodium hypochlorite and the removal of OM before chlorination are different ways to minimise the halophenol generation.

Sodium hypochlorite

Sodium hypochlorite has been shown to generate a high number of chloro/bromo species. The quality of this reagent has been evaluated with different products – high quality grade, commercial or collected on a plant. Tests were carried out with Evian water spiked with 2 mg/L HA. No difference was observable, neither qualitatively nor quantitatively. Figure 9 highlight these results for the 2 main compounds: 2,4,6-trichlorophenol and 2,4-dichloro-6-bromophenol.

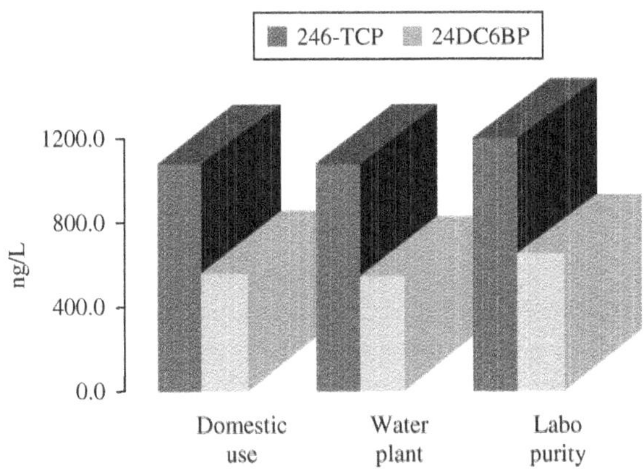

Figure 9. reagent quality – AH 2mg/L – TT 3 mg/L – 20°C – pH 7.3 – 180 min.

The effect of chlorine dose was also studied separately for lignin, fulvic acid and humic acid. Operating conditions were Evian water spiked with 2 mg/L NOM, pH 7.2, temperature 20°C and contact time varying from 1 to 180 minutes. For these conditions, the trials (Figure 10) with 1 mg/L chlorine show a rapid chlorine consumption. After 50 minutes, no more free chlorine was detectable. Chlorine being a limiting factor, as the chlorine demand was not satisfied, intermediate species – mono – 4 and 6-CP and 2-BP – and di-halophenol – 2,4 DCP and 2,6-DCP – were detected for 180 minutes contact time. For the same time, and a chlorine dose of 2 mg/L, the same species could be observed but at much lower concentration and were no more detectable for 3 mg/L. On the contrary and for these chlorine doses, the concentration of tri-halogenated species increased as shown in Figure 11 and 12. Humic acid for 3 mg/L chlorine (Figure 4) generates more tri-halophenol than fulvic acid, and a lower concentration was generated with lignin. It could also be noted that, for lignin, the distribution between these species was not the same as for the others and especially, 2,4-DC-6-BP, 2,6-DB-4-CP and 2,4-DB-6DP were more abundant relatively to 2,4,6-TCP and 2,6 DC-4BP.

As a conclusion, the same compounds appeared when applying NaOCl to the three different fractions of OM. However, the observed distribution was different. As a possible operational measure, the initial chlorine dose is an important parameter to be considered. As expected, an increase in the chlorine concentration dose leads to further transformation of phenolic compounds in highly substituted chloro- and bromo-derivatives. An increase in the chlorine concentration leads to a lower final concentration of the key compounds 2,6-dibromophenol and 2,6-dichlorophenol. However, formation of 2,4,6-trichloroanisole in the networks remains a risk which should be managed site by site.

Chlorine dioxide

Chlorine dioxide is commonly used in the United States to control tastes and odours associated with algae and decaying vegetation. Chlorine dioxide is also effective in removing taste and odour producing phenolic compounds. It is applied as pre-treatment or during the disinfection step. The application of chlorine dioxide does not produce THMs and produces only a small amount of total organic halide (TOX) (Werdehoff and Singer, 1987). However, by-products from the use of chlorine dioxide also include chlorite, chlorate, thereby limiting its use in France. Two kinds of tests were carried out, the first series aimed at comparing ClO_2 and NaOCl in terms of halophenol generated and the second series aimed at combining the effects of the 2 reagents. The first series was

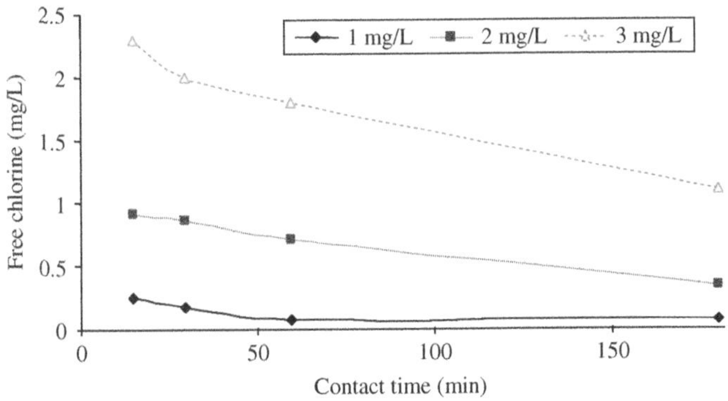

Figure 10. Free chlorine evolution – 2 mg/L HA or L or FA, 20°C, pH 7.2.

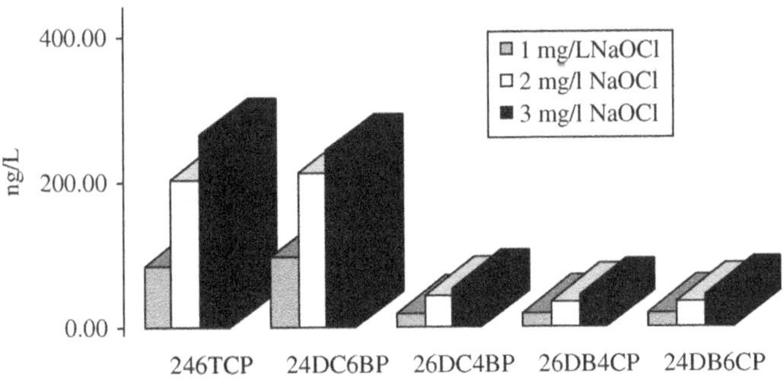

Figure 11. NaOCl dose effect on lignin – Lignin 2 mg/L – Evian – 180 min – 20°C – pH 7.2.

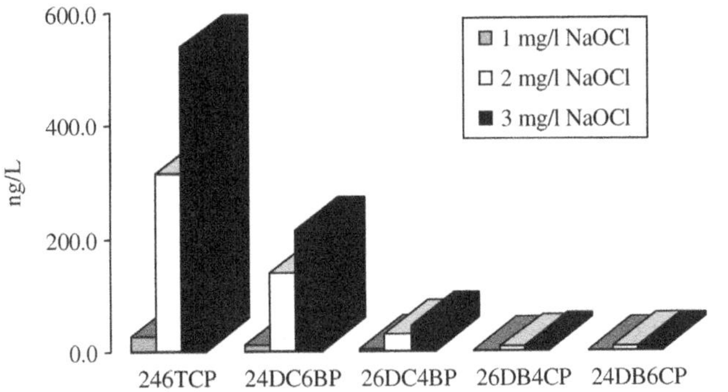

Figure 12. NaOCl dose effect on fulvic acid – FA 2 mg/L – Evian – 180 min – 20°C – pH 7.2.

conducted with HA, FA and lignin at 20°C, pH 7.2 in Evian water for contact times between 1 and 180 minutes.

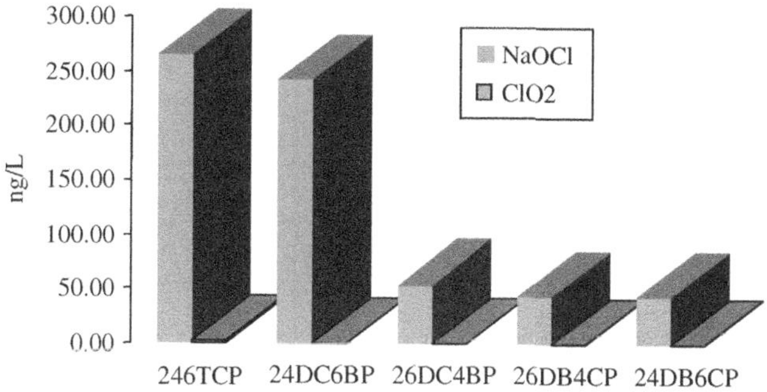

Figure 13. Effect of the reagent on the halophenol formation, Lignin 2 mg/L – 180 min – TT 3 mg/L – 20°C – pH 7.

The results presented in Figure 13 highlight the difference in terms of halophenol formation between the 2 reagents. More precisely and for the two main compounds, the concentrations were respectively 1.3 ng/L for 2,4,6-TCP and 0.1 ng/L for 2,4DC-6BP with ClO_2 and 267 ng/l and 243 ng/L with NaOCl. Qualitatively, the results were the same for HA, FA and lignin. The second series series aimed at satisfying partially the oxidant demand of the water with ClO_2 before introducing NaOCl. Tests were carried out as described above, temperature being 20°C, pH 7.2. ClO_2 was introduced at 0.4, 0.6 and 0.8 mg/L for 5 minutes contact time. This short contact time was chosen because on site time is often a limiting factor and it could not be too long before introduction of NaOCl. Free chlorine was not detectable for these conditions and 3 mg/L NaOCl were then introduced with a contact time of 180 minutes.

The results summarised in Table 2 show the difference in terms of halophenol concentrations. For these low concentrations of ClO_2, the chlorite concentration is negligible. The ratio dosage applied to oxidant demand is low and consequently the reactions between chlorine and chlorite if free chlorine is used for distribution system residual maintenance will be lowered leading to chlorate in the same proportions. Moreover, filtration onto granular activated carbon could also transform chlorite into chloride without apparent generation of chlorates (Suty *et al.*, 1995). These elements suggest an injection on line of small quantities of chlorine dioxide before the GAC filtration or disinfection step.

Table 2. Effect of "pre-chlorination" with ClO_2 on the formation of halophenol (ng/L), HA 1 mg/L, 20°C, pH 7.2, NaOCl 3 mg/L, TC 180 min.

		Evian + HA 3 mg/L NaOCL				
Halophenols	Evian + HA	Without ClO_2	CTime = 5 mn 0.4 ppm ClO_2	CTime = 5 mn 0.6 ppm ClO_2	CTime = 5 mn 0.8 ppm ClO_2	CTime = 10 mn 0.8 ppm ClO_2
246-TCP	1.0	562.4	297.5	148.9	71.7	74.5
24-DCl-6-BRP	0.1	304.0	186.8	102.8	42.3	48.3
26-DCl-4-BrP	0.0	84.3	40.5	18.9	9.9	10.7
26-DBr-4-ClP	0.1	19.2	16.5	10.0	3.8	4.6
24-DBr-6-ClP	0.1	26.4	17.6	8.9	4.3	5.0

Halophenol precursor removal

Conventional systems can lower the DBP formation potential of water prior to disinfection by removing precursors with enhanced coagulation, GAC adsorption, or membrane filtration prior to disinfection. Precursor removal efficiencies are site-specific and vary with different source waters and treatment techniques.

For our trials, waters were sampled on a plant at the inlet of each step and spiked with 1 mg/L with HA. Tests simulating treatments as described above were carried out in our laboratory. TOC, DO 254 nm and pH were measured at the beginning and the end of experiments. Furthermore, each matrix was chlorinated with 4 mg/L NaOCl before and after treatment in order to evaluate the removal of halophenol precursors. Before chlorination pH was adjusted to 7.5. There is not one universal treatment to eliminate NOM. Different treatments where successfully applied to remove halophenol precursors. Comparison between before and after chlorination (Figure 14) underlines the efficacy of the different steps with more than 95% precursor removal (99% for NF and RO membranes). Furthermore, these results suggest that if combined, these treatments would constitute a multibarrier with a greater safety. Alone, PAC for 10 and 20 mg/L dose and 10 minutes contact time show lower performances. These results also show higher removal for halophenol precursors than for THM precursors.

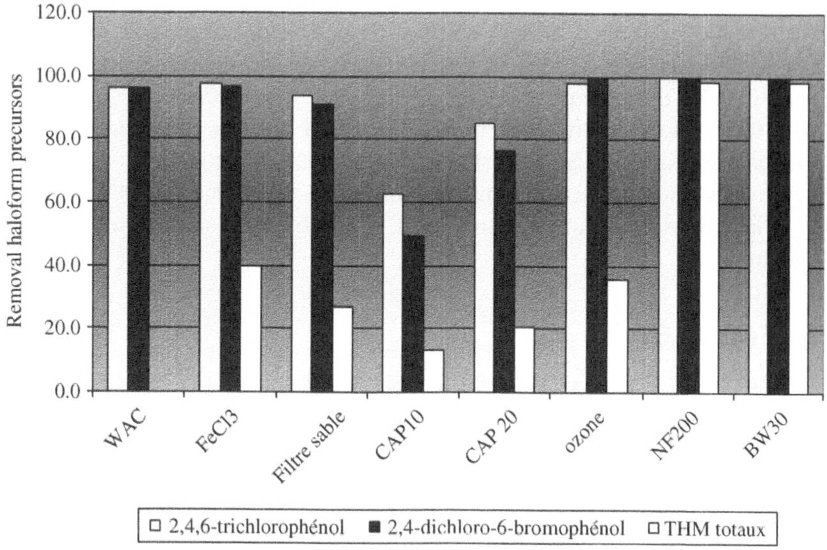

Figure 14. Removal of halophenol precursors on the different steps of a plant.

CONCLUSIONS

For drinking water production, many treatment processes exists that eliminate an important part of the raw water DOC. The ultimate disinfection step using disinfectants as chlorine guarantees the potability of the water from the station to the consumer's tap. However, the chlorination step involving reactions of chlorine with residual DOC could also lead to halophenols, which are involved in tastes and odours. The reactivity of chlorine depends on chlorine speciation as a function of pH. Among the different aqueous chlorine species, hypochlorous acid is the major reagent form during water treatment.

Halogenated phenols formation and speciation are directly linked to the disinfection parameters and physico-chemical properties of the matrix: pH, temperature, bromide concentration, chlorine level and disinfectant nature, contact time between disinfectant and NOM, and NOM concentration. But trials carried out with humic acids, fulvic acids or lignin lead to the same final trihalogenated compounds.

Chlorination of NOM (phenol) forms monochlorophenols or bromophenol if Br is present. Further halogenation leads to di-, then trihalophenol. Tetra and Pentachlorophenol have not been detected in our study. Brominated species (as dibromophenol) known to bring stronger T&O problems than chlorinated molecules can be formed in high levels if bromide ions are present. However, adapted chlorine dose should be applied to avoid the undesired mono- and di-substituted bromophenols.

As suggested by the literature removal of NOM is a key factor in removing halophenol precursors. The next step of this study will particularly be focused on characterization of the different fractions that can be found in NOM (surface water) in order to optimize the treatment of the fractions that are reactive to chlorination.

REFERENCES

Acero, J.L., Piriou, P. and Von Gunten, U. (2005). Kinetics and mechanisms of formation of bromophenols during drinking water chlorination: Assessment of taste and odor development. *Water Research*, **39**, 2979–2993.

Al-Samarrai, H., Matud, J., Wiesenthal, K., Atiyah, P., Bruchet, A. and Suffet, I.H. (2004). Sensory evaluation of the odors produced during bromophenol formation using a multi-level statistical model. *Water Science and Technology,* **49**(9), 241–248.

Amy, G.L., Tan, L. and Davis, M.K. (1991). The effects of ozonation and activated carbon adsorption on trihalomethane speciation, *Water Research,* **25**(2), 191–202.

Benanou, D., Acobas, F., de Roubin, M.R., David, F. and Sandra, P. (2003). Analysis of off-flavors in the aquatic environment by stir bar sorptive extraction – thermal desorption – Capillary/GC/MS/Olfactometry. *Anal Bioanal Chem.*, **376**, 69–77.

Bruchet, A., Prolongeau, C., Esperanza, M. and Coquery, M. (2008). Where do the odorous halogenated phenols in drinking water resources come from? *Water Science and Technology: Water Supply*, **8**(3).

Collins, M.R., Amy, G.L. and Steelink, C. (1986). *Environ. Sci. Technol.*, **20**, 1028–1032.

Corbi, E. (2006). *Potentialité de la technique Stir Bar Sorptive Extraction (SBSE) comme outil d'investigation et de caractérisation des composés odorants dans les eaux potables.* PhD Thesis.

Davidson, R. (1996). The photodegradation of some naturally occurring polymers. *J. Photoc. Photobiol. B*, **33**, 3–25.

Guay, C., Rodriguez, M. and Serodes, J. (2005). Using ozonation and chloramination to reduce the formation of trihalomethanes and haloacetic acids in drinking water. *Desalination*, **176**, 229–240.

Hirose, Y. and Okitsu, T. (1982). Formation of trihalomethanes by reaction of halogenated phenols or halogenated anilines with sodium hypochlorite. *Chemosphere*, **11**(1), 81–87.

Hua, G. and Reckow, D. (2007). A comparison of disinfection byproduct formation from chlorine and alternative disinfectants. *Water Research,* **41**, 1667–1678.

Jacangelo, J.G., DeMarco, J., Owen, D.M. and Randtke, S.J. (1995). Selected processes for removing NOM: an overview. *JAWWA*, **1**, 65–77.

Karlsson, S., Kaduare, S., Grimvall, A., Boren, H. and Savenhead, L. (1995). Formation of 2,4,6-trichlorophenol and 2,4,6-trichloroanisoles during treatment and distribution of drinking water. *Wat. Sci. Tech.*, **31**, 99–103.

Krasner, S.W., McGuire, M.J. and Jacangelo, J.G. (1989). The occurrence of disinfection byproducts in US drinking water. *JAWWA*, **81**(8), 41–53.

Liang, L. and Singer, P.C. (2003). Factors influencing the formation and relative distribution of haloacetic acids and trihalomethanes in drinking water. *Envir. Sci. Techn.*, **37**(13), 2920–8.

Lu, J., Zhang, T., Ma, J. and Chen, Z. (2009). Evaluation of disinfection by-products formation during chlorination and chloramination of dissolved natural organic matter fractions isolated from a filtered river water. *Journal of Hazardous Materials*, **162**, 140–145.

Michalowicz, J., Duda, W. and Stufka-Olczyk, J. (2007). Transformation of phenol, catechol, guaiacol and syringol exposed exposed to sodium hypochlorite. *Chemosphere,* **66**, 657–663.

Nyström, A., Grimvall, A., Krantz-Rülcker, C., Sävenhed, R. and Akerstrand, K. (1992). Drinking water off-flavour caused by 2,4,6 trichloroanisol. *Wat. Sci. Tech.*, **25**(2), 241–251.

Piriou, P., Soulet, C., Acero, J.L., Bruchet, A., Von Gunten, U. and Suffet, I.H. (2007). Understanding medicinal taste and odour formation in drinking waters. *Wat. Sci. Tech.*, **55**(5), 85–94.

Philipot, J.M. and Henriet, X. (1991). La préozonation, une étape de traitement en plein essor. Intérêt économique et technique. Exemple de Choisy-le-Roi. *TSM*, 175–179.

Rizzo, L., Belgiorno, V., Gallo, M. and Meriç, S. (2005). Removal of THM precursors from a high-alkaline surface water by enhanced coagulation and behaviour of THMFP toxicity on D. magna. *Desalination*, **176**, 177–188.

Roche, P., Volk, C., Carbonnier, F. and Paillard, H. (1994). Water oxidation by ozone or ozone/hydrogen peroxide using the "OZOTEST" or "PEROXOTEST" methods. O.S.E., Vol. 16, pp. 135–155.

Suty, H., Pouillot, M., Karpel, Vel leitner, de Laat J. and Doré, M. (1995). Traitement des eaux potables par le bioxide de chlore: incidence des traitements de potabilisation sur la concentration en ions chlorite et chlorate, *Conférence 11, JIE Poitiers*, 1–16.

Taylor, J.S., Thompson, D.M. and Carswell, J.K. (1987). Applying membrane processes to groundwater sources for trihalomethane precursor control, *JAWWA*, **79**(8), 72–82.

Tondelier, C., Thouvenot, T, Genin, A. and Benanou, D. (2009). On-tap passive enrichment, a new way to investigate off-flavor episodes in drinking water. *J. Chromatogr. A*, **1216**, 2854–2859.

Uyak, V. and Toroz, I. (2007). Investigation of bromide ion effects on disinfection by-products formation and speciation in an Istanbul water supply. *J. Hazardous Materials,* **149,** 445–451.

Werdehoff, K.S. and Singer, P.C. (1987). Chlorine dioxide effects on THMFP, TOXFP and the formation of inorganic by-products. *J. AWWA*, **79**(9), 107.

Whitfield, F.B., McBride, R.L. and Nguygen, T.H. (1987). Flavour perception of chloroanisoles in water and selected processed foods. *J. Sci. Food Agric.*, **40**, 357–365.

Young, W.F., Horth, H., Crane, R, Ogden, T. and Arnott, M. (1996). Taste and odour threshold concentrations of potential potable water contaminants. *Wat. Res.*, **30**(2), 331–340.

Embedding water quality changes in hydraulic modelling – the Techneau WP5.5 water quality modelling platform

S.T. Coelho, J. Menaia and H. Alegre

LNEC, Av. Brasil, 101 – 1700-066 Lisboa, Portugal (E-mail: stcoelho2@gamil.com; jmenaia@lnec.pt; halegre@lnec.pt)

Abstract The Techneau WP5.5 Water Quality Model (Coelho, 2008) is a water quality modeling platform based on an event-driven, hydrodynamic water supply network model. It provides a basis for the integration of the results of Techneau research being carried out into specific physical-chemical and microbiological processes taking place in the water and its environment, the network (Wricke *et al.*, 2007).

Physical-chemical and microbiological phenomena are developed as individual process models and incorporated into the water quality modelling framework as they become available. The processes being developed under Techneau research include particle movement and deposition, corrosion, and a certain number of microbiological processes related to biofilm regrowth, in addition to some of the most important background chemical processes such as chlorine decay. The process model development cycle is summarised in Figure 1.

The program is written in the Microsoft .NET framework, using mostly the C# programming language and the Epanet 2.0 network model library (Rossman, 2000) as a standard for hydraulic model specification. It uses Epanet –validated hydraulic model files as input. It includes its own hydrodynamic model and makes available to the user the Epanet MSX multi-species water quality programming library (Chang *et al.*, 2008). This allows the user to specify any water quality process at run time, with full flexibility to change and adjust the model at will and re-test as needed.

The use of the Techneau WQM requires a solid understanding of hydraulic modeling and the ability to prepare error-free hydraulic models using Epanet 2.0, as input files to the program.

The platform is also available to incorporate specific, *hard-wired* modelling formulations, particularly in the cases of more complex models that have been stabilized into a specific format for application by end-users.

The software provides a graphical and numerical user interface for result analysis (Figure 2). A network mapping facility, for spatial visualization of results, is under development.

Edited by Theo van den Hoven and Christian Kazner. ISBN: 9781843392750. Published by IWA Publishing, London, UK.

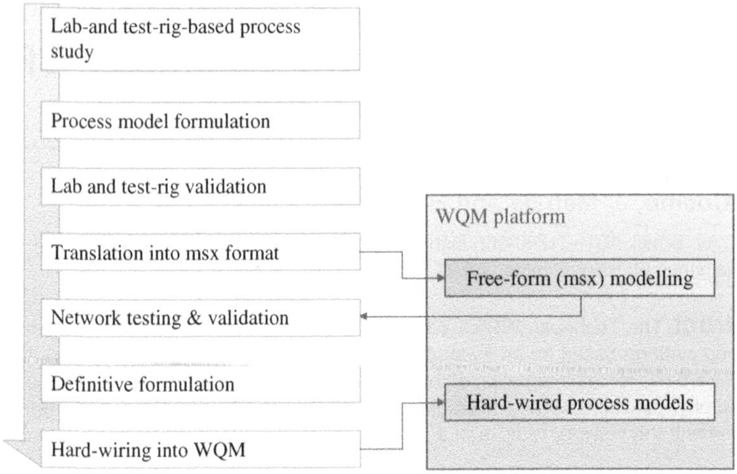

Figure 1. Process model development cycle.

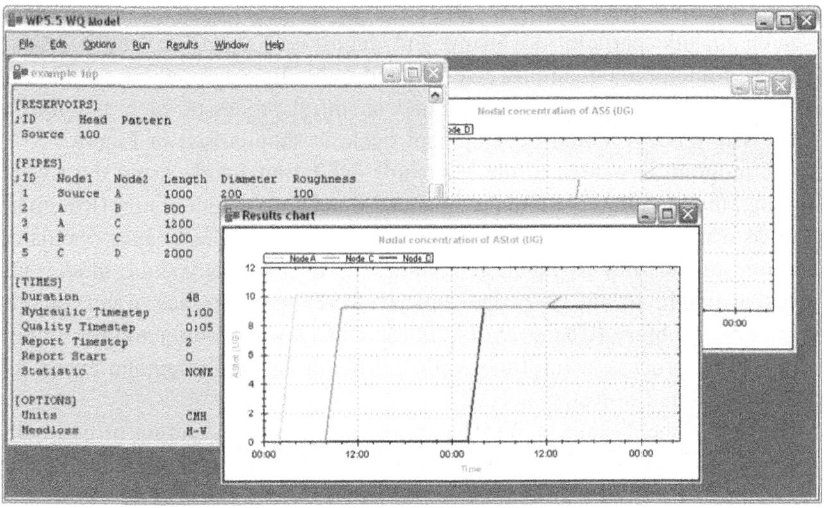

Figure 2. The Techneau WP5.5 Water Quality Model workspace.

At present, the WQM framework is available as a hydrodynamic model and includes one experimental process module, which simulates bacterial regrowth in the pipes. Other modules are expected to become available as a result of techneau WP5 research, namely to model corrosion and particle movement phenomena.

REFERENCES

Chang, F., Uber, J. and Rossman, L. (2008). Epanet *Multi-Species Extension User's Manual*, National Risk Management Research Laboratory, U.S. Environmental Protection Agency (http://www.epa.gov/nrmrl/wswrd/dw/epanet.html#extension).

Coelho, S.T. (2008). *Techneau WP 5.5 Water Quality Model – User's Manual, Release 1.1.0.2*. Techneau (www.techneau.org).

Rossman, L. (2000). Epanet user's manual, EPA/600/R-00/057, National Risk Management Research Laboratory, U.S. Environmental Protection Agency, September 2000.

Wricke, B., Henning, L., Korth, A., Vreeburg, J., Schaap, P., Osterhus, S., Juhna, T., Hammes, F. and Coelho, S. (2007). *Particles in Relation to Water Quality Deterioration and Problems in the Network*. Techneau Report D551, Techneau (www.techneau.org).

Small Scale Systems

Small-Scale Systems (SSS) for decentralised water supply and relevance of membrane technologies

B. Lesjean[a], W. Pronk[b] and E. Hoa[a]

[a]Berlin Centre of Competence for Water, Cicerost. 24, 10709 Berlin, Germany
(E-mail: boris.lesjean@kompetenz-wasser.de; eric.hoa@kompetenz-wasser.de)
[b]Eawag, Swiss Federal Institute of Aquatic Science and Technology, Überlandstr. 133, 8600 Dübendorf, Switzerland (E-mail: wouter.pronk@eawag.ch)

Abstract With respect to meeting the Millennium Development Goals (MDG), decentralised water supply stands as one major challenge in developing countries but also in developed areas (e.g. Europe). In order to raise awareness and provide advocacy for management and technological issues, WHO created two related networks whereas the EU Project TECHNEAU initiated the "Small-Scale System" (3S) Task Force. Membrane processes are identified as a promising technology to ensure a safe water supply at the community and the household levels. As shown in a survey performed in 2007, the market of membrane-based systems for decentralised applications has been growing and more standardised products are expected to appear within a few years. As the price of membranes has significantly decreased over the last years, such systems could be broadly applied in developed countries, but also for transition and developing countries. However, commercial point-of-use (POU) and small-scale systems (SSS) are not designed for long-term operation with low-energy and low-chemical requirements. Therefore, the R&D activities identified within TECHNEAU aims at responding this challenge.

Keywords decentralised water supply, low-energy, membrane technology, point-of-use, small-scale systems

INTRODUCTION

Universal access to safe water stands as one of the main priorities of the next decade. Indeed, the UN Millennium Declaration (UN, 2000) sets up the goal of *halving by 2015 the proportion of people without sustainable access to drinking water and basic sanitation* (target 10). As the construction and operation of water distribution networks are not technically or economically viable in many regions of the world, the use of decentralised small water systems has been developing. Point-of-Use (POU) technologies, which imply a drinking/cooking consumption of 20–200 L/d, have entered households and buildings (4–50 people) whereas Small-Scale Systems (for a full water use of 5–1000 m^3/d) have been installed in isolated communities (50–5000 people) (see Figure 1).

 TECHNEAU: Safe Drinking Water from Source to Tap. Edited by Theo van den Hoven and Christian Kazner. ISBN: 9781843392750. Published by IWA Publishing, London, UK.

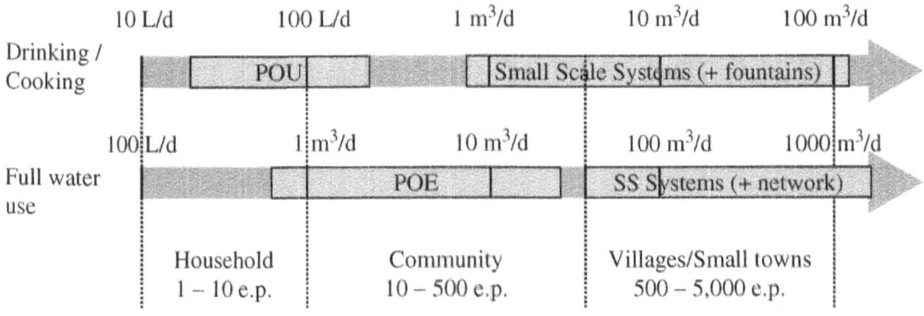

Figure 1. Ranges of the water facilities for small supply (based on Hulsmann, 2005; WHO, 2007).

GLOBAL PICTURE OF DECENTRALISED WATER SUPPLY

Reaching the Millennium Development Goals with small-scale systems

In 2005, 1.1 billion people lacked access to safe drinking water. That mostly explains the 4 billion cases of diarrhoea that occur every year and that cause 1.8 million deaths. Children under 5 are particularly affected by waterborne diseases which harm their nutrient uptakes and their growths (WHO, 2007). WHO estimates that 94% of those cases could be prevented with appropriate interventions to provide clean water and to improve sanitation and hygiene. Figure 2 shows the percentage of population who can use improved drinking water sources.

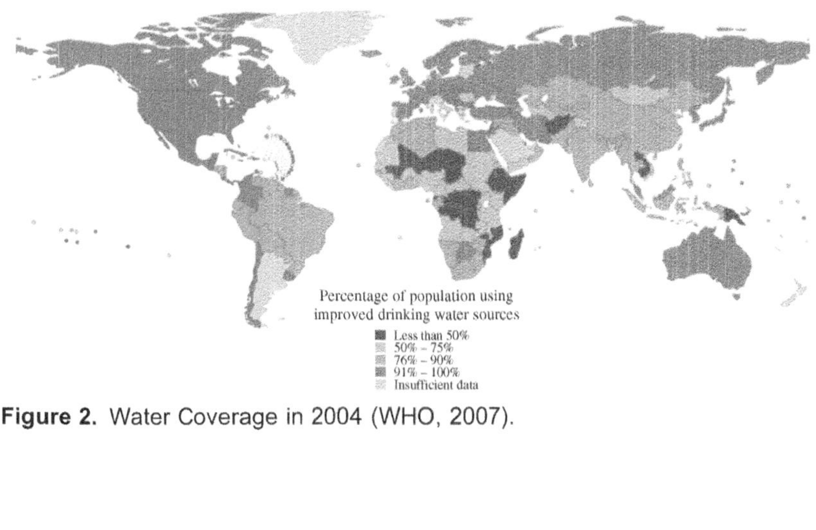

Figure 2. Water Coverage in 2004 (WHO, 2007).

Thus, when centralised water treatment systems are absent or inadequate, populations have to find their drinking water on their own. In many parts of the developing world, they depend on rivers, streams or other potentially contaminated surface sources. Moreover, the collected water is often held in unsafe storage facilities. That leads to an unsatisfying drinking water quality. Boiling the water can be used as a prevention measure but it cannot be considered as a sustainable and economical treatment method. The lack of safe water is often part of a problematic knot where populations suffer from poverty and live in precarious conditions. Thus, it is important to find simple and low-cost techniques to treat and store the water and by doing so, enhance health and contribute to better living conditions, development and productivity.

Working at the household and community levels enable targeted and immediate effects in order to reduce the risk of contaminated water. Self-sustaining and decentralized water schemes are clearly identified to be a priority for rapid implementation at global scale. Recent investigations reveal that improved point-of-use water quality alone would halve the diarrhoeal morbidity (WHO, 2007).

Overview in Europe

Within the framework of the project WEKNOW, a survey on (very) small water supplies in Europe was completed in 2003. Regarding the situation for 21 countries, at least 10% of the EU population receives their daily drinking water from small (<1000 m^3/d) and very small supplies (<10 m^3/d). The major problem with such water supplies is linked with microbial contamination (E.Coli, coliforms and faecal streptococci), especially in rural areas. As shown in Table 1, EUR-B subregion is particularly affected by diarrhoea and DALYs (Disability–adjusted life year). However, those are only estimated values: little precise information about the number of such supplies (especially private wells) and their users is available as very small supplies are not covered by national legislation and are not monitored. A better water safety approach within a legal framework is needed and it should be better addressed in the coming version of the European drinking water directive (Hoekstra, 2008).

WHO Networks and TECHNEAU 3S Task Force

With this emerging key issue, a clear call for time, money and intelligent capacities is highlighted. A WHO platform dedicated to small community water supply (SCWS, scwsm@who.int) was therefore set up to promote and accelerate improvements to the safety of SCWS via advocacy, the identification of research

Table 1. Estimates of rural supply basic data in Europe sub-regions (Hunter *et al.*, 2009).

	EUR-A	EUR-B	EUR-C
Total population (1000s)	428,195	216,039	236,976
Total rural population (1000s)	95,470	95,413	71,778
Estimated population relying on small community managed supply (1000s)	42,800	21,600	23,700
Unimproved Supply (1000s)	70	22,557	8,198
No tap in home (1000s)	3,101	42,701	36,927
Total mortality from diarrhoea (1000s)	2.35	12.95	1.44
Estimated deaths from diarrhoea per 100,000 population	0.55	6	0.61
Estimated total DALYs from diarrhoea (1000s)	114	492	96
Estimated total DALYs from diarrhoea per 100,000 population	27	228	40

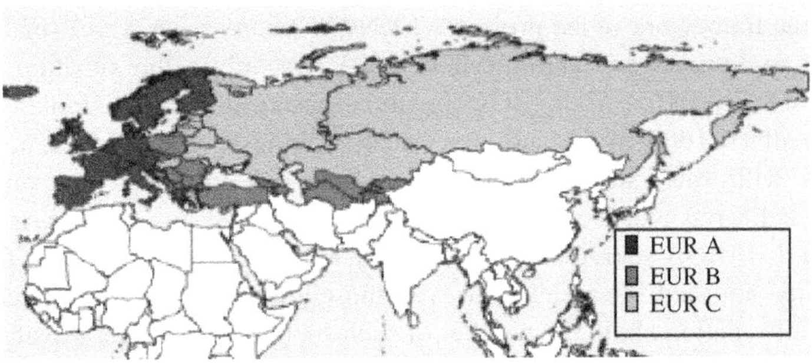

needs and the development of communication tools. Its objective is to provide strategic advice and recommendations to WHO, from a practitioner perspective and prior to and after implementation of guidelines and guidance on good practices. Thus, it aims at a better framework for safe drinking water with defined health-based targets, an appropriate water safety plan and independent surveillance.

WHO also funded another Network which focuses on household water treatment and safe storage (HWTS, hhwater@who.int). It has more than 120 member institutions – including NGOs (45%). Similarly to the SCWS network, its activities are based on advocacy, communication, research and implementation. It also collaborates with UNICEF in order to organize meetings and develop HWTS programs in priority countries (see Figure 3).

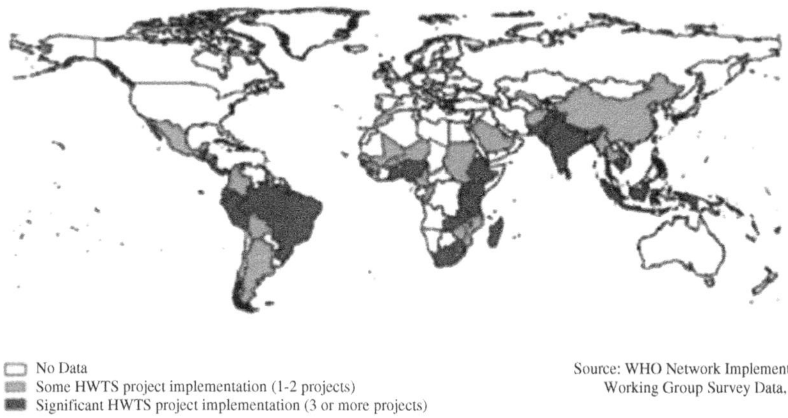

Figure 3. HWTS Project Implementation in 2005.

Within the framework of the project TECHNEAU, the Berlin Centre of Competence for Water (KWB) created the Techneau "Small Scale Systems" (3S) Task Force in order to enhance an effective integration and planning of the activities related to small-scale systems and decentralised water supply and to foster the positioning of TECHNEAU within the global picture. It currently involves more than 80 members, including, among others, representatives from TECHNEAU, WHO or the IWA group "Small-scale water and wastewater treatment". Around 30 scientists and professionals from Europe and South Africa attended annual workshops which were dedicated to the topics "Membrane Processes" (2007) and "International Networks and Monitoring and Risk Assessment for Small Communities" (2008). Minutes from those events are available on the Techneau website (www.techneau.eu).

MEMBRANE PROCESSES AS VIABLE SOLUTIONS TO DECENTRALISED WATER SUPPLY

Membrane responds to the decentralised needs

While POU technologies are widely used in developed countries as an additional treatment of tap water (EPA, 2006), small membrane-based systems stand also as alternative solutions for rural areas and in developing and transition countries where a centralised treatment would not be affordable. Indeed, membrane processes present many advantages: (i) a membrane is an absolute barrier which can guarantee the elimination of many components (e.g. high degree of pathogen reduction) through physical elimination, (ii) the production cost of membrane elements decreases significantly year after year – China now produces membranes at 10 USD/m^2, and (iii) the process appears flexible (with modular construction) and simple, as it only consists in one step of treatment (Peter-Verbanets *et al.*, 2009). An important development of those systems could then be expected in a close future with the apparition of competitive commercial offers.

Yet, in developing countries, membrane-based systems represent a pricey capital investment and they may face some specific regional problems regarding the quality of the raw water (e.g. nitrate, arsenic). Moreover, operation and maintenance may be inappropriate and fail to be performed by the local populations. In developing countries, such membrane-based systems should be designed to require as little energy and chemicals and maintenance as possible.

Membrane integrity

The membrane remains a barrier as long as the integrity is ensured (no leak or breakage). For long term operation (i.e. not necessarily for emergencies), it is observed that the integrity test is a necessity. If not available, it becomes necessary to add a second barrier, such as UV disinfection. It is observed that the system becomes then more complex and costly. In developing and transition countries, chemical disinfection such as chlorination or chloramination could be a cost effective post-treatment option. In any case, the performance of small membrane-based systems will rely upon appropriate and regular maintenance.

This observation raised a fundamental question: if a very high level of performance and quality is target, there is a risk of designing a system too complex and too expensive for sustainable application in developing countries. In other words, an issue is raised whether a less sophisticated system, with a lower level of control and warranty would be a good compromise for developing countries, to facilitate its use and acceptance in many areas. On short and medium term views, such solutions might be possible, if they lead to a considerable improvement of the current situation at reasonable costs. However, on long term view the cost of these technologies shall decrease, and the technology would continue to improve and get more reliable enabling high performance systems.

Considered applications fields

The difficulty to design a system adapted to each and every situation is also raised. An emergency case for instance will not demand the same level of performance that a long term system. Moreover, for a small scale system, an operator can be nominated from the community to take care of the system. This operator could be accepted and may be paid by the community. For a point of use system, each user is responsible for his system.

Respond to local needs

In addition to the above mentioned specificities, a regional specificity shall be added. For instance, in Eastern Europe, industrial and agricultural pollution generates a large part of the contamination of raw waters whereas in Madagascar (for example), natural organic pollution is more frequent. For the first case, elimination of arsenic or nitrate with ultrafiltration (UF) membrane alone is not feasible: a reverse osmosis membrane shall be used, which cannot be considered taking into account the high pressure required to feed such membranes, or additional physical-chemical treatment steps. For the second case, a pre-treatment may be required to prevent high fouling of the membrane, for instance a biosand filtration, or a carbon filter. The level of saturation of the carbon may then be monitored, which is difficult to consider in remote small scale systems in developing countries.

The autonomy in terms of energy is mentioned as a potential criterion to design the membrane system. Solar energy is a first solution, but a specific design shall be considered to avoid the peak power demand for the membrane backwash. A solution could be to slowly pressurize a vessel during the filtration phase, and to use this pressure for the backwash energy. Solar energy is also mentioned for the disinfection of the treated water. Wind energy is also a possibility, with the same kind of limitation in terms of maximum power delivered.

Market survey

A market survey for membrane-based compact systems for drinking water applications has been performed in 2007 within Techneau (Hoa and Lesjean, 2008) in order to identify the different market niches via Internet research and form requests to membrane companies during a 4 month-period. It is presumed that a part of the market was not consulted because the companies do not advertise on Internet and/or in English. Therefore, it cannot be regarded as an exhaustive survey but a good basis for development considerations.

Geographical market distribution

The market for membrane systems for small water supply seems mature in the US as many companies (36 identified companies) compete. In this large country, small-scale solutions have found customer recognition. Besides, the Asian market is significant with 22 companies identified in China and 8 in India. Due to their demographies and their development disparities, those countries face strong environmental issues which explain their interests for small decentralised water systems. Europe is also well-represented in the market for membrane-based systems for small supply. Germany (22) and the Netherlands (8) are particularly active in the water technology. Indeed, the populations are there known to be environmentally concerned and ensuring access to safe water stands as an essential right. Thus, the socio-economical background is here pointed out.

Current commercial applications

The survey highlighted different market niches (see Figure 4) which include: ceramic POU, organic POU, organic point-of-entries (POE), modular treatment units and emergency systems. The review of the marketed membrane modules reveals that ultrafiltration is currently the most available process. Besides, the survey also shows that the pre-treatment is a key parameter when considering options for decentralised water supply.

Figure 4. Identified market niches: a) ceramic POU: Clay pot by Potter for Peace; b) organic POU: Lifesaver® bottle; c) modular treatment unit: UFlex® by Krüger/Veolia; d) emergency system: SkyHydrant™ by Skyjuice Foundation.

Market development

The membrane market for water supply and sanitation is large and many products are commercially available. Yet, when looking more specifically at small units for drinking water uses, the market is still at a growing phase, and most of developments occurred in the past 3–5 years. Thus, many companies would claim that their membrane modules are effective for treating the water a small scale although the equipment was not directly designed or tested for that purpose. Companies adopt fast development strategies to fill empty market niches in regions where a socio-economical boom is expected. The market will be mature once few products make their proofs on the field site and start being standardised and produced in large quantities. Given the observed speed of development, this may turn soon into reality, and membrane technologies may become in the coming years one cost-effective option to address the issue of decentralised water supply in developed countries, but also in transition and developing countries.

The membrane market evolution for small systems can be summarized in Figure 5. Depending on the product niche, the membrane material and the filtration type (Microfiltration (MF), Ultrafiltration (UF), Nanofiltration (NF) and Reverse Osmosis (RO)), different degrees toward the market maturity were identified.

	Organic		Ceramic
POU	RO	UF	MF
POE	UF		/
Emergency Systems	MF/UF		MF
Small-scale Systems	MF/UF	NF/RO	MF/UF/NF

Established Market   Recent Market   Emerging Market

Figure 5. Market Overview for membrane-based systems regarding small water supply.

As only few simple UF POU and low-energy small-scale systems were identified, the EU research project TECHNEAU (www.techneau.eu) decided to contribute to the technological breakthrough with the development and the demonstration of new UF decentralised systems. Equipped with a biosand pre-filter, the gravity-driven prototypes are designed to be robust and run with limited energy supply (only to bring the water up to a 2 m height) and restricted chemical intervention (only chlorine for final disinfection and residual).

CONCLUSIONS

Decentralised water supply is a major issue on a worldwide scale and Europe can make significant contributions to its appropriate management and technological development. Indeed, there is a clear need to raise public and authorities awareness, to develop a suitable framework in order to manage and operate decentralised schemes and to design simple and low-cost systems with easy operation and maintenance. In that regard, membrane processes stands as a promising technology with a related market which is expected to keep growing.

ACKNOWLEDGEMENTS

This study was performed within the 6th framework of the European project "TECHNEAU" (Contract No 018320, www.techneau.eu). The authors would like to acknowledge the European Commission and Veolia Water for their financial contributions.

REFERENCES

EPA (2006). *Point-of-Use or Point-of-Entry Treatment Options for Small Drinking Water Systems*, EPA.

Hoa E. and Lesjean, B. (2008). *International Market Survey on Membrane-based Products for Decentralised Water supply (POU and SSS Units)*. EU Project TECHNEAU Report. (www.techneau.eu). KWB.

Hoekstra, E. (2008). *Status of the revision process of the Drinking Water Directive – sampling and monitoring*. Presentation 3S Task Force Workshop in Tabor in June 5th, JRC European Commission.

Hulsmann, A. (2005). *Small Systems Large Problems – A European inventory of small water systems and associated problems*, WEKNOW.

Hunter, P.R., Pond, K., Jagals, P. and Cameron, J. (2009). An assessment of the costs and benefits of interventions aimed at improving rural community water supplied in developed countries, *Sci. Total Environ* (2009), doi:10.1016/j.scitotenv2009.03.13

Peter-Varbanets, M., Zurbrügg, C., Swartz, C. and Pronk, W. (2009). Review: decentralized systems for potable water and the potential of membrane technology. *Water Research*, **43**(2), 2009, 245–265.

UN General Assembly. (2000). *Resolution 55/2 United Nations Millennium Declaration*, 8th Plenary Meeting, September 8.

WHO (2007). Combating Waterborne Disease at the Household Level, International Network to Promote Household.

Low-pressure UF and membrane fouling by polysaccharides

M. Peter-Varbanets[a,b], K. Borkmann[a] and W. Pronk[a]

[a]Eawag, Swiss Federal Institute of Aquatic Science and Technology, Überlandstrasse 133, 8600 Dübendorf, Switzerland (E-mail: maryna.peter@eawag.ch; wouter.pronk@eawag.ch)
[b]Institute of Environmental Engineering, ETH Zürich, 8093 Zürich, Switzerland

Abstract Low-pressure membrane systems have good perspectives for application in decentralized drinking water treatment. Fouling takes place but can be stabilized in dead-end mode at low pressures without membrane flushing or cleaning. In order to optimize such systems, it is important to investigate the fouling mechanisms. Polysaccharides are known to be relevant to fouling, but not much is known about the influence of polysaccharide structure and properties on fouling. Therefore, we systematically investigated the impact of polysaccharide and solution properties on UF membrane fouling. Regarding the initial stage of flux decline the polysaccharide structure, and particularly the availability of carboxyl groups, has a major impact on membrane fouling, while the molecular weights of polysaccharides does not play a significant role (in the studied range of Mw 5–250 kDa). The presence of metal ions and ionic strength also influences the fouling, which can be explained by the fact that both metal ions and ionic strength stabilize gel structure, and increasing the possibility of water trapping by hydrogen bonding, which leads to the higher permeability. However, independently of the initial solution conditions, flux stabilization on the level of approximately 10 $L \cdot h^{-1} \cdot m^{-2}$ was observed at higher permeate volumes. We suppose that the gel layer formed by polysaccharides play a role of a "second" membrane on the surface of the PES UF membrane, keeping permeability on the constant level, determined by the water retention properties of the gel structure.

INTRODUCTION

Access to microbiologically and chemically safe water is limited not only in developing countries, but also in transition countries and even in remote areas of developed countries (WHO, 2004). For these cases, decentralized water supply concepts such as point-of-use (POU), point-of-entry (POE) or small-scale system (SSS) technologies can be alternatives to centralized treatment concepts. In an extensive literature review (Peter-Varbanets *et al.*, 2009a) we showed that the choice of reliable and affordable systems for decentralized application is limited, and that membrane-based treatment systems provide opportunities. Furthermore, experiments were carried out with low-pressure, dead-end ultra-filtration without pretreatment and without using pumps, backflushing or cleaning (Peter-Varbanets *et al.*, 2009b). It was shown that such systems can be operated during extended periods of time (up to 5 months) with stable flux

Edited by Theo van den Hoven and Christian Kazner. ISBN: 9781843392750. Published by IWA Publishing, London, UK.

values using untreated river water and even using diluted waste water. However, also in these systems the flux is limited by membrane fouling and therefore, it is important to investigate the basic mechanisms of membrane fouling. Besides humic substances (Clark and Lucas, 1998), polysaccharides are recognized to be an important component of natural waters which cause membrane fouling (Kimura *et al.*, 2004), and it has been found that electrochemical interactions play an important role in the fouling process (Amy and Cho, 1999). However, relatively little knowledge is available on the influence of polysaccharide structure, composition and properties on fouling. In this paper, UF fouling by polysaccharides with different molecular weights and functional groups is investigated at varying water matrix compositions (ionic strength, metal ions). The data are interpreted and the consequences for decentralized water treatment are discussed.

METHODS

Several polysaccharides have been used in the study. Their main properties are shown in Table 1. The selection of these polysaccharides was based on the molecular structure of the polysaccharide, availability of the specific functional groups and the molecular weight.

Table 1. Characteristics of the polysaccharides.

Polysaccharide	Specific groups	Molecular weights of polymer, kDa	Molecular weight of single unit (approx.)
Alginate	$-COONa$	12–80	175
Hyaluronate	$-COONa$ $-NH\text{-}CO\text{-}CH_3$	68–340	189
Pectin	$-COONa$ $-COOCH_3$	30–100	158.5
Heparin	$-NH\text{-}SO_3Na$ $-O\text{-}SO_3Na$	4–6	195
N-acetylated-de-O-sulphated heparin	$-NH\text{-}CO\text{-}CH_3$ $-NH\text{-}SO_3Na$	12–15	184
Carboxymethylcellulose	$-O\text{-}CH_2\text{-}COONa$	250	189
Dextran	–	5 50 670	162
Starch	–		162

All reagents were obtained from Sigma-Aldrich and were of following characteristics: Carboxymethylcellulose sodium salt, medium viscosity, degree of substitution 0.6–0.95; Alginic acid, Sodium Salt, produced by brown algae in natural water; Dextran Standard, 5000, 50000, 630000, certified according to DIN for GPC; Hyaluronic acid, sodium salt, from human umbilical cords; Pectin from apples, degree of esterification 0.7–0.75; Heparin sodium salt, from porcine intestinal mucosa; Starch soluble.

In order to find a comparable concentration unit for polasaccharides with different chain lengths and different compositions, the same concentration of sugar-subunits was used for all solutions (0.002 mM). When a polasaccharide contained different types of sugars, the average molar mass of these sugars was taken. The molar concentration of the metal ions were determined in molar ratios to the polysaccharide concentrations and later plotted in the same way. The background solution for all experiments consisted of deionized water, NaCl (0.5 mM), $NaHCO_3$ (0.125 mM) and a buffer solution of 3-(N-Morpholino)-propanesulfonic acid (MOPS, 0.5 mM). The pH was adjusted by addition of HCl (0.1 M) or NaOH (0.1 M) and was kept constant at pH 7.6–7.8. The ionic strength was controlled by conductivity measurements and when necessary adjusted with NaCl.

UF membranes

We used a fresh flat sheet 100 kDa polyethersulfone membrane (Biomax, Millipore) for each experiment. Beforehand the membranes were stored for min. 24 h in deionized water, which has been changed three times, and then filtered with 5 L of deionized water at a transmembrane pressure of 150 mbar. This membrane pretreatment procedure was preliminary developed to remove conservation substances from the membrane. The efficiency of this procedure was assessed by DOC measurements of the deionized water permeate in the LC-OCD. The cleaning procedure was considered as prerequisite to avoid contamination of subsequent samples by membrane leaching and to perform reliable experiments with low NOM concentrations similar to natural water.

Membrane test units

The membrane test unit was a 400 ml stirring cell with removed stirrer (Amicon). The level in the stirring cell was maintained constant on 250 ml by a feed gear pump connected to a glass reservoir (10 L). The pressure in the cell was kept constant at 150 $\pm$ 5 mbar by a pressure control device, which steered the feed pump. The permeate flow was measured by a scale, which was connected to a PC. Data were logged every 2 min.

Fouling experiments

We performed all experiments at room temperature and constant transmembrane pressure (0.15 bar). Previous to every experiment the flux of the clean membrane was measured. The short term experiments consisted of three cycles. One cycle included filtration of 1 L of synthetic solution, backwash with 50 mL deionized water, filtration of 100 mL deionized water (used to determine the reversibility). The long-term experiments consisted of one cycle. At least 2 L of solution was filtered through the membrane and the flux was monitored as described above.

RESULTS AND DISCUSSION

Impact of different polysaccharides on the membrane fouling

The flux values of experiments with alginate, hyaluronate, carboxymethylcellulose and dextran of different molecular weights in the solutions of similar matrix on membrane fouling are depicted in Figure 1. As shown in Figure 1(a), the flux decreased much more severely during filtration of alginate, Hyaluronate (HA) and Carboxymethylcellulose (CMC) in comparison to Dextran independently of its MW. Relatively small differences in flux decline can be observed for Dextrans with different molecular weights. This result was not expected, considering the fact that membrane cut-off is equal to 100 kDa. Dextrans of 5 and 50 kDa were expected to pass through the membrane and lead to less fouling than the 630 kDa Dextran, which is completely retained.

In case of alginate, HA and CMC similar flux decline patterns were observed. Also the reversibility of fouling for these polymers was lower than in case of Dextran. The similarity of the flux decline should indicate the similarity of fouling mechanisms, and its difference to the case of Dextran. As can be see from Table 1, alginate, HA and CMC are independently of their structure negatively charged polysaccharides, while Dextran is neutral. This indicates the importance of electrostatic interactions in the membrane fouling mechanism.

However, as shown in Figure 1(b), the flux decreased for the same polysaccharides differently in the presence of Ca cations. While in case of Dextran, the difference is negligible, the charged polysaccharides react differently on the presence of Ca. In Figure 1(c), the influence of Pectin, Heparin, N-acetylated heparin and Starch in Ca containing solutions on flux is shown. Behavior of non charged starch is similar to one of dextran and pectin causes the similar fouling as alginate, however both Heparins have different impact. This difference may be caused by two factors – the lower molecular weights of both substances (12–15 kDa), and absence of the carboxyl functional group, which is present in all other charged polysaccharides considered in this study. While the

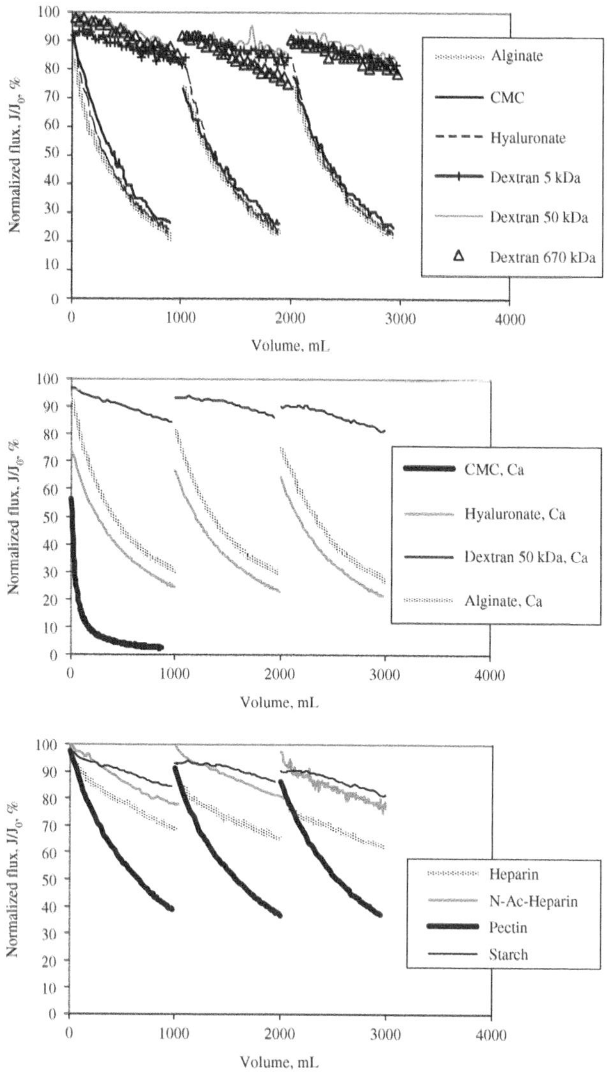

Figure 1. Impact of different polysaccharides on the flux decline: (a) polysaccharides in solutions without Ca; (b, c) in solutions with Ca; Polysaccharide/Ca = 1/1.

impact of the molecular weights on the fouling in case of Dextran has proven to be insignificant, we can assume that the presence of charged carboxyl group plays a significant role in fouling. The fact that alginate causes more severe flux decline than pectin, which has a similar structure, but partially methylated carboxyl groups, also supports this assumption. Summarizing the results and discussion above, we can conclude that the presence and availability of charged carboxyl groups and presence of Ca ions have a severe impact on the flux decline, while the molecular weights of the polysaccharides does not play a significant role in the conditions studied.

The role of metal ions

As we can see from the Figure 1(b) and 1(c), Ca plays an important role, and the character of the flux curves differs for different polysaccharides. It is generally accepted that metal ions are able to form networks with polysaccharides containing carboxyl groups through metal bridge formation and interlinking of polymers between each other, thus achieving organized structures. With increasing concentrations of polysaccharide and metals, gel structures may be formed. Alginate shows specific ion binding characteristics and the affinity for alkaline earth metals increases in the order $Mg \ll Ca < Sr < Ba$.

Thibault (Thibault and Rinaudo, 1985) found that under Na and Mg salt forms, the polymers behave as single chain polyelectrolytes, while with Ca, Sr and Ba a double chain structure is stabilized, also known as an egg-box model (Bryce *et al.*, 1975). It was found by these authors that with Na and Mg the polymers are water soluble while Ca, Ba and Sr are known to induce the gelation with an ionic selectivity corresponding to $Ba > Sr > Ca$ for the ability to form a gel (Rinaudo, 2006). In natural systems gel formation is considered as one of the mechanisms of membrane fouling. Also in studies of membrane fouling by alginate, the gel formation on the membrane surface has been considered as the main reason of flux decline (Jermann *et al.*, 2007).

Figure 2 indicates that fouling is more severe in case of alginate with Na and Mg ions, than the system with Ca and Ba in the same concentrations.

As shown in Figure 2, the fouling is more severe in the presence of Mg and Na than in the presence of Ca and Ba, which are able to form gels. It was mentioned before that in the presence of Mg and Na, alginate has the behavior of a single chain polysaccharide and therefore is not able to form organized networks and gels, while in the presence of Ca and Ba gels are formed. This can be explained by the fact that Ca/Ba containing gel layers are more hydrated, with water molecules trapped by hydrogen bonding in the gel matrix (de Kerchove and Elimelech, 2006). The higher rate of hydration leads to a higher permeability of the gel layer, and therefore to a higher flux.

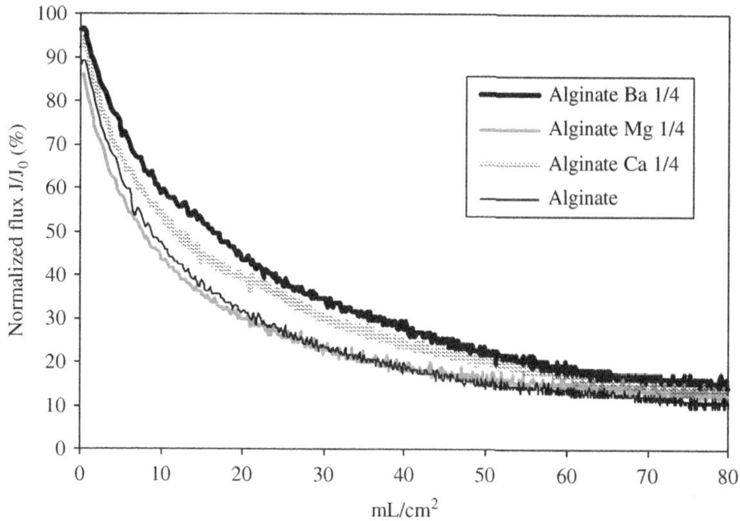

Figure 2. Impact of Ca, Mg, Ba and Na on flux in the presence of alginate.

Iron containing systems

Iron is often present in natural waters. The property of Fe(II) and Fe(III) ions to form complex substances with NOM and its fractions are well known. Also polysaccharides containing carboxyl functional groups are able to interact with Fe(III). Many studies have shown that Iron is one of the ions which is partly retained by the UF membranes and is often detected in the matrix of fouling layers (Xia *et al.*, 2004; Choo *et al.*, 2005). Therefore, it is necessary to consider the role of Fe ions in membrane fouling by polysaccharides.

We studied the impact of Fe(III) on the flux decline in alginate containing systems. While Fe(III) at the pH conditions of the experiments may form colloidal solution and therefore is retained by the membrane, also the impact of Fe(III) without alginate was investigated. The solutions were prepared in two different ways: preparing an acidic Fe(III) solutions and adding this to the alginate solution, and by slow oxidation of Fe(II) by increasing the pH in a system already containing alginate. The reason for that was to insure the complete reaction of Fe(III) with alginate. In the case where Fe(III) has been directly added to the system containing alginate, fast increase of pH may lead to formation of colloidal $Fe(OH)_3$ rather than to formation of Fe-alginate complexes, because the reactions with macromolecules are known to be slow. Therefore, slow oxidation of Fe(II), which does not react with alginate, to Fe(III) by air and slow pH increase ensures complete reaction and formation of the complex. However, in both cases we used an excess of Fe, and therefore, except of the Fe-alginate compound, colloidal iron hydroxides were also formed in the system.

Figure 3 represents the flux values of the experiments with Fe(III), with and without alginate. As shown in Figure 3, in both experiments without alginate under studied conditions (Fe concentration – 0.008 mM, pH 7.7) the flux is not influenced by the way the Fe(III) was prepared. There was no iron detected in the permeate in both cases, which meant that all iron was retained by the membrane, independently from the way the Fe(III) colloids have been obtained. Both systems containing alginate and Fe(III) showed significant flux decline, comparable to the system containing alginate and Ca. This indicates a similar mechanisms of fouling with Fe(III) as with Ca and Ba, suggesting that in the presence of Fe(III) structured polysaccharides complexes are formed.

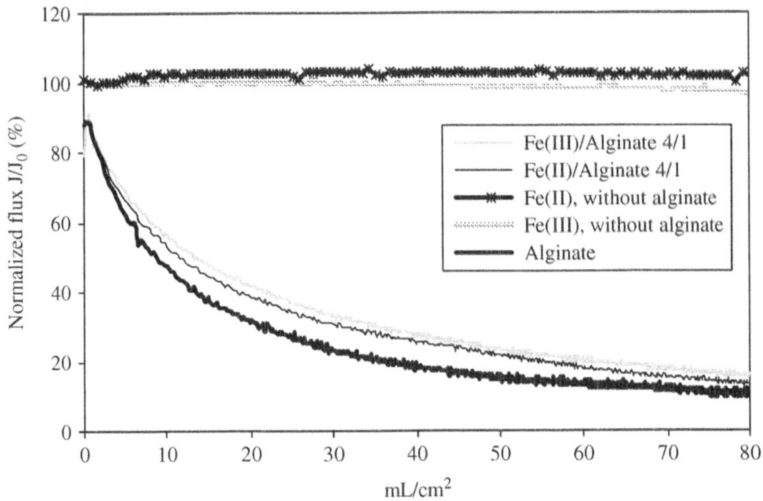

Figure 3. Impact of Fe(III) on flux in systems with and without alginate. Fe(III) was directly added to alginate solutions, while in the experiments marked "Fe(II)", the Fe(II) was added initially, and than slowly oxidized to Fe(III). The ionic strength of all solutions was adjusted by NaCl.

Furthermore, it can be concluded that the presence of (Fe-)colloids in the system does not influence the flux. As we can see on the Figure 4, the Fe(III) crystals obtained through oxidation of Fe(II) have crystalline structure, while in the other case, Fe(III) is amorphous. Apparently, both structures do not influence flux decline in the presence of alginate significantly.

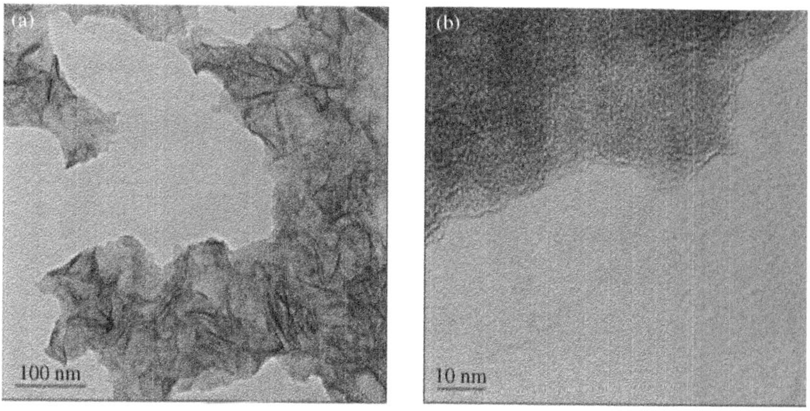

Figure 4. TEM pictures of particles obtained from Fe-alginate solution, where (a) $Fe(OH)_3$ was made by oxidation of Fe(II)-alginate solution with air and pH increase, (b) $-Fe(OH)_3$ was obtained by precipitation from Fe(III)-alginate solution.

The influence of ionic strength

The impact of the ionic strength on the polysaccharide behavior in solution and on the surface of the membrane is an important factor reflecting the role of electrostatic interactions in the fouling process.

In Figure 5, the impact of ionic strength on the flux decline caused by Hyaluronate is shown. As we can see from the flux decrease graphs in Figure 5, increase of the salt concentration 0.1 M leads to less fouling and higher flux values, and at higher concentrations the degree of fouling remains the same.

The impact of the ionic strength on the fouling by NOM has been considered in some studies before (Frank and Belfort, 2003). It is known that the cations of added salt neutralize the anionic carboxylic group of NOM by a charge shielding effect (Berriaud, 1998; Ise and Sogami, 2005). The charge shielding induces van der Waals interactions and formation of inter- and intra-molecular hydrogen bonds, stabilizing a certain structure of a macromolecule and increasing the water content of the gel formed. Thus, the gel acts as a looser and more fluid structure, resulting in an increase of flux.

The presence of the threshold NaCl concentration (0.1 M) after which the further increase of the ionic strength does not result in the increase of the flux also supports this theory: once the hydrogen bonds are formed, a further increase of the concentration of NaCl should not have an impact on the structure, and thus on the flux.

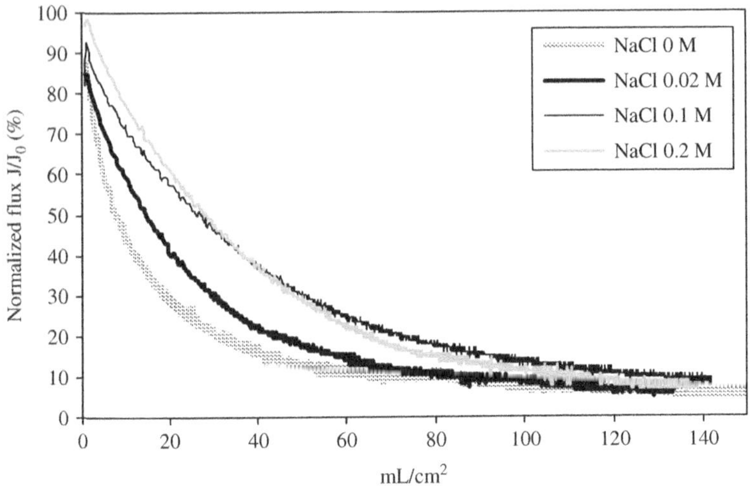

Figure 5. Impact of the ionic strength on the flux decline caused by hyaluronic acid.

However, another theory states that the increase of the ionic strength should result in compaction of the adsorbed layer and increase of the polysaccharide adsorption (De Kerchove and Elimelech, 2006). This theory is based on the reduction of the size of macromolecules with the increase of the ionic strength (Schaefer, 2001). Apparently, this effect is less significant for hyaluronate and the theory is not valid in the case of hyaluronic acid/NaCl.

Long-term experiments and flux stabilization

Figure 5 (and also Figures 2, 3) shows the flux behavior during the long-term filtration without backflushing. In the initial period of the flux curves (volume of the polysaccharide solution filtered through the membrane up to 80 mL/cm^2 or 800 L/m^2), a strong flux decline occurred, but with the increase of the volume of the filtered solution, the flux stabilized and the impact of different factors on fouling became negligible. The differences observed for different polysaccharides and the impact of solution conditions cannot be noticed anymore during this stage of ultrafiltration. These flux decline patterns indicate that the mechanisms of fouling layer formation at the beginning of the cycle differ from the situation where substantial fouling layer should already have been formed.

In this paragraph we will focus on phenomena of flux stabilization (the impact of different parameters on the first part of the cycle has been discussed before).

As mentioned above, the stabilization of the flux curve is being observed after approx. 80 mL of solution passed through 1 cm^2 of the membrane. This corresponds to the accumulation about 0.03 mg of alginate over 1 cm^2 of the membrane. Assuming an average polysaccharide concentration occurring in the gel structure of 2.9 g/L (Rolin *et al.*, 1998), the layer thickness can be calculated to be around 100 μm, which is a realistic value as seen by visual observation, but not confirmed by measurements. Being a network of randomly distributed chains, the gel can be expected to act as a filter for transport of other macromolecules (Laurent, 1995). This network, as any filter, can be characterized by a porosity and water permeability (Beeriaud *et al.*, 1998; Rolin, 1998). Therefore, the fouling layer formed on the membrane surface can be considered as a semi permeable membrane itself. Actually, alginate has been used for production of membranes in different fields of separation (Patil *et al.*, 2007). In order to investigate the structure of alginate gels, AFM pictures were made of an alginate gel obtained by centrifugation of an alginate solution of 0.365 μg/L, and compared to an UF membrane used in our investigations (Figures 6a and 6b). As shown, indeed membrane-like structures are observed in the case of alginate, with a similar surface-roughness as the UF membrane.

Relevance for practice

Stabilization of the flux on a level of around $7\ L \cdot h^{-1} \cdot m^{-2}$ has been observed in dead-end UF systems without backflushing operated during extended periods of time (Peter-Varbanets *et al.*, 2009b). As reported here, flux stabilization occurs with polysaccharide solutions at a level of around $10\ L \cdot h^{-1} \cdot m^{-2}$, which is in the upper range of the flux with natural waters. The difference between these values can probably be explained by the presence of more hydrophobic, smaller components in natural waters (e.g. humic substances) which can lead to pore blocking.

As shown in Figure 6, deposition of polysaccharides on the membrane can lead to the formation of a secondary membrane. It can be expected that the separation properties of this layer depends on the type of polysaccharide and the matrix composition. Also in the case of dead-end UF systems, stationary fouling layers are formed which are expected to contain bulk polysaccharides and EPS (Peter-Varbanets *et al.*, 2009b). Such may improve the separation characteristics of the membrane.

While biological activity exists in the systems operated with natural water, it can be assumed that a fraction of the polysaccharides present in natural waters is enzymatically hydrolysed when trapped in the fouling layer. Thus, the role of polysaccharides in dead-end low-pressure UF systems is considered less detrimental than in conventional membrane filtration. Actually, the hydrolysis of

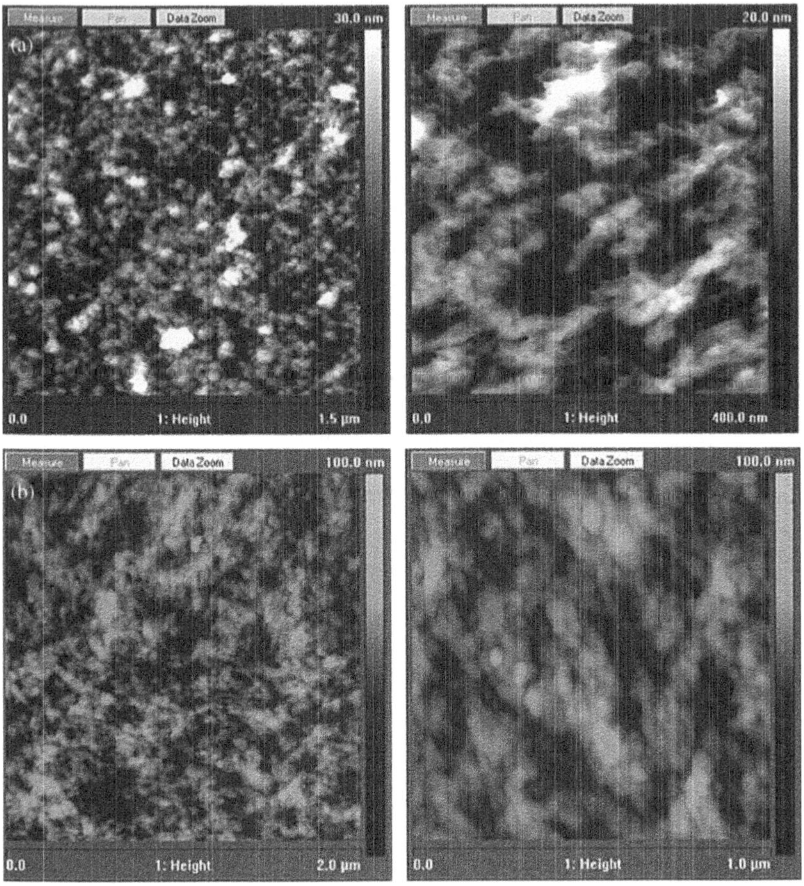

Figure 6. AFM Pictures: (a) centrifuged alginate gel; (b) clean UF membrane (PES).

polysaccharides in such fouling layers can have a positive effect in creating channels which facilitate flow and contribute to reducing the resistance in biologically active fouling layers.

CONCLUSIONS

We have systematically investigated the impact of polysaccharide structure, availability of functional groups and molecular weight on membrane fouling at conditions suitable for POU systems (hydrostatic pressure of 120–150 mbar and no backflushing). Our experimental results lead us to the following conclusions:

Regarding the initial stage of flux decline (0–800 L/m^2) the polysaccharide structure, and particularly availability of carboxyl groups, has a major impact on the membrane fouling, while the molecular weights of polysaccharides does not play a significant role (in the studied range of Mw 5–250 kDa).

The presence of metal ions has a strong impact on the fouling: Ca, Fe and Ba are found to stabilize a gel structure of the fouling layer. The metal containing gel layers are in general more hydrated than the gels without metals, with water molecules trapped by hydrogen bonding in the gel matrix. The higher rate of hydration results in a higher permeability of these metal containing gels. At elevated ionic strength the permeability is increased. This may be explained by the formation of hydrogen bonds, which can contribute to swelling of polysaccharides in the solution and on the surface in looser and more fluid gel structures.

Independently of the initial solution conditions, after approx. 800 L/m^2 has been filtered, the flux stabilizes on a level of approx. 10 $L \cdot h^{-1} \cdot m^{-2}$. The reasons of the long term flux stabilization are still not completely understood. However, we suppose that the gel layer formed by polysaccharides play a role of a "second" membrane on the surface of the PES UF membrane, keeping permeability on a certain level, determined by the water retention properties of the gel structure.

REFERENCES

Amy, G. and Cho, J. (1999). Interactions between natural organic matter and membranes: Rejection and fouling. *Water Science and Technology*, **40**(9), 131–139.

Beeriaud, N., Milas, M. and Rinaudo, M. (1998). Characterization and properties of hyaluronic acid. In: *Polysaccharides. Structural Diversity and Functional Versability*, S. Dumitriu (ed.), Marcel Dekker Inc., New York, pp. 313–335.

Bryce, T.A., McKinnon, A.A., Morris, E.R., Rees, D.A. and Thom, D. (1975). Chain conformations in sol-gel transitions for polysaccharide systems, and their characterization by spectroscopic methods. *Faraday Discussions,* **57**, 221–229.

Cho, J., Amy, G. and Pellegrino, J. (1999). Membrane filtration of natural organic matter: Initial comparison of rejection and flux decline characteristics with ultrafiltration and nanofiltration membranes. *Water Research*, **33**(11), 2517–2526.

Choo, K.H., Haebum, L. and Choi, S.J. (2005). Iron and manganese removal and membrane fouling during UF in conjunction with prechlorination for drinking water treatment. *Journal of Membrane Science,* **267**(1–2), 18–26.

Churchhouse, S. (2000). Membrane Bioreactors hit the big time – from lab to full scale application. In: *Membrantechnik, 3. Aachener Tagung*, R. Rautenbach, T. Melin and M. Dohmann (eds.), Klenkes Druck und Verlag GmbH, Aachen, pp. B12, 11–17.

Clark, M.M. and Lucas, P. (1998). Diffusion and partitioning of humic acid in a porous ultrafiltration membrane. *Journal of Membrane Science*, **143**(1–2), 13–25.

Ise, N. and Sogami, I.S. (2005). *Structure Formation in Solutions: Ionic Polymers and Colloidal Particles,* Springler Bln, Berlin, pp. 68–69.

Frank, B.P. and Belfort, G. (2003). Polysaccharides and sticky membrane surfaces: Critical ionic effects. *Journal of membrane science,* **212**(1–2), 205–212.

Jermann, D., Pronk, W., Meylan, S. and Boller, M. (2007). Interplay of different NOM fouling mechanisms during ultrafiltration fro drinking water production. *Water Research*, **41**, 1713–1722.

de Kerchove, A.J. and Elimelech, M. (2006). Structural growth and viscoelastic properties of adsorbed alginate layers in monovalent and divalent salts. *Macromolecules*, **39**, 6558–6564.

Kimura, K., Hane, Y., Watanabe, Y., Amy, G. and Ohkuma, N. (2004). Irreversible membrane fouling during ultrafiltration of surface water. *Water Research*, **38**, 3431–3441.

Laurent, T.C. (1995). An early look at macromolecular crowding. *Biophysical Chemistry,* **57**, 7–14.

Morris, E.R. (1990). Mixed polymer gels. In: *Food Gels*, P. Harris (ed.), Elsevier Applied Science, London.

Patil, M.B., Veerapur, R.S., Patil, S.A., Madhusoodana, C.D. and Aminabhavi, T.M. (2007). Preparation and characterization of filled matrix membranes of sodium alginate icorporated with aluminium-containing mesoporous silica for pervaporation dehydration of alchohols. *Separation and Purification Technology*, 54(1), 34–43.

Peter-Varbanets, M., Zurbrügg, C., Swartz, C. and Pronk, W. (2009a). Review: Decentralized systems for potable water and the potential of membrane technology. *Water Research*, **43**(2), 245–265.

Peter-Varbanets, M., Hammes, F., Gujer, W. and Pronk, W. (2009b). Biofouling of Ultrafiltration Membranes at ultra-low Pressures and its Relevance for Decentralized Water Supply, Submitted for publication.

Rinaudo, M. (2006). Non-covalent interactions in polysaccharide systems. *Macromolecular Bioscience*, **6**, 590–610.

Rolin, C., Nielsen, B.U. and Glahn, P.E. (1998). Pectin. In: *Polysaccharides. Structural Diversity and Functional Versability*, S. Dumitriu (ed.), Marcel Dekker Inc., New York, pp. 377–431.

Schaefer, A.I. (2001). *Natural Organics Removal using Membranes: Principles, Performance and Cost.* Lancaster, Penn, Technomic, p. 406.

Thibault, J.F. and Rinaudo, M. (1985). Interactions of mono-valent and divalent counterions with alkali-deesterified and enzyme-deesterified pectins in salt-free solutions. *Biopolymers,* **24**(11), 2131–2143.

Van der Berg, L., van Vliet, T., van der Linden, E., van Boekel, M.A.J.S. and van de Velde, F. (2007). Breakdown properties and sensory perception of whey proteins/polysaccharide mixed gels as a function of microstructure. *Food Hydrocolloids*, **21**, 961–976.

WHO (2004). *Report: Meeting the MDG Drinking Water and Sanitation Target; A Midterm Assessment of Progress.* World Health Organization and United Nations Children's Fund.

Xia, S., Nan, J., Ruiping, L. and Li, G. (2004). Study of drinking water treatment by ultrafiltration of surface water and its application to China. *Desalination*, **170**, 41–47.

Ye, Y., Le Glech, P., Chen, V., Fane, A. and Jefferson, B. (2005). Fouling mechanisms of alginate solutions as model extracellular polymeric substances. *Desalination*, **175**(1), 7–20.

First results of a 5 m³/d gravity-driven ultrafiltration unit for decentralised water supply

E. Hoa[a], M. Peter[b], L. Pred'homme[c], B. Lesjean[a] and A. Tazi-Pain[d]

[a]Berlin Centre of Competence for Water (KWB), Cicerost.24, 10709 Berlin, Germany
(E-mail: eric.hoa@kompetenz-wasser.de; boris.lesjean@kompetenz-wasser.de)
[b]Eawag, Swiss Federal Institute of Aquatic Science and Technology, Überlandstr. 133, 8600 Dübendorf, Switzerland (E-mail: maryna.peter@eawag.ch)
[c]Opalium, 178, avenue du Mar. de Lattre de Tassigny, 94120 Fontenay-sous-Bois, France
(E-mail: laurent.pred'homme@otv.fr)
[d]Anjou Recherche, Chemin de la Digue, 78600 Maisons-Laffitte, France
(E-mail: annie.tazi-pain@veolia.com)

Abstract The study aims at validating the point-of-use investigations on long-term gravity-driven ultrafiltration for a 5 m^3/d scaled-up system, which is fed with Marne surface water in France. Equipped with a biosand filter as pretreatment, the membrane-based unit is designed to be robust, energy-sufficient (gravity-driven) and run with restricted chemical intervention (only residual chlorine). Tested in winter conditions, the unit shows a stabilized flux of 2.5 lmh within 20 operating days. This value is lower than the 7–10 lmh observed at lab scale at 20°C but it could increase to 4–5 lmh when a weekly drainage of the membrane reactor is performed, which is consistent with the design specification in terms of water production. With warmer temperatures, a flux enhancement is expected as the membrane permeability and the performance of the biological activity of the system would be improved and is to be confirmed with further tests–notably in South Africa.

Keywords decentralised water supply, low-energy, small-scale system (SSS), ultrafiltration

INTRODUCTION

As it may be neither economically nor technically viable to set up a reliable water network in developing countries or in rural areas, decentralised water supply stands as one of the greatest challenges in the forthcoming years. In these regards, membrane processes seem promising as they efficiently remove pathogens and offer a modular design that enables flexibility in terms of flow capacity reduction. With regards to the Millennium Development Goals, novel decentralised water systems should be robust, low-cost and as independent as possible from chemical and energy requirements and they are expected to enter the market within the next years (Hoa and Lesjean, 2008).

 TECHNEAU: Safe Drinking Water from Source to Tap. Edited by Theo van den Hoven and Christian Kazner. ISBN: 9781843392750. Published by IWA Publishing, London, UK.

Within the European project TECHNEAU (www.techneau.eu), a research group aims to develop a low-energy ultrafiltration (UF) unit for small drinking water applications. The Swiss Federal Institute of Aquatic Science and Technology – Eawag – performs lab work on long-term gravity-driven membrane filtration at a point-of-use (POU) scale. These investigations have enabled to design and build a scaled-up unit (dimensioned for 5 m^3/d) to be tested in real environments in France and in South Africa.

This study presents the first results of the tests which are performed under controlled conditions in Annet-sur-Marne, France, on Marne river water in order to ensure the technological reliability of the compact unit. The focus is set on the process features of the unit (flow capacity) in function of periodicities of mechanical cleaning (drainage) and intermittent operation. These investigations will demonstrate if UF membrane systems can be operated without chemicals and energy, and stand as (cost)-effective options for decentralised water supply.

MATERIALS AND METHODS

Description of the unit

This small-scale system (SSS) is based on a gravity-driven UF process developed by Eawag, which enables operation without crossflow, backflush, aeration or chemical cleaning. Hence, Eawag, KWB and Opalium conceived a membrane-based SSS, which could treat up to 5 m^3/d of natural surface water – enough to satisfy drinking water needs for a community of 100–200 inhabitants. The unit only needs to be fed with raw water at a 2 m-height.

Compacted in a 10 feet-long maritime container, the unit is composed of (see Figure 1):

- A biological sand filter, as pre-treatment, which enables an elimination of the turbidity and significantly improves the membrane permeability as the biological layer ("Schmutzdecke") on the sand surface consumes some of the raw water constituents and thereby contributes to the reduction of the organic fraction (Varbanets and Pronk, 2006).
- a flat-sheet UF module (area : 40 m^2), whose long-term operation can produce a stable flux (about 7–10 lmh was observed at POU scale) over months, although no maintenance is performed (Varbanets and Pronk, 2006).
- a storage tank for residual chlorination to avoid recontamination of the treated water.

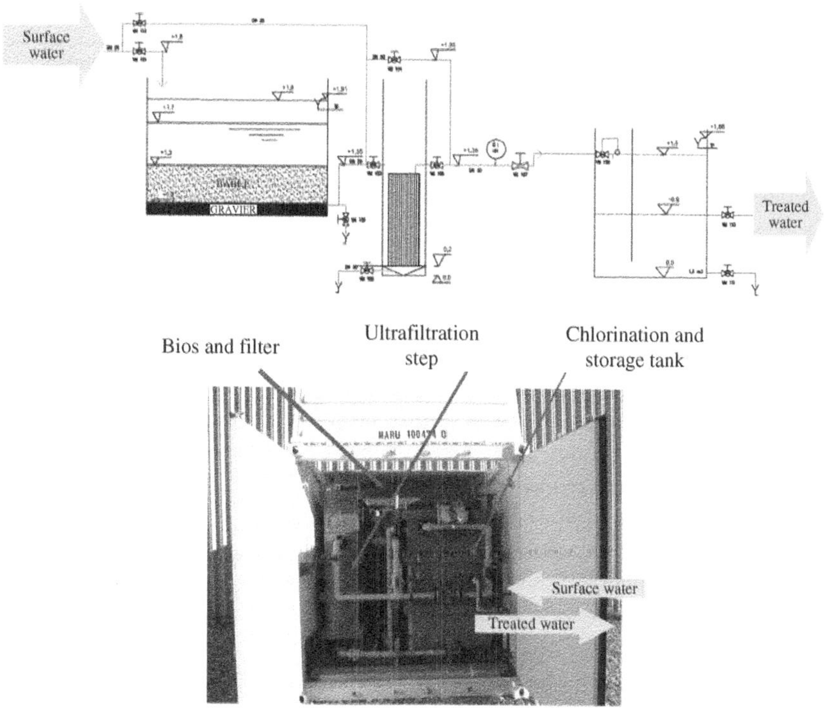

Figure 1. Process Instrument Diagram and Photo of the compact unit.

As sole operational requirement, the membrane reactor is simply drained (i.e. emptied) on daily to weekly basis to remove the material retained by the membrane and accumulated in the module. Moreover, the operating rate (number of hours/days) can also vary. Apart from those two control parameters, the system, which is only driven by a 40 cm differential pressure head (i.e. 40 mbar), is totally self-determined and autonomous.

Site for trials

The unit is installed at the Anjou Recherche – Veolia Water Research – facilities in Annet-sur-Marne (France) and is fed with Marne river water. As presented in Table 1, the water quality of the Marne River is consistent with the ranges of water qualities on which Eawag made its lab tests (Chriesbach water and Chriesbach water mixed with wastewater (Peter-Verbanets and Pronk, 2008)), in regards to organic content and turbidity. Therefore, the "scale-up" challenge is relevant and results with the Eawag POU investigations with a membrane area of 25 cm^2 could be comparable.

Table 1. Marne river and Chriesbach water qualities.

Marne River (at Annet-sur-Marne) Average in 2008 (Min–Max)	Chriesbach water	Chriesbach water mixed with 15% wastewater
TOC: 2.7 mg/L (0.9–7.7)	DOC: 2–3 mg/L	DOC: 10–15 mg/L
Turbidity: 23 NTU (3.3–258)	Turbidity: 0.3–1 NTU (peak 300)	Turbidity: 30–40 NTU

Action plan

The article here presents the trials performed in Annet-sur-Marne from November 2008 to early March 2009. Initially, different operating conditions were to be tested in order to put progressively the unit under more constraints (i.e. increase the operating rate – 12, 18 and 24 hours/day – and decrease the drainage frequency of the membrane reactor).

The initial phase (from Day 1 to Day 21) consisted in by-passing the UF step in order to let the biological activity build in the slow biosand filter. After a 3 week-operation, the UF step was sequenced back in the process train. However, preliminary values of outlet flux were very low (1–2 lmh) and showed that the unit presented some unexpected head loss. Due to these hydraulic setbacks which required constructive modifications of the system and the hostile/freezing winter conditions, the unit could only be run properly from Mid-January 2009 (Day 62) and a reschedule of the phases with different operating conditions took place:

- From Day 62 to Day 75, the unit run 12h/d and no drainage was performed.
- From Day 76 to Day 91, the unit run continuously and no drainage was performed.
- Since Days 92, the unit has run continuously and a weekly drainage was organized.

The chlorination step was not implemented in this study. As the unit is autonomous, the monitoring tasks simply include the general visual control of the unit, the recordings of the temperature and the volume flow rate (a mechanical flow meter is included in the unit) and the measurements of the oxygen content and the turbidity in the inlet, after the sand filter and after the UF step. From Day 62 to Day 110, data were collected 2 to 3 times/day every weekday, but after the apparent stabilization of the system, the measurements took place only 3 days/week. Data for the raw water were directly read from on-line instrumentation of the water plant, whereas data for the waters sampled after the sand filter and after the UF step were measured by portable probes (for oxygen content and turbidity).

If not specified as corrected values, the flux/permeability results that are presented in the study are actual flux measurements at ambient (winter) temperature. A correction at 20°C, based on the variation of the water viscosity (Trussell *et al.*, 2005) was also considered for complementary discussions.

Commissioning and identification of hydraulic issues

Some permeability tests with clean water were made in order to identify the causes of the initial head loss (Day 21). Indeed, the low flux (1–2 lmh) measured in the outlet may be explained by an insufficient pre-conditioning of the new membrane (possible need of a chemical wetting to enhance the membrane porosity) or a head loss due to the connectors and the piping of the unit. In a first step, it was decided to let the membrane filtration run continuously on longer periods (several hours), at the highest head pressure (1m in this case, as it is a gravity system) in order to ensure that the whole membrane surface was wet. As observed in Figure 2, the permeability (corrected at 20°C) did not evolve significantly although an increase from 80 lmh/bar to close to 250 l/(m^2.h.bar) was observed in the first 20 operating hours.

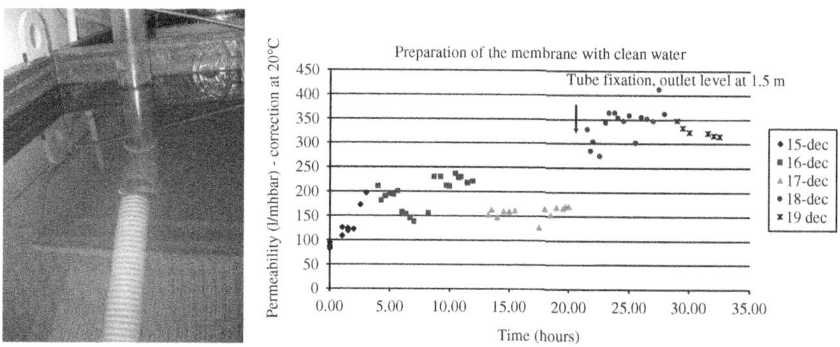

Figure 2. Permeablity tests with the membrane before and after the addition of an open-air tube on the permeate pipe.

Before suggesting any chemical wetting of the membrane, the next option was to chase any air that would be stuck in the module (without the use of any pumps). Therefore, a tube connected to the air was added on the permeate tube, in the membrane reactor (see Figure 2), in order to ensure that any air bubbles could be released on the permeate side. Moreover, the addition of this pipe removes a pinch on the flexible permeate tube which could have caused significant head loss as well. Once the tube was fixed, a corrected permeability of 300–350 l/(m^2.h.bar) was observed, which is consistent with tests made at lab scale with a membrane sample from the same manufacturer (A3 Water Solutions GmbH).

RESULTS AND DISCUSSION

Flux stabilization

Influence of the intermittent operation and of the turbidity feed

From Day 62, the process unit was running with the biosand filter and the UF step in winter conditions (temperature range: 3.1–10.5°C). Figure 3 shows that the flux stabilizes to around 2.5 lmh (compared to 7–10 lmh for Eawag lab conditions at 20°C ± 2°C), in 20 days. The stabilization curve seems here slightly different that the one expected from the Eawag lab tests. Indeed, the decrease of the flux is slower in the first 10 days than in the next 10 days.

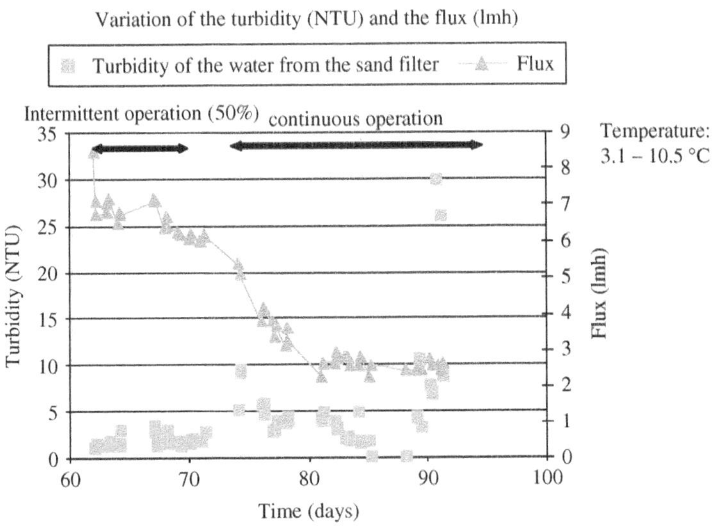

Figure 3. Flux variation in regards to intermittent operation and turbidity feed.

This is probably due to the fact that the unit was only running half-time on the first 10 days and thereby the flux was enhanced (also observed at Eawag lab scale). Another option is that the membrane reactor had to cope with higher turbidities (>5 NTU) in the next 10 days, which could also have lead to some flux reduction.

Influence of the drainage frequency

Once the flux has stabilized for more than 10 days, a weekly drainage (D) took place. This should lead to the removal of the accumulated material which is not attached to the membrane surface, facilitating the stabilization of the fouling layer and preventing the increase of its thickness. From Figure 4, the first drainage event (D1) enables a flux enhancement from 2.5 to 6 lmh.

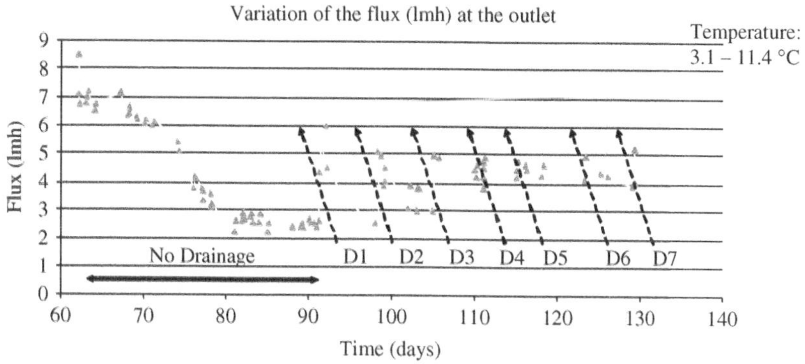

Figure 4. Flux variation in regards to the drainage frequency.

However, the following drainage events have progressively less impact on the flux enhancement. Indeed, Figure 5 illustrates a linear decrease of the flux enhancement for the 4 first drainage events (transition period of 1 month) and then a stabilization at around 20% is observed (to be confirmed by more measurements).

The flux stabilization is however not reached during the 1-week period of the drainage routine. Instead, a pattern seems to be created with flux values varying between 4–5 lmh after the 4th drainage event (D4). With these operating conditions (continuous operation and weekly drainage), between 3.8 and 4.8 m^3/d of treated water is produced. That is consistent with the 5 m^3/d target that was established when scaling-up the Eawag lab investigations.

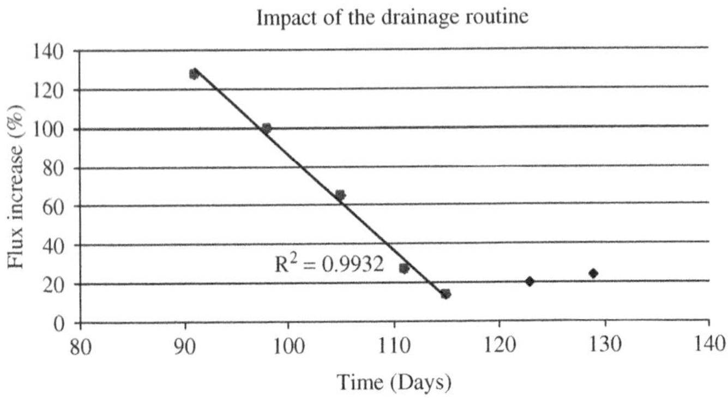

Figure 5. Flux enhancement due to the drainage routine.

Influence of the temperature

The current results in regards to flux values are promising. However, it should be mentioned that these investigations were performed in winter, with the temperatures of the raw water varying from 3.1°C to 11.4°C. That range of temperatures is not well-adapted for the biological activity which is needed in the process unit: in the biosand filter and in the fouling layer of the membrane. Figure 6 shows the variations of the oxygen contents in the raw water and the water from the sand filter and illustrates the slow activation of the biology in the sand filter through the consumption of the organic matter. Spring times will surely enhance biological activity and flux patterns.

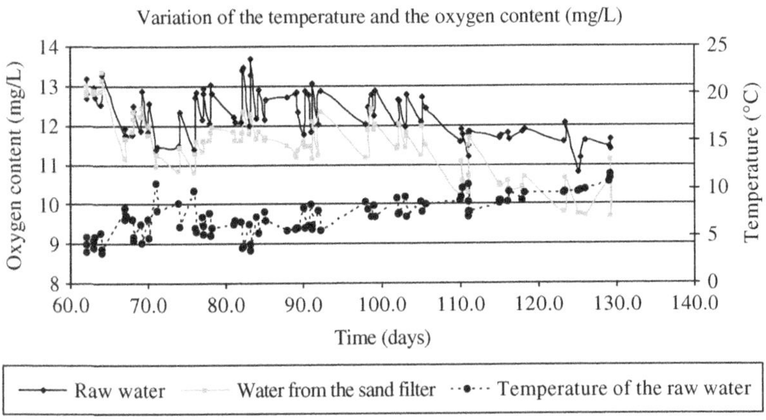

Figure 6. Effect of temperature on the oxygen content.

Expectations for the South African conditions

As the unit will be also tested in South Africa, it is interesting to transpose the current results to South African-like conditions. Figure 7 simulates the flux variation which is corrected at 20°C based on water viscosity considerations (Trussell *et al.*, 2005) and it shows a flux stabilization between 5 and 6.5 lmh with similar operating conditions. In that virtual case, between 4.8 and 6.2 m^3/d of treated water would be produced, satisfying the design specification of the system. Besides, an increase of the temperature would also lead a better biological performance in the sand filter and the membrane reactor, and thereby better outputs.

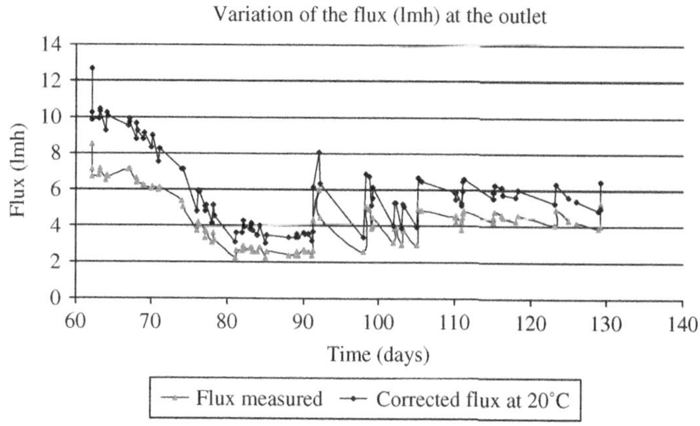

Figure 7. Flux enhancement with a 20°C-correction.

However, in South Africa, coping with high turbidity peaks (>300 NTU) could impact the present flux pattern. In that regard, more investigations on the pre-treatment options are therefore planned in the rest of the study.

CONCLUSIONS

The gravity-driven UF compact unit has shown promising results in regards to flow capacity. Although these preliminary investigations occur in winter, a flux stabilization of 2.5 lmh was observed, which is below the results from Eawag lab tests (i.e. 7–10 lmh, at 20 ± 2°C). However, the "scaled-up" system can benefit from a weekly drainage which seems to enhance the flux to 4–5 lmh, and thereby, the unit is to produce more than 4 m^3/d, which is consistent with the design target of 5 m^3/d. Moreover, it can be expected that with warmer temperatures – leading to a better membrane permeability and an enhanced biological activity, the outputs would improve to the expected lab values. This is particularly relevant for South Africa, where the unit is to be further tested from Summer 2009. The capacity of the system to cope with high turbidity peaks would then be relevant. In that regards, capital and operational costs of the pre-treatment option will be further assessed.

ACKNOWLEDGEMENTS

This study was performed within the 6th framework of the European project "TECHNEAU" (Contract No 018320, www.techneau.eu). The authors would like to acknowledge the European Commission and VEOLIA for their financial contributions and Anjou Recherche for offering facilities and technical support.

REFERENCES

Hoa, E. and Lesjean, B. (2008). *International Market Survey on Membrane-based Products for Decentralised Water supply (POU and SSS Units)*. EU Project TECHNEAU Report. D2.5.3. Berlin Centre of Competence for Water. Available at www.techneau.eu

Peter-Verbanets, M. and Pronk, W. (2008). *Mechanisms of biofouling of UF membranes and evaluation of pre-treatment on fouling of UF membranes*. EU Project TECHNEAU report.D2.5.6/8/10. Eawag, CH.

Trussell, S., Adham, S. and Trussell, R. (2005). Process limits of municipal wastewater treatment with the submerged membrane bioreactor. *Journal of Environmental Engineering*.

Varbanets, M. and Pronk, W. (2006). *Point-of-use Membrane Systems: Place in the World of Supply*, EU Project TECHNEAU Report D2.5.2. Eawag, CH.

Simulation

Modelling with the European drinking water treatment simulator

P.S. Ross[a], K.M. van Schagen[b,c], A.W.C. van der Helm[a,d], J. Dudley[e] and L.C. Rietveld[a]

[a]Delft University of Technology, PO Box 5048, 2600 GA Delft, The Netherlands (E-mail: p.s.ross@tudelft.nl)
[b]DHV BV, PO Box 1132, 3800 BC Amersfoort, The Netherlands
[c]Delft Center for Systems and Control, Delft University of Technology, Mekelweg 2, 2628 CD Delft, The Netherlands
[d]Waternet, PO Box 8169, 1005 AD Amsterdam, The Netherlands
[e]WRc plc, Frankland Road, Swindon, SN5 8YF, United Kingdom

Abstract By combining validated measurements or estimated process parameters (basic control) with mathematical models (model-based monitoring), a model-based optimization can be done. When the optimisation criteria are deterimined a model-based control can be implemented such that a more stable and robust operation may be achieved. To achieve this, research needs on modelling and on-line measurements were determined based on a plant wide optimisation of the treatment plant. An example is given for the treatment step ozonation. It was concluded that the mathematical model used for ozonation gives a reliable prediction of the water quality produced, however, with respect to the prediction of the formation of easily assimilable organic carbon (AOC) still more research is needed. The applicability of an on-line measurement for the estimation of the AOC concentration gives promising results, but the amount of data on which the algorithm was developed is still limited, so additional experiments should be carried out to enhance the algorithm and applicability in different treatment plants.

Keywords modelling, model-based control, water treatment simulator

INTRODUCTION

When designing a water treatment plant the focus, in general, is on the individual treatment steps, whether it is the improvement of an existing step or the development of new technologies; interactions with other treatment processes present are usually ignored. During operation, the performance of the drinking water treatment plant is monitored on a regular basis by treatment operators and water technologists. If process changes occur the operators are expected to react appropriately, mostly based upon experience. Bosklopper *et al.* (2004) reported that more efficient operation can be achieved when interactions between the individual processes are taken into account. But those decisions regarding a plant wide approach can be difficult to make.

A sub-optimum operation and maintenance of a water treatment plant can result in water quality problems, increasing costs and environmental impact issues. By using mathematical models to represent individual treatment steps and by connecting these processes, the entire treatment train can be simulated, making a plant wide approach easier. A plant wide model can predict the water quality parameters that are influenced by preceding processes or influence subsequent processes (van der Helm, 2007). This may result in a more stable and robust operation which will lead to less maintenance. In addition the models can play a vital role in retaining the process knowledge, since the different people and specialisms involved in operating and supporting a water works gain knowledge about the drinking water treatment processes under different circumstances. Information about the full-scale plant is obtained by treatment operators, during the design of a process, the designer collects information from pilot plant research and detailed information about a process is gained by researchers using laboratory scale tests (Rietveld, 2005).

After performing a water quality assessment, environmental assessment and a financial assessment of the operation of drinking water treatment plants van der Helm (2007) concluded that the objective for plant wide optimization of the operation of drinking water treatment plants should be the improvement of water quality. To maintain a good water quality in the distribution system, the drinking water produced should be free of particles, biologically stable and chemically stable (Vreeburg *et al.*, 2008; van der Kooij *et al.*, 2000). Recently, special focus is on the removal of natural organic matter (NOM), since the biological stability in the distribution system is one of the major concerns as well as the formation of disinfection by-products. Optimisation efforts should focus on the processes affecting these properties, such as ozonation. (Kruithof *et al.*, 1999; Muller *et al.*, 2003; Juhna *et al.*, 2004; Rosa *et al.*, 2004; and Fabris *et al.*, 2008).

To be able to achieve good drinking water quality with the aid of mathematical modelling van Schagen (2009) developed a model-based approach to meet the challenges in drinking water production which is illustrated in Figure 1.

The model-based approach consists of three layers: basic control, advanced control (consisting of model-based monitoring and model-based control) and the process optimisation level. In the basic control of the treatment processes on-line measurements are used for direct control. During model-based monitoring mathematical models combined with laboratory measurements give an estimate of the actual state of the treatment process; if these results differ from the basic control values this can be an indication that the on-line measurements are failing. When using the mathematical models off-line, optimisation strategies can be calculated, after which these strategies can be applied during the model-based control.

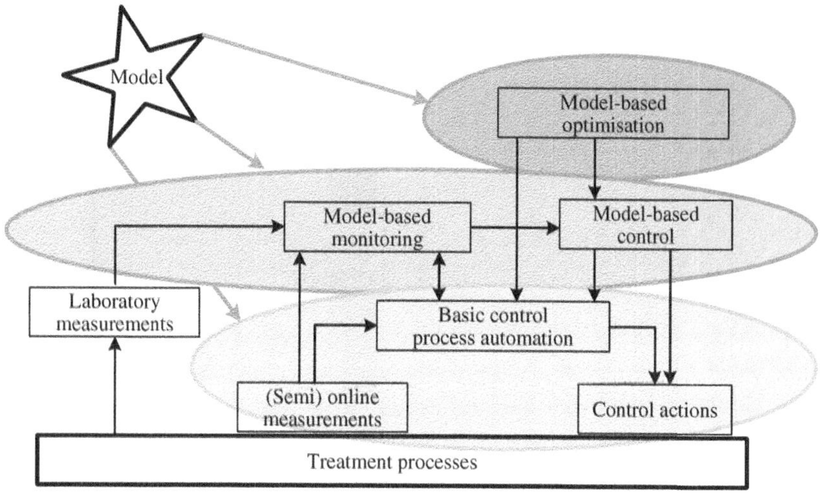

Figure 1. Model-based approach drinking water treatment plant. (source: van Schagen, 2009)

In this paper the model-based approach is discussed as the basis for mathematical modelling and on-line monitoring in European treatment plants for improved operation and maintenance of drinking water treatment, in particular during ozonation.

Materials and methods

Mathematical models

Within the EU 6th Framework Project TECHNEAU a water treatment plant simulator for improving operation and maintenance of existing treatment plants (the 'European Water Treatment Simulator (SimEau)') is developed. The simulator was build upon the experiences of existing modelling platforms and models (Dudley *et al.*, 2007). From this review a list of programme requirements was constructed. Basically, the programme is freely available (this implies no third party commercial packages are involved), easy handling of import and export of water quality measurements, possibility for communication with other platforms and the programme is easily extendable with new water quality parameters and process models. In Figure 2, a screen-shot of SimEau is given.

During ozonation, NOM is oxidized, transforming higher molecular weight compounds into lower molecular weight compounds, measured as assimilable organic carbon (AOC). Besides the oxidation of NOM, disinfection, oxidation of

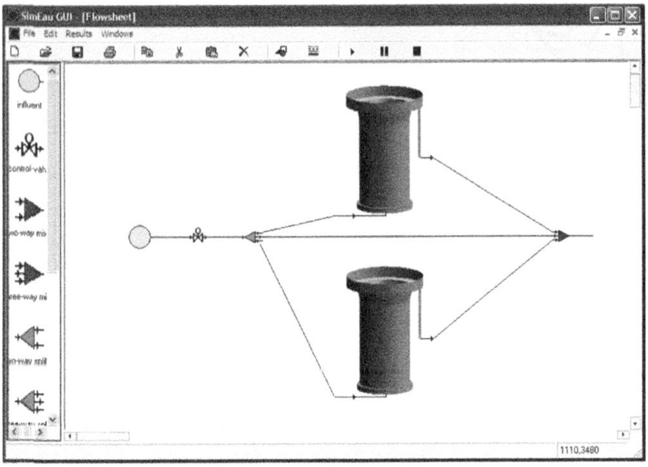

Figure 2. Screen shot water treatment simulator (SimEau).

micro pollutants takes place and bromate can be formed influencing the toxicity of the water produced. A measure for disinfection is the CT-value. The CT-value is determined by multiplying the ozone concentration with the residence time. In this paper the mathematical ozone model developed by van der Helm (2007) was used:

$$\frac{\partial c_{O3}}{\partial t} = -u\frac{\partial c_{O3}}{\partial x} + k_L RQ \frac{u}{u_g}\frac{6}{d_b}(\alpha k_D c_{O3_g} - c_{O3}) - k_{UVA}(UVA - UVA_0)Y - k_{O3}c_{O3}$$

where u is the water velocity (m/s), x is the length of the reactor (m), k_L is the gas transfer coefficient (m/s), RQ is the gas to water flow ratio (Q_g/Q) (Nm^3/m^3), Q and Q_g are the water flow (m^3/h) and gas flow (Nm^3/h), respectively, u_g is the gas velocity (m/s), d_b is the bubble diameter (m), α is the temperature and pressure correction factor, k_D is the distribution coefficient (-), c_{O3g} is the ozone in gas concentration (g-O_3/Nm^3), k_{UVA} is UVA_{254} decay rate (1/s), UVA is UVA_{254} in water (1/m), UVA_0 is the stable UVA_{254} after completion of the ozonation process (1/m), Y is the yield for ozone consumed per UVA_{254} decrease ((mg-O_3/l)/(1/m)) and k_{O3} is the slow ozone decay rate (1/s). The ozone concentration in the gas phase is described by:

$$\frac{\partial c_{O3g}}{\partial t} = -u_g\frac{\partial c_{O3g}}{\partial x} + k_L\frac{6}{d_b}(\alpha k_D c_{O3g} - c_{O3})$$

Measurements

Generally the flow and temperature are parameters that influence all drinking water treatment processes. The parameters influenced by ozonation are bromate, UV absorbance at a wavelength of 254 nm (UV254), dissolved organic carbon (DOC), AOC and the ozone concentration in water for the determination of the CT-value. Besides laboratory measurements, on-line measurements are necessary to monitor the treatment plant on a continuous interval. Van der Broeke *et al.* (2008) developed an on-line measurement tool for the estimation of dissolved ozone and AOC concentrations in drinking water by means of an UV/Vis-spectrometer.

Drinking water treatment plant Weesperkarspel

At the drinking water treatment plant Weesperkarspel of Waternet, the water cycle company for Amsterdam and surrounding areas, situated in the Netherlands, surface water rich in humic acids is pre-treated by coagulation and sedimentation followed by 100 days retention in a reservoir and subsequently filtered over rapid sand filtration. At Weesperkarspel it is then further treated by ozonation, pellet softening, biological activated carbon (BAC) filtration and slow sand filtration. The DOC levels before ozonation vary between 4 and 6 mg-C/L. Because the origin is surface water the temperature of the water varies between 3 and 24°C.

RESULTS AND DISCUSSION

Basic control

The basic control option for ozonation is the ozone dosage. The ozone dosage needs to be adjusted such that the desired disinfection is reached. The disinfection can be checked by measuring the dissolved ozone concentration and calculating the CT-value. On-line determination of the dissolved ozone concentration is derived from changes in UV/Vis-spectrum, see Figure 3.

Depending on the raw water quality the amount of AOC formed during ozonation may vary. Since the AOC concentration is an important parameter regarding the biological stability of the water it is desirable to determine the AOC concentration on-line. By applying Principal Component Analysis (PCA) and Partial Least Squares regression (PLS) on differential spectra for individual treatment steps a good relation was identified between a selection of wavelengths in the differential spectrum and laboratory AOC measurements (Van den Broeke *et al.,* 2008), see Figure 4.

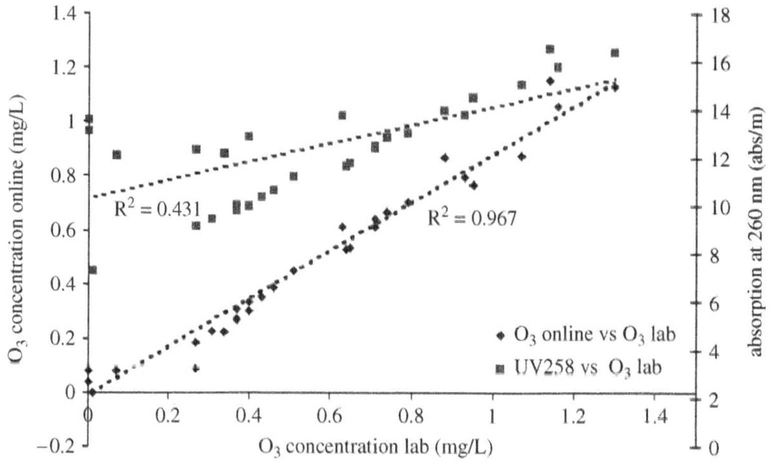

Figure 3. Correlation of on-line and laboratory ozone measurements and correlation of laboratory ozone measurements against single wavelength measurements at 258 nm. (source: van den Broeke *et al.*, 2008)

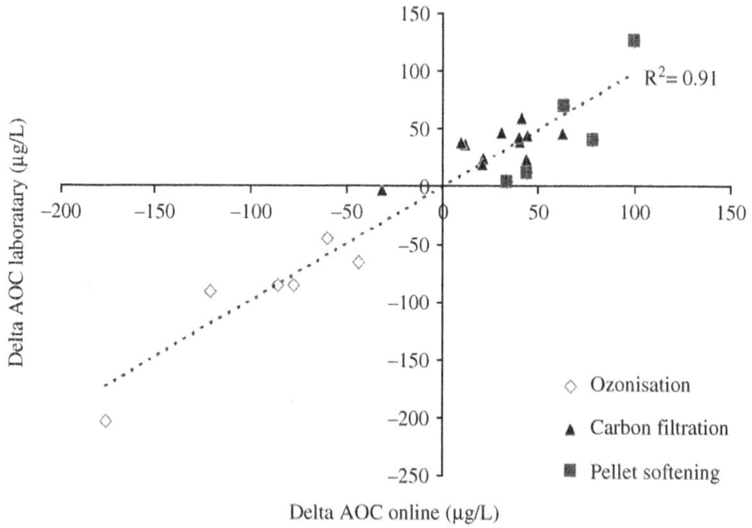

Figure 4. Relation between determined delta AOC laboratory vs delta AOC on-line. (source: van den Broeke *et al.*, 2008)

The correlation found is very good, but further work is needed to establish what correlations are appropriate for other water types.

Model-based monitoring

By feeding the model for ozonation with laboratory measurements of the influent, the effluent water quality is calculated. In Figure 5 the measured effluent water quality parameters and the model predictions of UVA_{254}, AOC concentration, bromate formation and the disinfection by means of the E. coli concentration are given.

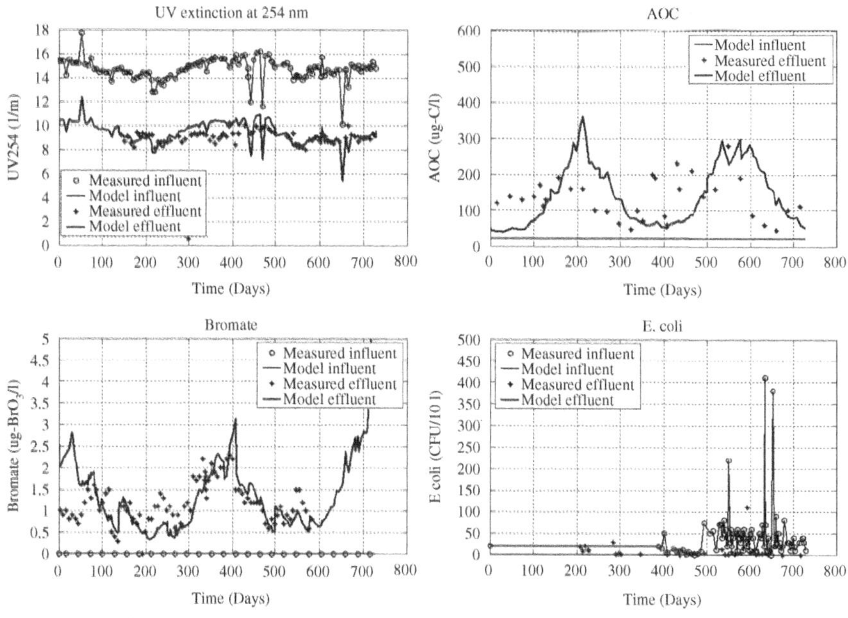

Figure 5. Values of ozonation influent and effluent water quality parameters. (source: van der Helm *et al.*, 2009)

The model results show that it accurately predicts the values for the UV_{254}, bromate formation and *E. coli*, while the prediction of AOC needs further research.

Model-based optimisation

During ozonation the optimised setting could be to get a maximum CT-value under the constraint that the AOC and bromate concentrations should not exceed a maximum allowable value. The CT-value is dependant on the water temperature, see Table 1.

Table 1. CT-values for inactivation of *Giardia* by ozone (USEPA, 1989).

Giardia	Temperature [°C]					
log inactivation	0.5	5	10	15	20	25
0.5	0.48	0.32	0.23	0.16	0.12	0.08
1.0	0.97	0.63	0.48	0.32	0.24	0.16
1.5	1.5	0.95	0.72	0.48	0.36	0.24
2.0	1.9	1.3	0.95	0.63	0.48	0.32
2.5	2.4	1.6	1.2	0.79	0.60	0.40
3.0	2.9	1.9	1.4	0.95	0.72	0.48

Model-based control

Based on the optimisation criteria to get a maximum CT-value without exceeding AOC concentration of 110 μg-C/L and bromate concentrations of 3 μg-BrO_3/L van der Helm (2007) determined the control of the ozone dosage, see Figure 6.

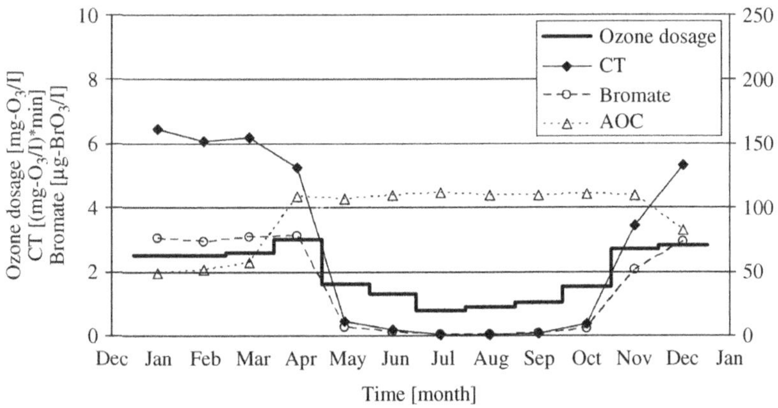

Figure 6. Control strategy for maximal CT limited by maximal allowed bromate concentration and AOC concentration. (source: van der Helm, 2007)

CONCLUSIONS

During ozonation disinfection takes place as well as the oxidation of NOM. Oxidation of NOM leads to AOC which influences the biological stability.

By combining validated measurements or estimated process parameters (basic control) with mathematical models (model-based monitoring), a model-based optimization can be done. In order to use model-based optimization and implement model-based control, mathematical process models are necessary combined with on-line measurements. By applying the mathematical models off-line optimal settings (model-based optimisation) can be determined after which the controller can be optimised. The mathematical model used for ozonation gives a good prediction of the water quality produced. However, with respect to the prediction of the formation of AOC still more research is needed.

Preliminary results on the use of differential UV/V is measurements for the on-line estimation of the amount of AOC formed during ozonation (regardless of the raw water quality) showed that this can be a good tool for on-line monitoring of the biological stability. However the amount of data on which the algorithm was developed is still limited, so additional experiments should be carried out to enhance the algorithm and applicability in different treatment plants. With the availability of on-line AOC estimation combined with an enhanced mathematical model for ozonation and in future for biological activated carbon filtration a plant wide water treatment control system can be designed for the production of water with a higher biological stability.

ACKNOWLEDGEMENTS

This research was financially supported by the European Commission through the project TECHNEAU.

REFERENCES

Bosklopper, Th. G. J., Rietveld, L.C., Babuska, R., Smaal, B. and Timmer, J. (2004). Integrated operation of drinking water treatment plant at Amsterdam water supply, *Water Science and Technology: Water Supply*, **4**(5–6), 263–270.

Dudley, J., Dillon, G., Ross, P.S., Rietveld, L.C. and van Schagen, K.M. (2007). The European Water Treatment Simulator. In *Proceedings of Water Management Challenges in Global Change*, Leicester, United Kingdom, September 2007.

Fabris, R., Chow, C.W.K., Drikas, M. and Eikebrokk, B. (2008). Comparison of NOM character in selected Australian and Norwegian drinking waters. *Water Research*, **42**, 4188–4196.

Juhna, T. and Rubulis, J. (2004). Problem of DOC removal during biological treatment of surface water with a high amount of humic substances. *Water Science & Technology: Water Supply*, **4**(4), 183–187.

Kruithof, J.C. and Masschelein, W.J. (1999). State-of-the-Art of the Application of Ozonation in BENELUX Drinking Water Treatment. *Ozone: Science & Engineering*, **21**(2), 139–152.

Muller, K.C., Forster, R., Gammeter, S. and Hambsch, B. (2003). Influence of ozonated cyanobacteria on bacterial growth in rapid sand filters. *Journal of Water Supply: Research and Technology – AQUA*, **52**(5), 333–340.

Rietveld, L.C. (2005). *Improving operation of drinking water treatment through modelling*. PhD thesis, Faculty of Civil Engineering and Geosciences. Delft University of Technology, the Netherlands.

Rosa, M.J., Cecilio, T, Ribau Teixeira, M., Viriato, M., Coelho, R. and Lucas, H. (2004). Monitoring of hazardous substances at Alchantarilha's water treatment plant, Portugal. *Water Science and Technology: Water Supply*, **4**(5), 343–353.

Ross, P.S., van der Helm, A.W.C., van den Broeke, J., van der Aa, L.T.J. and Rietveld, L.C. (2008). Effect of raw water quality on performance of ozone and biofiltration based on modeling and on-line monitoring. In *Proceedings of WISA*, Sun City, South Africa.

USEPA (1989). Guidance manual for compliance with the filtration and disinfection requirements for public water systems using surface water supplies. USEPA, Washington, D.C.

van den Broeke, J., Ross, P.S., van der Helm, A.W.C., Baars, E.T. and Rietveld, L.C. (2007). Use of on-line UV/Vis-spectrometry in the measurement of dissolved ozone and AOC concentrations in drinking water treatment. *Water Science and Technology*, **57**(8), 1169–1175.

van der Kooij, D. (2000). Biological stability: a multidimensional quality aspect of treated water. *Water, Air and Soil Pollution*, **123**, 25–34.

van der Helm, A.W.C. (2007). *Integraded modeling of ozonation for optimization of drinking water treatment*. PhD thesis. Faculty of Civil Engineering and Geosciences, Delft University of Technology, the Netherlands.

van der Helm, A.W.C., van Schagen, K.M. and Rietveld, L.C. (2009). Modeling of Full-scale Drinking Water Treatment Plants with Embedded Plant Control within Stimela. In *Proceedings of Instrumentation Control and Automation*, Cairns, Australia.

van Schagen, K.M. (2009). *Model-Based Control of Drinking-Water Treatment Plants*. PhD thesis, Delft Center for Systems and Control, Delft University of Technology, Delft, the Netherlands.

Vreeburg, J.H.G., Schippers, D., Verberk, J.Q.J.C. and van Dijk, J.C. (2008). Impact of particles on sediment accumulation in a drinking water distribution system. *Water Research*, **42**, 4233–4242.

Adsorption and biodegradation of natural organic matter in biological granular activated carbon filters: model validation

L.T.J. van der Aa[a], L.C. Rietveld[b] and J.C. van Dijk[b]

[a]Waternet, P.O. Box 94370, 1090 GJ Amsterdam, The Netherlands
(E-mail: rene.van.der.aa@waternet.nl)
[b]Delft University of Technology, P.O. Box 5048, 2600 GA Delft, The Netherlands
(E-mail: l.c.rietveld@tudelft.nl; j.c.vandijk@tudelft.nl)

Abstract A dynamic simulation model was developed to simulate adsorption and biodegradation of natural organic matter (NOM) in (biological) granular activated carbon ((B)GAC) filters. The calibration parameters were derived from batch-wise determined NOM adsorption isotherms and kinetic experiments. The microbiological parameters were derived from dominant bacterial species and from literature. Model results were compared with three (B)GAC filters and a tailor-made biological granular non activated carbon (BGNAC) filter. The model was able to predict the S-shape NOM breakthrough curves from the pilot filters roughly. However, the model was not accurate enough to predict the effect of pre-ozonation dosage on NOM removal in the adsorption phase. At filter run times longer than 200 days, the model predicted a reduction of adsorbed NOM at increasing pre-ozonation dosages. This results in less competition, pre-loading and pore blocking and therefore in more effective removal of micro-pollutants. The model was found suitable for such qualitative predictions of the effects of changing raw water conditions and operations and for design considerations in (B)GAC filters.

Keywords adsorption, biodegradation, biological granular activated carbon filtration, model, natural organic matter

INTRODUCTION

In drinking water treatment activated carbon (AC) is used for the removal of different kinds of organic compounds, including natural organic matter (NOM), disinfection by-products, organic micro-pollutants and substances causing color, taste and odor. Oxidation, especially ozonation, prior to granular activated carbon (GAC) filtration enhances biological processes, resulting in biological granular activated carbon (BGAC) filtration (Sontheimer *et al.*, 1988; Graveland, 1994). In BGAC filters adsorption and biodegradation occur simultaneously. NOM significantly hinders the adsorption of other target compounds due to pore blocking, competition and pre-loading (Carter *et al.*, 1992; Carter and Weber, 1994; Knappe *et al.*, 1997). Therefore, NOM is a key parameter in

(B)GAC filtration. Many different mathematical models were developed to predict either adsorption in GAC filters or biodegradation of NOM and/or organic micro-pollutants in biofilters (Sontheimer *et al.*, 1988). Models for (B)GAC filters filled with exhausted AC, which is not regenerated, exist, but only include biodegradation (Billen *et al.*, 1992). These (B)GAC models usually were developed to optimize the design of these filters (Hand *et al.*, 1997). Nowadays many drinking water treatment plants (DWTPs) have installed (B)GAC filters. Because influent water quality and operations vary in time, the need for dynamic models for optimization of operational conditions increases (Bosklopper *et al.*, 2004; van der Helm *et al.*, 2008).

The objective of this study was to validate a developed dynamic simulation model for simultaneous NOM adsorption and biodegradation in (B)GAC filters. Model parameters were calibrated from laboratory experiments, from pilot experiments with exhausted AC and from literature data. Four pilot (B)GAC filters were operated with different pre-ozone dosages to generate different NOM breakthrough curves with simultaneous adsorption and biodegradation for model validation.

MATERIALS AND METHODS

Model description

A dynamic white-box model was developed to simulate NOM adsorption and biodegradation in (B)GAC filters. In the model the NOM is splitted into 3 fractions: non adsorbable non biodegradable NOM ($NOM_{NA\text{-}NB}$), adsorbable non biodegradable NOM ($NOM_{A\text{-}NB}$) and adsorbable biodegradable NOM ($NOM_{A\text{-}B}$). The non adsorbable fraction was assumed not biodegradable. In the model influent water is led into the (B)GAC filter. Convection transports the dissolved substances through the filter bed pores. The AC grains are surrounded by a laminar liquid film. The dissolved substances diffuse through this film from the bulk liquid towards the surface of the AC grains (Sontheimer *et al.*, 1988). Biomass is located on the outer surface of the AC grains. $NOM_{A\text{-}B}$ is biodegraded according to Monod growth kinetics based on the locally available concentrations $NOM_{A\text{-}B}$, oxygen and phosphate. Parts of the substrates convert into biomass. Biomass growth rate increases with water temperature. Suspended biomass, attachment and detachment of bacteria are neglected. Biomass death rate is linear to the amount of fixed biomass. Dead biomass is assumed to be washed out during backwashing. Biomass maintenance is neglected. $NOM_{A\text{-}NB}$ and the remaining $NOM_{A\text{-}B}$ diffuse into the pores of the AC grain via surface diffusion. Internal pore diffusion is neglected.

Inside the pores of the AC non reversible adsorption of $NOM_{A\text{-}NB}$ and $NOM_{A\text{-}B}$ takes place according to Freundlich adsorption isotherms combined with the ideal adsorption solution theory (IAST) (Sontheimer *et al.*, 1988; Kümmel and Worch, 1990).

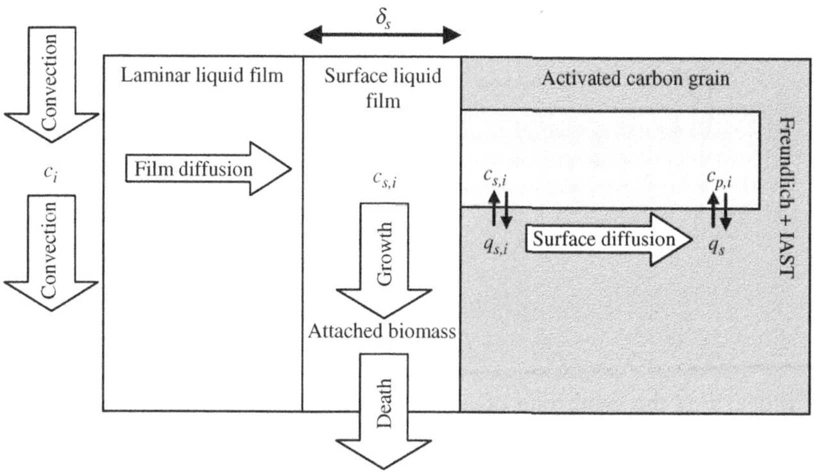

Figure 1. Model scheme (for explanation of symbols see Equations 1 to 7).

The model was implemented in "Stimela", a Matlab-Simulink tool for modeling drinking water processes (van der Helm and Rietveld, 2002; Rietveld, 2005). In the model the (B)GAC filter is divided into a number of completely stirred tank reactors (CSTRs) in series. The model scheme for each CSTR is shown in Figure 1. For each CSTR mass balances are made for the bulk liquid, the surface liquid film and the AC grain. From these mass balances the variables are calculated dynamically, see Equations 1 to 4. Equation 5 is used to calculate the specific growth rate of the biomass (Billen *et al.*, 1992). Equation 6 is used to calculate the surface diffusion mass transfer coefficient (Glueckauf, 1955).

$$\frac{\partial c_i}{\partial t} = -\frac{v}{\varepsilon}\frac{\partial c_i}{\partial x} - \frac{6(1-\varepsilon)\beta_{f,i}}{\varepsilon \cdot d_{\text{part}}}\left(c_i - c_{s,i}\right) \tag{1}$$

$$\frac{dc_{s,i}}{dt} = \frac{\beta_{f,i}}{\delta_s}(c_i - c_{s,i}) - \frac{\rho_{\text{filt}} \cdot \beta_{s,i}}{\delta_s(1-\varepsilon)}\left(q_{s,i} - \overline{q_i}\right) - \frac{\mu_{\text{spec}}}{Y_i} \cdot X_s \cdot \frac{\rho_{\text{filt}} \cdot d_{\text{part}}}{6 \cdot \delta_s(1-\varepsilon)} \tag{2}$$

$$\frac{d\overline{q_i}}{dt} = \frac{6\beta_{s,i}}{d_{\text{part}}}\left(q_{s,i} - \overline{q_i}\right) \tag{3}$$

$$\frac{dX_s}{dt} = (\mu_{\text{spec}} - b) \cdot X_s \tag{4}$$

$$\mu_{\text{spec}} = \min\left(\left(\frac{c_{s,1}}{K_{S,i}+c_{s,1}} \cdot \frac{e^{-\left(\frac{60-T}{44}\right)^2}}{e^{-\left(\frac{45}{44}\right)^2}} \cdot \mu_{\max,15^\circ\text{C}}\right), \left(\frac{c_{s,2}}{K_{S,i}+c_{s,2}} \cdot \frac{e^{-\left(\frac{60-T}{44}\right)^2}}{e^{-\left(\frac{45}{44}\right)^2}} \cdot \mu_{\max,15^\circ\text{C}}\right), \text{etc.}\right) \tag{5}$$

$$\beta_{s,i} = 10\frac{D_{s,i}}{d_{\text{part}}} \tag{6}$$

$$q_i = K_{F,i} \cdot c_{p,i}^{n_i} \tag{7}$$

in which, b is the death rate of biomass (s^{-1}), $\beta_{f,i}$ is the film diffusion mass transfer coefficient of component i ($m \cdot s^{-1}$), $\beta_{s,i}$ is the surface diffusion mass transfer coefficient of component i ($m \cdot s^{-1}$), c_i is the liquid phase concentration of component i in the bulk liquid (g $C \cdot m^{-3}$), $c_{s,i}$ is the liquid phase concentration of component i in the surface liquid film (g $C \cdot m^{-3}$), $c_{p,i}$ is the liquid phase concentration of component i in the pore liquid (g $C \cdot m^{-3}$), d_{part} is the AC particle diameter (m), $D_{s,i}$ is the surface diffusion coefficient of component i ($m^2 \cdot s^{-1}$), δ_s is thickness of the surface liquid film (m), ε is the filter bed porosity (−), $K_{F,i}$ is the mass based Freundlich constant of component i (g $C \cdot$(g AC)$^{-1} \cdot$(g $C \cdot m^{-3})^{-n}$), $K_{S,i}$ is the Monod half velocity constant for component i (g $C \cdot m^{-3}$), n_i is the Freundlich exponent of component i (−), q_i is the solid phase concentration of component i (g $C \cdot$(g AC^{-1})), $q_{s,i}$ is the solid phase concentration of component i at the outer surface of the AC particle (g $C \cdot$(g AC^{-1})), ρ_{filt} is the AC mass density of a packed filter bed ($g \cdot m^{-3}$), t is the time (s), T is the temperature (°C), $\mu_{\max,15^\circ\text{C}}$ is the maximum growth rate of biomass at 15°C (s^{-1}), μ_{spec} is the specific growth rate of the biomass (s^{-1}), v is the surface load of the filter ($m \cdot s^{-1}$), x is the distance in direction of flow (m), X_s is the total activity from biomass attached onto the AC (g ATP$\cdot$(g AC)$^{-1}$) and Y_i is the yield of biomass on component i (g ATP$\cdot$(g C)$^{-1}$).

In Equations 2 and 3 $q_{s,i}$ is based on adsorption equilibrium according to Freundlich (Equation 7), combined with IAST (Equations 8, 9 and 10). IAST is valid for molar liquid and solid phase concentrations. Therefore, in the IAST module Equations 11 and 12 are used instead of Equation 7. Note that NOM usually is expressed in g $C \cdot m^{-3}$. Therefore, in Equation 12 for NOM only the molar weight of the C-atoms is used.

$$\sum_{i=1}^{N} z_i = \sum_{i=1}^{N} \frac{c_{p,i,M}}{\left(\frac{\Theta \cdot n_i}{K_{i,M}}\right)^{1/n_i}} = \sum_{i=1}^{N} \frac{q_{p,i,M}}{\Theta \cdot n_i} = 1 \tag{8}$$

$$q_{t,M} = \left(\sum_{i=1}^{N} \frac{z_i}{\Theta \cdot n_i}\right)^{-1} \tag{9}$$

$$q_{i,M} = z_i \cdot q_{t,M} \tag{10}$$

$$q_{i,M} = K_{F,i,M} \cdot c_{p,i,M}^{n_i} \tag{11}$$

$$K_{F,i,M} = K_{F,i} \cdot MW_i^{n_i - 1} \tag{12}$$

in which, $c_{p,i,M}$ is the molar liquid phase concentration of component i in the pore liquid ($mol \cdot m^{-3}$), Θ is the spreading pressure ($mol \cdot (g\ AC^{-1})$), $K_{F,i,M}$ is the molar Freundlich constant of component i ($g\ C \cdot (g\ AC)^{-1} \cdot (g\ C \cdot m^{-3})^{-n}$), MW_i is the molar weight of the C-atoms of component i ($g \cdot mol^{-1}$), $q_{i,M}$ is the molar solid phase concentration of component i ($mol \cdot (g\ AC^{-1})$), $q_{t,M}$ is the total molar solid phase concentration of all components ($mol \cdot (g\ AC^{-1})$) and z_i is the mol fraction of component i adsorbed (−).

Model parameters

NOM isotherms for ozonated and non ozonated (B)GAC influents were determined in batch experiments according the bottle-point method using entire AC grains (van der Aa *et al.*, 2006a). Fictive component analysis (Sontheimer *et al.*, 1988) and IAST were used to describe the different NOM fractions. The ratio between biodegradable and non biodegradable adsorbable fractions was estimated from other pilot plant experiments (data not included). The results are summarized in Table 1.

External mass transport was calculated from film diffusion coefficients (Sontheimer *et al.*, 1988). In a separate experiment the internal kinetic parameters were determined in ozonated BGAC influent (data not shown). All NOM fractions were assumed to have the same kinetic characteristics. Biomass

Table 1. NOM isotherms.

Ozone dosage	NOM	$NOM_{NA\text{-}NB}$			$NOM_{A\text{-}NB}$			$NOM_{A\text{-}B}$		
($g\ O_3 \cdot g\ C^{-1}$)	$c_{i,0}$	$c_{i,0}$	$K_{F,i}$	n_i	$c_{i,0}$	$K_{F,i}$	n_i	$c_{i,0}$	$K_{F,i}$	n_i
0 (measured)	6.2	0.88	0	–	4.26	14.2	1.047	1.06	14.2	1.047
1.5 (interpolated)	5.79	0.84	0	–	3.71	14.82	0.934	1.24	14.82	0.934
2.0 (measured)	5.65	0.83	0	–	3.49	15.02	0.896	1.33	15.02	0.896
2.5 (extrapolated)	5.51	0.82	0	–	3.28	15.23	0.858	1.41	15.23	0.858

In which, $c_{i,0}$ is the initial concentration of component i (g C · m-3), $K_{F,i}$ is the Freundlich constant of component i (103 · g C · (g AC)-1 · (g C · m-3)-n), n_i is the Freundlich component of component i (–).

characteristics were taken from measurements on *Polaromonas*, which was one of the dominant bacterial species in the (B)GAC filter in the pilot plant (Magic-Knezev and Van der Kooij, 2006). Other biomass parameters were estimated from literature values. Table 2 shows an overview of the used model parameters.

Pilot plant

At DWTP Weesperkarspel of Waternet, the water cycle company of Amsterdam and surrounding areas, a pilot plant was operated for 176 days. The raw water from the Bethunepolder water was pre-treated in the full-scale plant with coagulation, sedimentation, a reservoir and rapid sand filtration. This pre-treatment reduced the dissolved organic carbon (DOC) concentrations from 8 to 10 $g\ C \cdot m^{-3}$ to 6 $g\ C \cdot m^{-3}$. The pilot plant consisted of 2 ozonation systems, followed by 4 identical gravity (B)GAC filters operated in down flow mode. The filter media was Norit GAC830 for the filters W-GAC8, W-BGAC6 and W-BGAC7. Filter W-BGNAC5 was filled with a tailor-made granular non activated carbon (GNAC): the production procedure was similar to that of GAC830, only it was not activated. The influent waters of filters W-GAC8, W-BGNAC5, W-BGAC6, and W-BGAC7, received ozone dosages of 0, 1.5, 1.5 and 2.5 $g\ O_3 \cdot m^{-3}$, respectively. In all filters flow rates were kept constant at $0.88 \cdot 10^{-3}\ m \cdot s^{-1}$ and EBCT was 2394 s. During operation the (B)GAC filters were backwashed with air and water every 4 to 20 days, depending on the increase rate of filter headloss. The columns were monitored on influent and effluent, once per two or once per four weeks. Measured parameters included DOC, UV-extinction at a wavelength of 254 nm, AOC, pH and temperature. Ozone was analyzed according to the DPD method (Gilbert, 1981). AOC was measured in duplicate, using the simultaneous incubation of

Table 2. Model parameters.

Parameter	Model value	Reference values / notes
b (s^{-1})	$1.6 \cdot 10^{-6}$	$1.2 \cdot 10^{-6}$ (Rittmann and McCarty, 1980) $2.1 \cdot 10^{-6}$ (van der Kooij, 1984) $2.8 \cdot 10^{-6}$–$2.2 \cdot 10^{-5}$ (Billen *et al.*, 1992) $2.3 \cdot 10^{-6}$ (Hozalski, 1996) $1.6 \cdot 10^{-6}$ (Magic-Knezev, personal comm.)
$D_{f,NOM}$ ($m \cdot s^{-1}$)	$1.2 \cdot 10^{-9}$	$9.3 \cdot 10^{-10}$ (Rittmann and McCarty, 1980) $6 \cdot 10^{-11}$–$2.5 \cdot 10^{-10}$ (Sontheimer *et al.*, 1988) $4 \cdot 10^{-10}$–$10 \cdot 10^{-10}$ (van Lier, 1989) $1.2 \cdot 10^{-9}$ (Hozalski, 1996)
$D_{s,NOM}$ ($m \cdot s^{-1}$)	$2.4 \cdot 10^{-15}$	$3.4 \cdot 10^{-15}$–$2.9 \cdot 10^{-11}$ (Sontheimer *et al.*, 1988) $8.0 \cdot 10^{-16}$–$1.2 \cdot 10^{-12}$ (van Lier, 1989) $2.4 \cdot 10^{-15}$ (Experimental Work, unpublished)
d_{part} (m)	0.0012	Measured d_{50} value was used
δ_s (m)	0.00001	Assumption, no impact on model results
ε (–)	0.4	Measured value
$K_{S,NOM\_A\text{-}B}$ ($g\ C \cdot m^{-3}$)	$1.5 \cdot 10^{-3}$	10 (Rittmann and McCarty, 1980) $2.8 \cdot 10^{-3}$–0.372 (van der Kooij *et al.*, 1982c) $0.05 \cdot 10^{-3}$–0.25 (Billen *et al.*, 1992) 0.19–54.7 (Hozalski, 1996) $1.5 \cdot 10^{-3}$ (Magic-Knezev and Van der Kooij, 2006)
MW_{NOM} ($g \cdot mol^{-1}$)	712	712 (Grefte *et al.*, 2007)
ρ_{filt} ($g \cdot m^{-3}$)	$445 \cdot 10^3$	Supplier information
$\mu_{max,\ 15°C}$ (s^{-1})	$1.9 \cdot 10^{-5}$	$9.3 \cdot 10^{-5}$ (Rittmann and McCarty, 1980) $5.6 \cdot 10^{-5}$–$2.7 \cdot 10^{-4}$ (Billen *et al.*, 1992) $1.5 \cdot 10^{-6}$–$6.9 \cdot 10^{-5}$ (van der Kooij *et al.*, 1982c) $1.9 \cdot 10^{-5}$ (Magic-Knezev and Van der Kooij, 2006)
Y_l ($g\ ATP \cdot (g\ C)^{-1}$)	$0.8 \cdot 10^{-6}$	$0.8 \cdot 10^{-6}$–$1.0 \cdot 10^{-6}$ (Magic-Knezev and Van der Kooij, 2006)

strains P17 and NOX (van der Kooij *et al.*, 1982b). Other water quality parameters were determined according to standard procedures (Standard Methods, 2005). Table 3 shows an overview of the pilot plant characteristics and influent water quality.

Table 3. Characteristics and influent water quality of pilot plant Weesperkarpsel.

Column	W-GAC8 Run 2	W-BGNAC5 Run 2	W-BGAC6 Run 2	W-BGAC7 Run 2
Column diameter (m)	0.25	0.25	0.25	0.25
Bed height (m)	2.10	2.10	2.1	2.1
Carbon type (–)	GAC830	GNAC830	GAC830	GAC830
Operation time (s)	$176.24 \cdot 3600$	$176 \cdot 24 \cdot 3600$	$176 \cdot 24 \cdot 3600$	$176 \cdot 24 \cdot 3600$
Flow ($m^3 \cdot s^{-1}$)	$4.3 \cdot 10^{-5}$	$4.3 \cdot 10^{-5}$	$4.3 \cdot 10^{-5}$	$4.3 \cdot 10^{-5}$
Surface load ($m \cdot s^{-1}$)	$8.8 \cdot 10^{-4}$	$8.8 \cdot 10^{-4}$	$8.8 \cdot 10^{-4}$	$8.8 \cdot 10^{-4}$
EBCT (s)	2394	2394	2394	2394
Ozone dosage (g $O_3 \cdot m^{-3}$)	–	0.5	1.5	2.5
Temp (°C)	9.7 (5.8–17.1)	11.1 (5.9–18.8)		11.0 (5.8–18.9)
pH (–)	7.6 (7.6–7.7)	7.7 (7.6–7.8)		7.6 (7.6–7.7)
DOC (g C $\cdot m^{-3}$)	6.0 (5.8–6.2)	5.8 (5.5–6.0)		5.7 (5.3–6.0)
UV254 (m^{-1})	14.5 (13.6–15.8)	10.3 (9.5–11.5)		8.2 (7.4–8.9)
Oxygen (g $O_2 \cdot m^{-3}$)	9.1 (7.3–10.4)	10.9 (9.2–12.3)		11.4 (9.7–12.7)
AOC (10^{-3} g Acetate-C $\cdot m^{-3}$)	9.9 (7.5–13.6)	132 (90–185)		176 (147–242)

Note water quality parameters: table values are average concentrations; values between parentheses are minimum and maximum concentrations.

RESULTS AND DISCUSSION

NOM adsorption and biodegradation

Figure 2 shows both measured and simulated DOC breakthrough curves and the simulated concentrations of adsorbed DOC. Initial breakthrough in the model simulations was higher than in the pilot (B)GAC filters. Apparently the amount of $NOM_{NA\text{-}NB}$ at the start of the filter runs was not the same as in the water sample taken for the isotherm experiments. The S-shape breakthrough curves predicted by the model corresponded roughly with the pilot plant data. However, the observed differences between the three (B)GAC pilot filters were not predicted by the model. At filter run times longer than 200 days the model predicted less adsorption, more biodegradation and in total more effective DOC removal at increasing pre-ozonation dosages. The decrease in q_{DOC}, see Figure 2 bottom graph, results in less competition, pre-loading and pore blocking by NOM, making the removal of micro-pollutants more effective. The pilot plant was not operated long enough to confirm these model results. However, such observations were made in other pilot experiments (van der Aa *et al.*, 2006b).

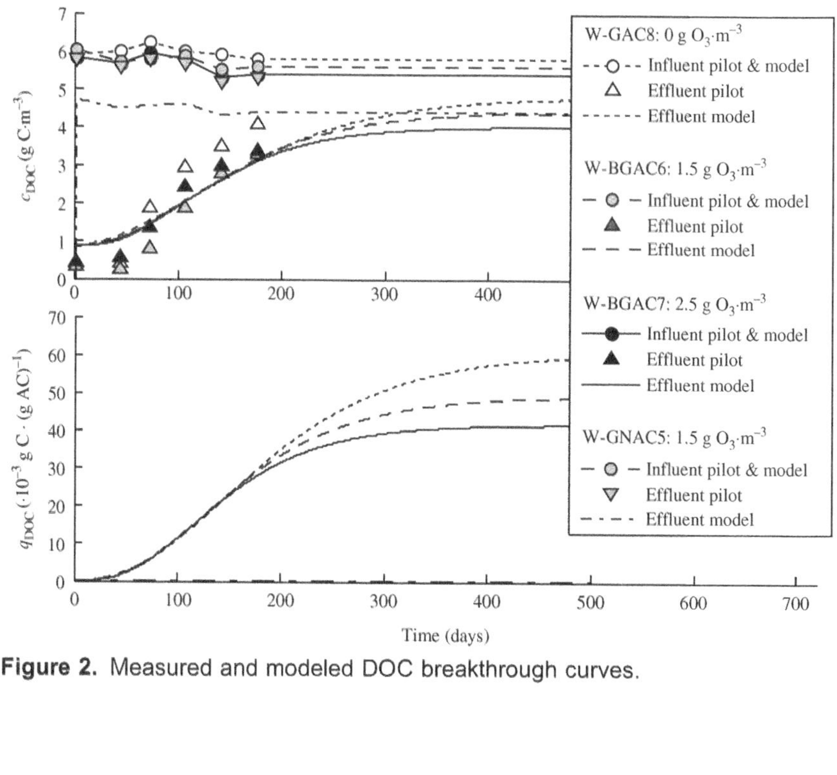

Figure 2. Measured and modeled DOC breakthrough curves.

During the experiment (B)GAC was more suitable for removal of biodegradable NOM than GNAC. This was not predicted by the model. The model predicted the removal of almost all biodegradable NOM in each filter.

Biomass development

After one week the model predicted biodegradation of more than 99% of the amount of biodegradable NOM and biomass activity reached values of approximately $5 \cdot 10^{-9}$ g ATP $\cdot$ (g AC)$^{-1}$. In the pilot filters biomass activity developed slowly, and increased until the end of the experiment, to maximum values of $215 \cdot 10^{-9}$ to $328 \cdot 10^{-9}$ g ATP $\cdot$ (g AC)$^{-1}$ for the (B)GAC filters, while not all the biodegradable NOM was removed (data not included).

Evaluation

The model can be used for qualitative predictions of the effects of changing raw water conditions and operations and for design considerations in (B)GAC filters. However, the model was not accurate enough to make quantitative predictions of these effects. The adsorption isotherm and the kinetic adsorption parameters were determined only once and twice respectively. Combined with the error in the analytical method for DOC this made the calibration values for the adsorption process not very accurate. Improvement of the biodegradation aspects of the model is necessary to make the model suitable for prediction of NOM biodegradation in biofilters filled with other media than AC. The differences between modeled and observed biomass amounts could be explained by a 1000 times larger yield (van der Kooij *et al.*, 1982a; Velten *et al.*, 2007) and/or a 10 times larger death rate due to predation (Billen *et al.*, 1992). Usually the ATP concentration in the influent was higher than in the effluent (data not included). This means that besides growth and death, attachment and detachment of suspended biomass should be included in the model (Billen *et al.*, 1992; Uhl, 2000). By "tuning" the biological parameters the ATP profile as measured in the pilot plant could be approached.

To improve the accuracy of the model predictions of NOM adsorption and biodegradation following suggestions are made: a) increase accuracy of NOM isotherms and b) include attachment and detachment of biomass. It should be kept in mind that the model remains a simplification of the complex reality, in which NOM is a heterogeneous mixture of hundreds of different substances, consequently resulting in a heterogeneous biomass in the filters. Even when the simulation models improve, especially for surface water they have to cope with variations in raw water quality and operations. In such cases either faster and less labor intense model calibration procedures are needed, or the simulation model should be used for qualitative purposes only.

CONCLUSIONS

The developed simulation model combined biodegradation and adsorption of NOM in one model. The model was able to describe the S-shaped NOM breakthrough curves in (B)GAC filters roughly. However, minor changes in raw water quality and the effects of pre-ozonation on NOM removal were not predicted accurately. The model can be used for qualitative predictions of the effects of changing raw water conditions and operations and for design considerations in (B)GAC filters. The most interesting observation was that the model predicted a reduction of q_{DOC} at increasing ozone dosages, especially at long filter run times. This results in less competition, pre-loading and pore blocking by NOM and therefore in more effective removal of micro-pollutants.

ACKNOWLEDGEMENTS

This research was conducted by Waternet, Vitens, KWR, Norit, Delft University of Technology and Wageningen University and Research. The research was co-funded by the SenterNovem, agency of the Dutch Ministry of Economic Affairs.

REFERENCES

Billen, G., Servais, P., Bouillot, P. and Ventresque, C. (1992). Functioning of biological filters used in drinking-water treatment – the Chabrol model. *Journal of Water Supply: Research and Technology – AQUA*, **41**(4), 231–241.

Bosklopper, T.G.J., Rietveld, L.C., Babuska, R., Smaal, B. and Timmer, J. (2004). Integrated operation of drinking water treatment plant at Amsterdam water supply. *Water Supply*, **4**(5–6), 263–270.

Carter, M.C. and Weber, W.J., Jr. (1994). Modeling adsorption of TCE by activated carbon preloaded by background organic matter. *Environmental Science & Technology*, **28**(4), 614–623.

Carter, M.C., Weber, W.J., Jr. and Olmstead, K.P. (1992). Effects of background dissolved organic matter on TCE adsorption by GAC. *Journal American Water Works Association*, **84**(8), 81–91.

Gilbert, E. (1981). Photometrische Bestimmung niedriger Ozonkonzentrationen in Wasser mit Hilfe von Diaëthyl-p-phenylendiamin (DPD) (Photometric determination of low ozone concentrations in water dusing Diethyl-p-phenylenediamine (DPD)). *GWF Wasser Abwasser*, **122**(9), 410–416.

Glueckauf, E. (1955). Theory of chromatography. Part 10. Formulæ for diffusion into spheres and their application to chromatography. *Transactions of the Faraday Society*, **51**, 1540–1551.

Graveland, A. (1994). Application of biological activated carbon filtration at Amsterdam Water Supply. *Water Supply*, **14**, 233–241.

Grefte, A., Dignum, M., Rietveld, L.C., Cornelissen, E.R. and Amy, G.L. (2007). Removal of NOM fractions by different ion exchange (IEX) and adsorbent resins and effect on the performance of activated carbon. *Water Quality and Technology Conference*, Charlotte, North Carolina, USA.

Hand, D.W., Crittenden, J.C., Hokanson, D.R. and Bulloch, J.L. (1997). Predicting the performance of fixed-bed granular activated carbon filters. *Water Science & Technology*, **35**(7), 235–241.

Hozalski, R.M. (1996). Removal of biodegradable organic matter in drinking water biofilters: Experimental studies and model development, PhD, John Hopkins University, Ann Arbor, Michigan, USA.

Knappe, D.R.U., Snoeyink, V.L., Roche, P., Prados, M.J. and Bourbigot, M.-M. (1997). The effect of preloading on rapid small-scale column test predictions of atrazine removal by GAC adsorbers. *Water Research*, **31**(11), 2899–2909.

Kümmel, R. and Worch, E. (1990). Adsorption aus wäßrigen Lösungen (Adsorption from aqueous solutions), VEB Deutscher Verlag für Grundstoffindustrie, Leipzig, Germany.

Magic-Knezev, A. and Van der Kooij, D. (2006). Nutritional versatility of two Polaromonas related bacteria isolated from biological granular activated carbon filters. In: *Recent Progress in Slow Sand and Alternative Biofiltration Processes*, R. Gimbel, N. Graham and R. Collins (eds.), IWA Publishing.

Rietveld, L.C. (2005). Improving operation of drinking water treatment through modeling, PhD, Delft University of Technology, Delft, The Netherlands.

Rittmann, B.E. and McCarty, P.L. (1980). Model of steady-state-biofilm kinetics. *Biotechnology and Bioengineering*, **22**(11), 2343–2357.

Sontheimer, H., Crittenden, J.C. and Summers, R.S. (1988). Activated carbon for water treatment, AWWA – DVGW Forschungsssteile Engler Bunte Institut, Karlsruhe, Germany.

Standard Methods. (2005). Standard Methods for the examination of water & wastewater 2005, American Public Health Association/American Water Works Association/ Water Environment Federation, Baltimore, USA.

Uhl, W. (2000). Einfluss von Schüttungsmaterial und Prozessparmetern auf die Leistung von Bioreaktoren bei der Trinkwasseraufbereitung (Influence of filter material and process parameters on the performance of bioreactors in the drinking water treatment), PhD, IWW Rheinisch-Westfällisches Institut für Wasserforschung gemeinnützige GmbH, Institut an der Gerhard-Mercator-Universität Duisburg, Mülheim an der Ruhr, Germany.

van der Aa, L.T.J., Magic-Knezev, A., Rietveld, L.C. and van Dijk, J.C. (2006a). Biomass development in biological activated carbon filters. In: *Recent Progress in Slow Sand and Alternative Biofiltration Processes*, R. Gimbel, N. Graham and R. Collins (eds.), IWA Publishing.

van der Aa, L.T.J., Rietveld, L.C., Siegers, W.G. and van Dijk, J.C. (2006b). Adsorption of natural organic matter and atrazin on granular activated carbon filters. Workshop Delft University of Technology/IHE-UNESCO "Developments in drinking water treatment modelling", Delft, The Netherlands.

van der Helm, A.W.C. and Rietveld, L.C. (2002). Modelling of drinking water treatment processes within the Stimela environment. *Water Supply*, **2**(1), 87–93.

van der Helm, A.W.C., Rietveld, L.C., Bosklopper, T.G.J., Kappelhof, J.W.N.M. and van Dijk, J.C. (2008). Objectives for optimization and consequences for operation, design and concept of drinking water treatment plants. *Water Science & Technology: Water Supply*, **8**(3), 297–304.

van der Kooij, D. (1984). The growth of bacteria on organic compounds in drinking water, PhD, Landbouwhogeschool/KIWA N.V., Wageningen the Netherlands.

van der Kooij, D., Oranje, J.P. and Hijnen, W.A.M. (1982a). Growth of Pseudomonas aeruginosa in tap water in relation to utilization of substrates at concentrations of a few micrograms per liter. *Applied and Environmental Microbiology*, **44**(5), 1086–1095.

van der Kooij, D., Visser, A. and Hijnen, W.A.M. (1982b). Determining the concentration of easily assimilable organic carbon in drinking water. *Journal American Water Works Association*, **74**(10), 540–545.

van der Kooij, D., Visser, A. and Oranje, J.P. (1982c). Multiplication of fluorescent pseudomonads at low substrate concentrations in tap water. *Antonie van Leeuwenhoek*, **48**(3), 229–243.

van Lier, W. (1989). Mass transfer to activated carbon in aqueous solutions, PhD, Delft University of Technology, Delft, The Netherlands.

Velten, S., Hammes, F., Boller, M. and Egli, T. (2007). Rapid and direct estimation of active biomass on granular activated carbon through adenosine tri-phosphate (ATP) determination. *Water Research*, **41**(9), 1973–1983.

Modelling of NOM removal by coagulation

B. Eikebrokk, S.W. Østerhus and G. Thorvaldsen

SINTEF Building and Infrastructure, N-7465 Trondheim, Norway
(E-mail: Bjornar.Eikebrokk@sintef.no)

Abstract Natural organic matter (NOM) affects most water treatment and distribution processes, and can completely control coagulation. In order to remove NOM, enhanced coagulation involving elevated coagulant doses and strict pH-control is widely used. In order to optimize design and operation of complex enhanced coagulation processes, modelling and simulations can be very valuable. Within the EU-financed R&D project Techneau significant efforts are made to develop an applicable water treatment simulator including various unit process models. This paper discusses NOM removal by enhanced coagulation processes: 1) good reasons for removing NOM (DOC) from drinking water, 2) NOM coagulation mechanisms, and 3) models for the prediction of DOC residuals and coagulant demand. Priority models are suggested for further testing and calibration within Techneau.

Keywords DOC removal; enhanced coagulation; modelling; optimization; water treatment

INTRODUCTION

Natural organic matter (NOM) plays a major role in water treatment and distribution systems and 12 good reasons why NOM should be removed from drinking water are presented below:

1. Affect colour, taste and odour.
2. Increase coagulant demand and control coagulation processes (high SUVA).
3. Increase disinfectant demands and form DBPs.
4. Affects UV-disinfection processes (decrease UV transmittance and can contribute to fouling of lamp sleeves and UV-sensor windows).
5. Affect stability and removal of inorganic substances and turbidity.
6. Affect stability and removal of pathogens and treatment barrier efficiency.
7. Foul membranes.
8. Block AC pores/compete for adsorption sites.
9. Show great seasonal and spatial variability in concentration and composition, thereby imposing significant challenges to process control systems.
10. Affect corrosion processes.
11. Affect biostability, biofilm formation and regrowth.
12. Increase mobility and human exposure to micro pollutants.

 TECHNEAU: Safe Drinking Water from Source to Tap. Edited by Theo van den Hoven and Christian Kazner. ISBN: 9781843392750. Published by IWA Publishing, London, UK.

Coagulation is a vital part of most water treatment processes. *Enhanced coagulation* is a specific treatment strategy and among the best available and most widely used technologies for removing NOM (Total Organic Carbon, TOC). Compared to conventional coagulation processes where turbidity removal was the major treatment objective, enhanced coagulation processes employ elevated coagulant doses and strict pH control. In modern water treatment practice, coagulation has at least seven objectives: 1) to maximize removal of turbidity and particles, 2) to maximize removal of pathogens, i.e. to maximize treatment barrier efficiency and stability, 3) to maximize removal of NOM (TOC) and disinfection by-product (DBP) precursors, i.e. "enhanced coagulation", 4) to minimize coagulant residuals, 5) to minimize production of sludge and waste products, and 6) to minimize use of resources like chemicals and energy, and 7) to minimize operating costs.

Optimum coagulation can be defined as the coagulation conditions of dosage and pH that achieve maximum compliance with the given objectives (Edzwald and Tobiason, 1999). Hence, sub-optimum coagulation can lead to poor operation performance, e.g. reduced quality of treated water and reduced treatment barrier efficiency as well as reduced overall water supply safety. Because NOM affects and controls coagulation processes in many cases, knowledge of the seasonal and spatial variability in NOM concentration and composition is important in order to be able to control and optimize the treatment process adequately.

A failure in coagulation can have severe consequences for the water treatment and for the entire water supply system, e.g. poor removal of particles, pathogens and natural organic matter (NOM) and poor disinfection efficiency. This can significantly decrease water safety and result in distribution problems as well, e.g. solids deposition, corrosion, formation of disinfection by-products (DBPs), regrowth and biofilm formation.

On this background, the development of models for various purposes can be very useful. The models can be applied for the a number of purposes, e.g. improved operation performance and better process control, better overview and understanding of complex processes and relationships, for process simulations and education/training. From the models listed below, this paper will however address residual DOC and coagulant demand prediction models only:

- Mathematical TOC/DOC removal prediction models.
- Semi-empiric operation models for predictions of coagulant demand, sludge production, filter run length, net water production, etc.
- Models for the decay of disinfectants (e.g. chlorine, chloramines, ozone).
- Models for DBP formation (e.g. THMs, HAAs).

The Techneau project intends to develop a new-generation water treatment simulator which will also address shortcomings regarding coagulation/flocculation processes where the models of today are too empirical, too data intensive, and have limited applicability range outside the calibration area (Dudley *et al.*, 2008).

NOM COAGULATION MECHANISMS

Over the years, a number of different coagulation mechanisms have been suggested, e.g. electric double layer compression, adsorption and charge neutralization, bridging, enmeshment in precipitate, complexation, and adsorption to metal precipitates.

As an example, Edzwald and Tobiason (1999) proposed two dominating mechanisms for coagulation of natural organic matter (NOM), using fulvic acid (FA) as a model for coagulation of aquatic NOM. FA has a pH-dependent negative charge of about 7.5 μeq/mg C at pH 5.5 and 10 μeq/mg C or greater at pH 7. The first coagulation mechanism involves reactions between aluminium hydrolysis species formed very quickly upon alum addition and FA. The negative charge of FA creates a coagulant demand. Once satisfied by positively charges Al hydrolysis species, there is direct precipitation of the FA.

$$Al(OH)_n^{3-n} + FA^{x-} \rightarrow \underset{\text{(soluble complex)}}{Al(OH)_nFA} \rightarrow \underset{\text{(solid)}}{Al(OH)_nFA_{(s)}} \quad (1)$$

The second mechanism involves reactions between Al hydrolysis species and FA in which the charge of FA is neutralized or reduced followed by adsorption of the complex onto precipitated aluminium hydroxide solids.

$$Al(OH)_n^{3-n} + FA^{x-} \rightarrow \underset{\text{(soluble complex)}}{Al(OH)_nFA} + \underset{\text{(solid)}}{Al(OH)_{3(s)}} \rightarrow \underset{\text{(adsorbed)}}{Al(OH)_nFA \cdot Al(OH)_{3(s)}} \quad (2)$$

Edzwald and Tobiason (1999) also presented aluminium equilibrium constants for 4 and 25°C (Table 1).

More than 45 years ago, Black *et al.* (1963) reported stoichiometric coagulation reactions between coagulant and natural colour. The negative charge associated with the NOM creates a coagulant demand. This demand is pH-dependent, because the NOM charge itself and the positively charged Al coagulation species available for reaction with NOM are both pH dependent. The average charge on the Al hydrolysis species at pH 5.5 is about +1.5 ($Al(OH)_{1.5}^{+1.5}$) decreasing to +0.5 at pH 6.5 ($Al(OH)_{2.5}^{+0.5}$) . Above pH 7, the

Table 1 Aluminium equilibrium constants for 4 and 25°C (Edzwald and Tobiason, 1999)

	4°C	25°C
Hydrolysis	Log β	
$Al^{3+} + H_2O = Al(OH)^{2+} + H^+$	−5.77	−4.99
$Al^{3+} + 2\ H_2O = Al(OH)_2^+ + 2\ H^+$	−11.7	−10.1
$Al^{3+} + 3\ H_2O = Al(OH)_3 + 3\ H^+$	−17.3	−15.0
$Al^{3+} + 4H_2O = Al(OH)_4^- + 4H^+$	−26.1	−23.0
Dissociation of water	Log K_w	
$H_2O = H^+ + OH^-$	−14.78	−14.0

fraction of positively charged Al hydrolysis species is low and decrease with increasing temperature. At pH 6.5 for example, 1 mg/L of soluble Al would provide about 18 μeq/L of positive charge while the negative charge of FA is about 9 μeq/mg C. This yields a stoichiometry for coagulation of about 0.5 mg Al/mg C at pH 6.5, increasing to about 1 mg Al/mg C at pH 7. Shin *et al.* (2007) reported values of 0.47 mg Al/mg DOC at pH 7 and only 0.30 mg Al/mg DOC at pH 6. The lower values reported by Chin *et al.* (2007) are supported by those reported by Eikebrokk (1999) who performed coagulation-contact filtration experiments on soft and humic surface waters with low levels of turbidity and alkalinity. He reported alum and poly-aluminium chloride coagulant dosage requirements of 0.3 to 0.6 mg Al/mg DOC for raw water DOC-levels in the range of 2 to 5 mg/L. The residual coagulant concentration standard value (0.1 mg Al/L) appeared to be the dose determining parameter in this study.

The above analysis by Edzwald and Tobiason (1999) illustrates well why metal based coagulation can be done at lower dosages at low pH, and why coagulation at pH above 7 is difficult and costly, especially with higher water temperatures. Cold water coagulation with alum at pH 5.5 has a higher concentration of $Al(OH)^{2+}$ and Al^{3+} which lowers the dosage for charge neutralization, but coagulation at this low pH may cause problems with residual Al and particle stability resulting in poor clarification and filtration (Edzwald and Tobiason, 1999).

Removal of NOM by coagulation (enhanced coagulation) depends on: 1) the DOC concentration, 2) the nature and composition of DOC (charge, size or molecular weight, hydrophobicity/hydrophilicity), 3) the type of coagulant, 4) the coagulant dosage, and 5) the pH.

Specific ultraviolet absorbance (SUVA), i.e. the absorbance at 254 nm per unit DOC and expressed as m^{-1} of absorbance per mg/L DOC, is used as an operational indicator of the nature of NOM and the effectiveness of coagulation (Edzwald and van Benschoten, 1990). In the past few years, increased efforts are made in order to link NOM characteristics to treatability by coagulation (Chow *et al.*, 2004; Chow *et al.*, 2005; Bèle *et al.*, 2006; Sharp *et al.*, 2006). Rapid resin fractionation, size exclusion chromatography (HPSEC) with UV, fluorescence and/or DOC detectors, and analysis of excitation-emission fluorescence spectra (EEM) are among the most valuable NOM characterization methods applied. Sharp *et al.* (2006) found that the hydrophobic NOM content controls the coagulant demand, and that the raw water hydrophilic NOM content, and specifically the non-acid fraction, can be used as an indicator of the residual DOC. In addition, a clear relationship was found between zeta potential and residual DOC, and an operational zeta potential range between −10 and +3 mV was suggested.

Similar to Edzwald and Tobiason (1999), Shin *et al.* (2007) suggested two pathways as the controlling processes for the removal of NOM (DOC): 1) formation of soluble Al-NOM complexes followed by Al-NOM precipitation, and 2) formation of $Al(OH)_3(s)$ followed by adsorption of NOM and/or Al-NOM complexes on $Al(OH)_3(s)$.

Tseng and Edwards (1999) however, showed that the most accurate prediction of final DOC by enhanced coagulation was demonstrated in a model developed by Edwards (1997) that assumed that DOC removal was mainly the result of adsorption onto metal hydroxide flocs formed during coagulation. The adsorption model is also supported by results from Jekel (1986), and by results from contact filtration experiments (Eikebrokk, 2000) where NOM removal appeared to persist for several hours in a dual media filter bed after a complete termination of the coagulant dosage after some hours of normal operation. In addition, NOM removal efficiency was relatively insensitive to mixing conditions. Thus, adsorption of DOC to metal hydroxide deposits seems to be a major coagulation mechanism in contact filtration systems.

DOC REMOVAL MODELS

DOC removal models presented in the literature reflect the different coagulation mechanisms presented above. Although the coagulation mechanisms and the factors that influence NOM removal by coagulation are now relatively well understood, their complex interplay has hindered formulation of quantitative, applicable and conceptually valid models.

Several empirical models predicting DOC or TOC removal by coagulation can be found in the literature. Some of these are summarized in Table 2.

Table 2 Models to predict DOC or TOC removal during coagulation.

Coag	Equation	Reference
Al[a)]	$TOC_{final} = 0.405TOC_{raw} - 0.55dose + 0.688pH_{coag} - 3.32$	Moomaw *et al.* (1992)
Fe[a)]	$TOC_{final} = 0.52TOC_{raw} - 1.49dose + 0.740pH_{coag} - 3.50$	Moomaw *et al.* (1992)
Al[b)]	$TOC_{final} = 1.076[TOC_{raw}]^{0.923}[dose]^{-0.298}[Alk_{raw}]^{0.173}$	Zhu (1995)
Fe[b)]	$TOC_{final} = 0.839[TOC_{raw}]^{1.094}[dose]^{-0.355}[Alk_{raw}]^{0.150}$	Zhu (1995)
Al[c)]	$\ln(TOC_{final}) = -0.33 + 1.31\ln(TOC_{raw}) - 0.54\ln(dose) - 0.08\ln(TOC_{raw})\ln(dose) + 0.077\,pH_{coag}\ln(dose)$	Harrington *et al.* (1992)
Fe[c)]	$\ln(TOC_{final}) = -0.10 + 1.25\ln(TOC_{raw}) - 0.71\ln(dose) - 0.05\ln(TOC_{raw})\ln(dose) + 0.092\,pH_{coag}\ln(dose)$	Harrington *et al.* (1992)
Al/Fe[d)]	$TOC = a + b\,(2.717)^{-c(dose)}$	Chowdhury and Owen (1996)

a) dose in meq/L Al or Fe;
b) alk. in mg/L $CaCO_3$ and dose in mg/L as $Al_2(SO_4)_3 \cdot 18H_2O$ or $FeCl_3 \cdot 6H_2O$
c) dose in mg/L as $Al_2(SO_4)_3 \cdot 18H_2O$ or $FeCl_3 \cdot 6H_2O$
d) a, b and c are different empirical constants depending on TOC, alkalinity and coagulant.

The Langmuir isotherm-based adsorption model

The accuracy of the models presented in Table 2 is however lower than that of the Langmuir adsorption model presented by Edwards (1997). This model assumes the following (Tseng and Edwards (1999):

- DOC removal is mainly the result of adsorption onto metal hydroxide flocs formed during coagulation.
- There exists a non-sorbable fraction of DOC that is not amendable to removal by the metal coagulant considered.
- The adsorption can be described by the Langmuir isotherm.
- The adsorption capacity of the floc is a function of pH.
- The characteristics of NOM do not change with pH.
- The DOC variation of source waters is due to concentration or dilution, i.e. the removable fraction remains constant despite changes in DOC values of the raw water.

Equation (3) describes the adsorption model proposed by Edwards (1997).

$$(DOC_{sorbable} - DOC_{eq})/D = ab\ DOC_{eq}/(1 + b\ DOC_{eq}) \qquad (3)$$

where

DOC_{eq} = Equilibrium (final) concentration of sorbable DOC in water
$DOC_{sorbable}$ = Sorbable DOC (mg/L) in water before coagulant addition
D = Dose of alum or ferric coagulant (meq/L)
a = maximum DOC sorption capacity (mg DOC/meq metal)
b = the sorption coefficient (L/mg).

The fraction of non-sorbable DOC that is not removed by coagulation can be calculated as:

$$\text{Fraction of non} - \text{sorbable DOC} = K_1(SUVA)_{raw\ water} + K_2 \qquad (4)$$

where K_1 and K_2 are empirical fitting constants. The sorbable DOC can then be calculated by:

$$\text{Sorbable DOC} = \{1 - \text{fraction non-sorbable DOC}\}\ DOC_{initial} \qquad (5)$$

The model is based on the Langmuir equation:

$$x/M = (ab\ [DOC]_{eq})/(1 + b[DOC]_{eq}) \qquad (6)$$

where x is DOC removed (mg/L DOC), M is coagulant added and metal hydroxide formed (mmole/L), DOC_{eq} is sorbable DOC in solution at equilibrium, and a and b are sorption constants. The constant a, i.e. the maximum sorption capacity of a metal hydroxide floc, is supposed to be controlled by pH and can be determined by:

$$a = x_3\ pH^3 + x_2\ pH^2 + x_1\ pH \qquad (7)$$

where x_1, x_2 and x_3 are fitting constants. Combining these equations gives:

$$\{(1 - SUVA\ K_1 - K_2)\ DOC_{initial} - [DOC]_{eq}\}/M$$
$$= (x_3\ pH^3 + x_2\ pH^2 + x_1\ pH)\ b\ [DOC]_{eq}\}/\{1 + b\ [DOC]_{eq}\} \qquad (8)$$

This equation can be solved in an Excel sheet if the six empirical constants (K_1, K_2, x_1, x_2, x_3, b) are known. The DOC concentration remaining after coagulation is then:

$$DOC\ (mg/L) = DOC_{eq}\ (mg/L) + \text{non} - \text{sorbable DOC}\ (mg/L) \qquad (9)$$

Edwards (1997) also determined the values for the empirical constants, still keeping the standard error below 10%. These general values for the empirical constants are listed in Table 3 below.

Table 3 Parameter values used in the model for predicting DOC removal (Edwards, 1997).

Parameter	Fe coagulant	Al coagulant	General DOC < 10 mg/L
Standard error, %	9.3	9.5	10
x_3	4.96	4.91	6.44
x_2	−73.9	−74.2	−99.2
x_1	280	284	387
K_1	−0.028	−0.075	−0.053
K_2	0.23	0.56	0.54
b	0.068	0.147	0.107

Modifications to the adsorption model

Kastl *et al.* (2004, 2008) proposed a modification to the Langmuir isotherm-based adsorption model developed by Edwards (1997). The modification assumes that DOC present in water is assumed to comprise threes major groups of compounds: 1) non-sorbable compounds ($DOC_{nonsorb}$) which are not removed by metal coagulants, 2) humic acid compounds (DOC_{HA}) that are adsorbed depending on the concentration of their neutral (associated) form and thus on pH, and 3) non-polar compounds ($DOC_{nonpolar}$). If initial DOC is denoted DOC_0, then:

$$DOC_{HA,0} + DOC_{nonpolar,0} + DOC_{nonsorb} = DOC_0 \tag{10}$$

where the subscript 0 refers to initial conditions.

Because only DOC_{HA} and $DOC_{nonpolar}$ can be adsorbed, the initial sorbable DOC ($DOC_{sorbable,0}$) can be defined as:

$$DOC_{sorbable,0} = DOC_{nonpolar,0} + DOC_{HA,0} \tag{11}$$

The initial fractions of can be defined as follows:

$$DOC_{nonsorb} = f_{nonsorb} \cdot DOC_0 \tag{12}$$

$$DOC_{HA,0} = f_{HA} \cdot DOC_0 \tag{13}$$

$$DOC_{nonpolar,0} = f_{nonpolar} \cdot DOC_0 \tag{14}$$

It is assumed that the fractions are constant for a given source water. The equilibrium between associated and dissociated forms of humic acids in water is described by:

$$K = ([H^+] \cdot [A^-])/[HA] \tag{15}$$

where K is the apparent dissociation constant.

If $[HA]_s$ and $DOC_{nonpolar,s}$ are the amounts of humic and nonpolar compounds adsorbed onto the metal hydroxide and $DOC_{nonpolar,w}$ is the equilibrium concentration in water after metal coagulant addition, then equations 16–18 hold:

$$[HA]_s + [HA] + [A^-] = DOC_{HA,0} \tag{16}$$

$$[HA]_s/D = a \cdot b \cdot [HA]/(1 + b \cdot [HA] + k \cdot DOC_{nonpolar,w}) \tag{17}$$

$$\begin{aligned}(DOC_{nonpolar,0} - DOC_{nonpolar,w})/D \\ = a \cdot b \cdot k \cdot DOC_{nonpolar,w}/(1 + b \cdot [HA] + k \cdot DOC_{nonpolar,w})\end{aligned} \tag{18}$$

where k is the relative adsorption constant for the nonpolar fraction.

The predicted residual DOC can then be determined:

$$DOC_{res} = DOC_{nonsorb} + [HA] + [A^-] + DOC_{nonpolar,w} \tag{19}$$

The model, which requires determination of five coefficients, i.e. maximum surface density (a), saturation constant (b), humic acid dissociation constant (K), fraction of humic acid (f_{HA}), and non-polar fraction of DOC ($f_{nonpolar}$), was fitted to jar test data on two Australian waters and 14 US waters (Figure 1). It predicted the final DOC for a given metal coagulant and water within a standard deviation of the error of either 5.5% or 0.15 mg/L of DOC over the valid combinations of coagulant dose and pH. The model prediction was within the measurement error of the instrument of both pH and DOC. pH ranges of 4–9 and 5–9 was chosen to avoid issues of floc solubility for ferric hydroxide and aluminium hydroxide, respectively (Kastl *et al.*, 2004, 2008).

ENHANCED COAGULATION OPERATION MODELS

Several empirical model equations for operational application within enhanced coagulation and filtration processes can be found in the literature (Eikebrokk *et al.*, 2007). Among them are guidance models to predict coagulant dose requirements, sludge production, filter run time, net unit water production from filters, backwash requirements, etc. Due to space limitations, only coagulant dose prediction models are presented in this paper.

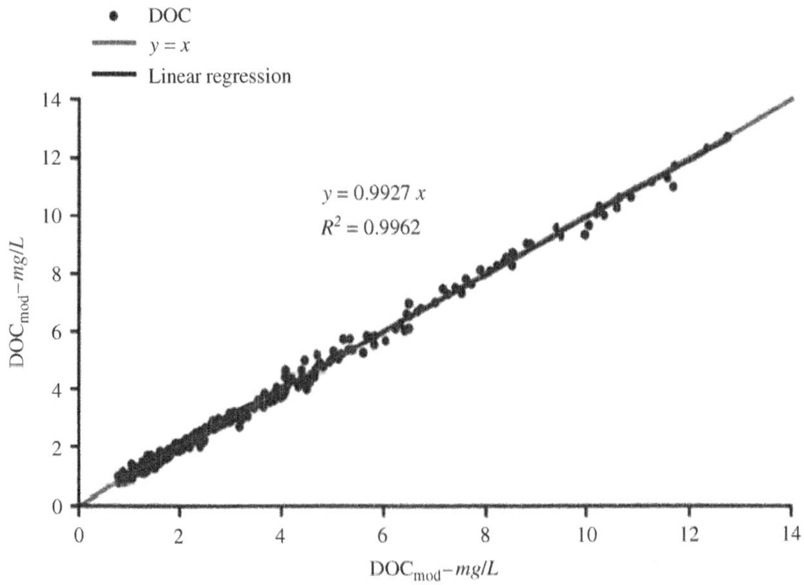

Figure 1 Comparison of measured and model-predicted residual DOC concentration for 14 US waters (Kastl *et al.*, 2004, 2008)

Coagulant dose demand

Some empirical models for predicting enhanced coagulant dose demand are summarized in Table 4.

Table 4 Model equations for coagulant dosage requirements.

Coag	Dose demand equation	References
Alum	$Al = 10.9 + 2.63\,(TOC_o) + 0.17\,(Alk) + 0.74\,(pH)$	Rodriguez *et al.*, 2000
Alum[1)]	$Al_{min} = 0.043 \cdot Colour + 0.30$	Eikebrokk, 2004
	$Al_{pract} = 0.054 \cdot Colour + 0.37$	
Ferric[1)]	$Fe_{min} = 0.107 \cdot Colour + 0.58$	Eikebrokk, 2004
	$Fe_{pract} = 0.134 \cdot Colour + 0.72$	Soviet Build. Norms, 1984
Alum	$Al = 4 \times Colour^{-1/2}$	

TOC_0 : concentration of TOC before treatment (mg/l); Al : coagulation dose (mg/l);
Alk: water alkalinity (mg/l $CaCO_3$); 1) Absolute and practical minimum dosages determined by the residual coagulant concentration standard (<0.15 mg Me/L).

Figure 2 shows a comparison of the absolute and practical minimum coagulant dose predictions from the model developed by Eikebrokk (2004) and the seasonal variability in raw water colour and applied coagulant dose levels at five WTPs in Bergen, Norway using alum or ferric. In a major part of the year some facilities seem to be over-dosing, and the dose levels have now been reduced with good results. However, some WTPs are very close to the minimum coagulant dose level during periods of high raw water colour. This may compromise treated water quality and safety barrier efficiency.

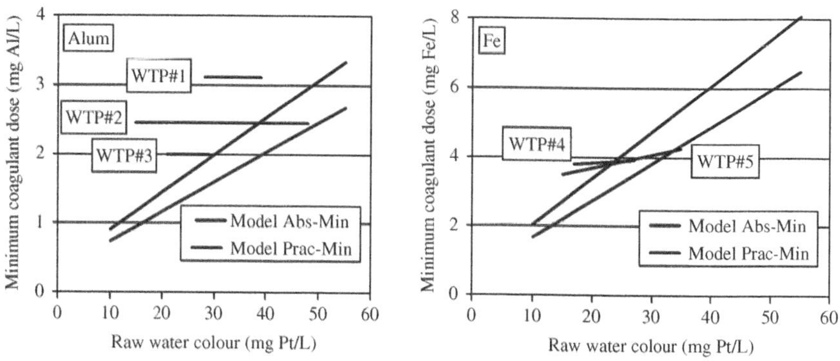

Figure 2 Variability in raw water colour levels, coagulant dose predictions and actual dosages applied at five WTPs in Bergen, Norway in 2007.

Enhanced coagulation software models

A software enhanced coagulation model (mEnCo©) has been developed by the Australian Cooperative Research Centre for Water Quality and Treatment that enables accurate predictions of residual DOC in treated water based on the model presented above (Holmes *et al.*, 2006; van Leeuwen *et al.*, 2003, 2005; Kastl *et al.*, 2004). The software allows WTP operators to rapidly and accurately determine the optimum coagulation pH as well as the dose rates for coagulants and/or pH-controlling chemicals. Enhanced coagulation is defined in mEnCo© as the point where increasing the coagulant dose provides a reduction of only 0.15 mg/L DOC per 10 mg/L increase in coagulant dose. The model is used by United Water International as an operational tool to optimize WTPs in metropolitan Adelaide.

Examples of the applied relationships between raw water quality (DOC, UV254 and colour) and coagulant dose requirements to maximize DOC removal are presented below (van Leeuwen *et al.*, 2005):

$$\text{Dose}(\text{mg/L}) = A_1 + B_1(\text{UV254})^{0.5} + C_1^{(-\text{UV254})} \tag{20}$$

where A_1, B_1, and C_1 are constants.

$$\text{Dose}(\text{mg/L}) = A_2 + B_2(\text{UVCol})^{0.5} + C_2^{(-\text{UVCol})} \tag{21}$$

where A_2, B_2, and C_2 are constants and UVCol is UV254 • log (Colour • 10).

A picture of the mEnCo© main user interface screen is shown in Figure 3.

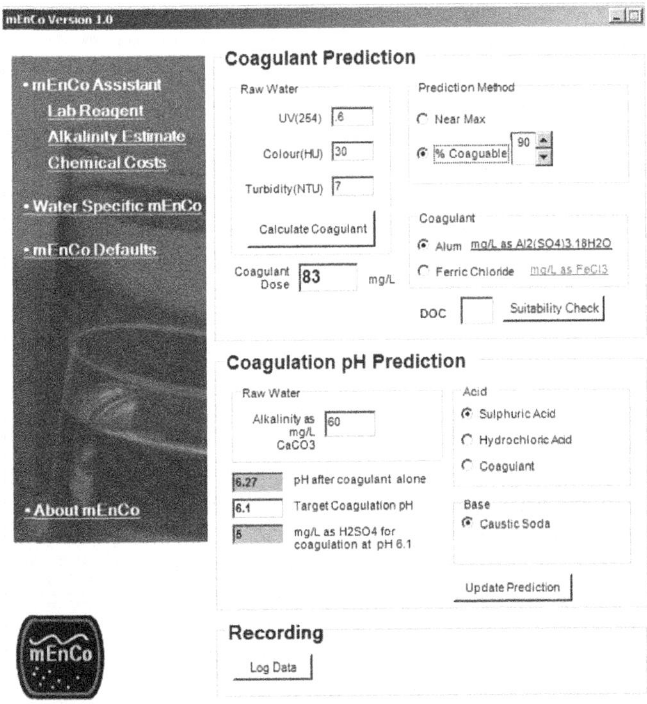

Figure 3 Picture of the mEnCo© main user interface screen (Holmes *et al.*, 2006).

SUMMARY AND CONCLUSIONS

There are several models available for the prediction of residual DOC and coagulant demand for enhanced coagulation processes. However, the Langmuir isotherm-based adsorption model developed by Edwards (1997) and later modified by Kastl *et al.* (2004, 2008) is suggested for further testing and verification within the EU-funded project Techneau. The adsorption model is considered as a good basis for making the Techneau water treatment simulator a powerful tool also for the design, operation and optimization of enhanced coagulation processes. Input data from Techneau optimization trials will be applied for the purpose of testing and validating the model.

ACKNOWLEDGEMENT

This paper was prepared as part of the EU-funded R&D-project Techneau. The authors are grateful for the contributions from co-workers within Techneau, i.e. J. Dudley and G. Dillon, WRc plc, L. Rietveld, Delft Univ. of Technology, and T. Juhna, Riga Technical University.

REFERENCES

Bèle, C., Habarou, H., Ambonguilat, S., Croué, J-P. amd Heim, V. (2006). How to optimize NOM removal on drinking water plants? First step: Making a good diagnosis thanks to HSEP. *Water Practice and Technology*, **1**(4).

Black, A.P., Singley, J.E., Whittle, G.P. and Maulding, J.S. (1963). Stoichiometry of the coagulation of color-causing organic compounds with ferric sulphate. *Journal AWWA*, **55**(10), 1347–1366.

Chow, C., Fabris, R., and Drikas, M. (2004). A rapid fractionation technique to characterise natural organic matter for the optimisation of water treatment processes. *J. Water Supply: Research and Technology – AQUA*, **53**(2), 85–92.

Chow, C., Fabris, R., Drikas, M. and Holmes, M. (2005). A case study of treatment performance and organic character. *J. Water Supply: Research and Technology – AQUA*, **54**(6), 385–395.

Chowdhury, Z. and Owen, D. (1996). Final report to Water Industry Technical Action Fund, USA.

Dudley, J., Dillon, G. and Rietveld, L. (2008). Water treatment simulators. *J. Water Supply: Research and Technology – AQUA*, **57**(1), 13–21.

Edwards, M. (1997). Predicting DOC removal during enhanced coagulation. *Jour. AWWA*, 89:5:78.

Edzwald, J.K. and Tobiason, J.E. (1999). Enhanced coagulation: US requirements and a broader view. *Water Science and Technology*, **40**(9), 63–70.

Edzwald, J.K. and Van Benschoten, J.B. (1990). Aluminium coagulation of natural organic matter. In: *Chemical Water and Wastewater Treatment*, H. Hahn and R. Klute (eds), Springer Verlag, NY, 341–359.

Eikebrokk, B. (1999). Coagulation-direct filtration of soft, low alkalinity humic waters. *Water Science and Technology*, **40**(9), 55–62.

Eikebrokk, B. (2000). Effects of coagulant type and coagulation conditions on NOM removal from drinking water. Proc. 9th Gothenburg Symposium on Chemical Water and Wastewater Treatment, Istanbul, Turkey, 2–4 October 2000. In: *Chemical Water and Wastewater Treatment VI*, H. H. Hahn, E. Hoffmann and H. Ødegaard (eds), Springer-Verlag, Berlin, Heidelberg 2000, pp 211–220. ISBN 3-540-67574-4.

Eikebrokk, B., Vogt, R.D and Liltved, H. (2004). NOM increase in Northern European Source Waters: Impacts on coagulation/contact filtration processes. Proc NOM Innovations and Applications, Adelaide, Australia, March 2–5.

Eikebrokk, B. Juhna, T., Melin, E. and Osterhus, S.W. (2007). Water treatment by enhanced coagulation and ozonation-biofiltration. Techneau-Report D5.3.2A, Dec. http://www.techneau.eu.

Harrington, G.W., Chowdhury, Z.K., and Owen, D.M. (1992). Developing a computer model to simulate DBP formation during water treatment. *Journal AWWA*, **84**(11), 78.

Holmes, M., Kaeding, U., van Leeuwen, J., Daly, R., Soenarjo, M., Crossing, N., Haskard, C. and Bursill, D. (2006). The application of mEnCo© by United Water at Metropolitan Adelaide WTPs. Proc.: NZWWA 48th Annual Conf. and Expo, Christchurch, New Zealand, Oct. 11–13

Jekel, M. (1986). Interactions of humic acids and aluminium salts in the flocculation process. *Water Research*, **20**(12), 1535.

Kastl, G., Sathasivan, A. Fisher, I. and van Leeuwen, J. (2004). Modelling DOC removal by enhanced coagulation. *Jour. AWWA*, **96**(2), 79–89.

Kastl, G., Sathasivan, A. and Fisher, I. (2008). Modelling DOC removal by enhanced coagulation. Research report No. 57, CRC for Water Quality and Treatment, Salisbury, Australia, ISBN 18766 16830.

Moomaw, C., Amy, G., Krasner, S. and Najm, I. (1992). Predictive models for coagulation efficiency in DBP precursor removal. Proc. AWWA Annual Conference, Vancouver, B.C.

Rodriguez, M.J., Serodes, J. and Morin, M., (2000). Estimation of water utilities compliance with trihalomethane regulation using a modeling approach. *Jour. Wat. Suppl. Resear. Tech.-AQUA*, **49**(2), 57–73.

Sharp, E.L., Parsons S.A. and Jefferson, B. (2006). Coagulation of NOM: linking character to treatment. *Water Science and Technology*, **53**(7), 67–76.

Shin, J.Y., Spinette, R.F. and O'Melia, C.R. (2008). Stoichiometry of coagulation revisited. *Env. Sci. Technol.*, **42**, 2582–2589.

Tseng, T. and Edwards, M. (1999). Predicting full-scale TOC removal. *Jour. AWWA*, **91**(4), 159–170.

Zhu, H.W. (1995). Modeling the effects of coagulation on chlorination by-product formation. Doctoral dissertation, University of Colorado, Boulder.

van Leeuwen, J., Daly, R. and Holmes, M. (2005). Modelling the treatment of drinking water to maximize dissolved organic matter removal and minimize disinfection by-product formation. *Desalination*, **177**, 81–89.

van Leeuwen, J., Holmes, M., Heidenreich, C., Daly, R., Fischer, I., Kastl, G., Sathasvian, A. and Bursill, D. (2003). Modelling the application of inorganic coagulants and pH control reagents for removal of organic matter from drinking waters. Proc.: MODSIM integrative modelling of biophysical, social and economic systems for resource management solutions. Townsville, Australia, 14–17th July, 1835–1840.

Adaptive strategies for drinking water production in The Netherlands

J.J.G. Zwolsman, G.A. Van den Berg and J. Frijns

KWR Watercycle Research Institute, P.O. Box 1072, 3430 BB Nieuwegein, The Netherlands (E-mail: gertjan.zwolsman@kwrwater.nl; gerard.van.den.berg@kwrwater.nl; jos.frijns@kwrwater.nl)

Abstract This paper describes the results of a feasibility study on adaptive strategies for drinking water production in The Netherlands. The currently applied strategies to deal with risk and uncertainty in water resources management will generally be insufficient for dealing with the anticipated increased variability induced by climate change. Climate change will result in different types of water stress: droughts, floodings, rainstorms and decline in water quality. New adaptive strategies for exploitation and management of natural resources of drinking water production are needed and must be developed. Examples are 1) use of alternative raw water sources such as precipitation, brackish groundwater and domestic wastewater, 2) temporary storage of excess run off or pretreated water, and 3) flexible production strategies, using various water resources and treatment procedures.

In selecting appropriate adaptive strategies, also the contribution of drinking water production and distribution to climate change should be taken into account. Although the carbon footprint of drinking water production and distribution is limited, the sector is putting an effort into reducing its climate change impact.

Keywords adaptation, climate change, flexible strategies, mitigation, water supply

INTRODUCTION

Within the EU funded TECHNEAU project ten major global trends have been recognised that will affect the water industry. Relevant trends for drinking water supply include climate change, urbanisation and energy use and costs (Segrave *et al.*, 2007). Concern for the impact of climate change on water systems and associated ecosystems is increasing. Climate change affects water system dynamics through temperature changes, changes in precipitation patterns (more rainstorms, floods), an increase in summer evaporation and a decrease in snow-fed river run off (extended droughts). These changes will have major impacts on the management of river deltas and river catchments in general (EEA, 2001). The capacity of water systems to cope with changing climate dynamics depends strongly on catchment soil characteristics, the presence of inundation areas and run-off management.

Climate change impact assessments generally focus on risks of flooding of heavily populated low-lying areas and hydrodynamic effects on navigation, agriculture and energy production (EEA, 2007). Climate change impacts on droughts and associated water quality effects (e.g. Van Vliet and Zwolsman, 2008) are lesser recognized so far, although they may have huge economic effects on agricultural water use and drinking water production.

Production of reliable drinking water is of major importance for society as drinking water is the number one requirement for life. Water utility companies and governments are mainly focused on solving the problems of today, i.e. renovating existing water production facilities. However, the vulnerability of drinking water production demands a careful evaluation of future threats to our natural water resources. In this paper, the results of a feasibility study on adaptive strategies for drinking water production in The Netherlands are described.

WATER STRESS AND DRINKING WATER PRODUCTION

Drinking water supply companies have always been challenged by increased variations in water availability, water quality, production, distribution and water demand. Climate change makes the situation more dynamic, however. The tendency towards increasing hydrological extremes will result in e.g. lowering of groundwater tables, reduction of the capacity of water reservoirs, as well as water quality effects.

Groundwater and surface water dynamics are expected to change significantly due to climate change. In some eastern parts of The Netherlands the groundwater table may be lowered by tens of centimetres during dry periods (Cirkel *et al.*, 2006). This may have serious consequences for the use of groundwater as a source for drinking water production, given the effects on terrestrial ecosystems. The anticipated effects are even more pronounced when the increase in agricultural and drinking water demand during periods of water scarcity is taken into account.

The trend towards more frequent and extended droughts is expected to cause problems for drinking water supply companies with respect to contamination of surface water systems. This may ultimately result in additional and elongated intake stops. Low river discharges are generally accompanied by less dilution of the pollution load (Van Vliet and Zwolsman, 2008), resulting in higher concentrations of contaminants. This is clearly illustrated in Figure 1. Methyl tert-butyl ether (MTBE), the most widely used oxygenate in gasoline, shows a seasonal pattern in the river Meuse with high concentrations during low discharge, frequently excceding the indicator value of 1 μg/l.

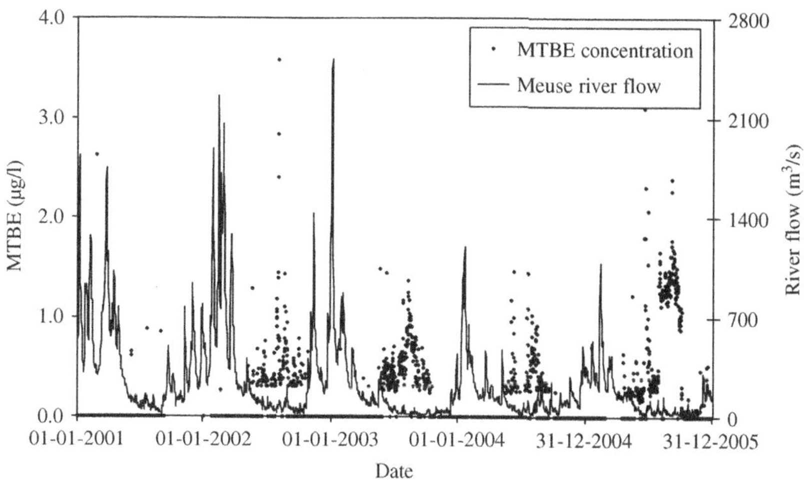

Figure 1. MTBE concentrations and river discharge of the river Meuse at measuring station Eijsden during the period 2001–2005 (data from www.aqualarm.nl and www.waterbase.nl).

Similarly, reduced dilution of pollution loads is expected for compounds that are emitted to the surface water at a constant rate, e.g. via effluents of industrial or domestic waste water treatment plants (WWTP). Several emerging compounds, like pharmaceutical residues and endocrine disruptive compounds, may consequently be observed in higher concentrations in surface water during hydrological droughts. Therefore, quality standards should be set to actively control their emissions to surface water. Other relevant parameters, like oxygen and algae levels (including the toxic cyanobacteria), are also significantly influenced by extended droughts and increased temperatures (Mooij *et al.*, 2005). Cyanobacteria are considered a potential threat to drinking water production in The Netherlands (Hoogenboezem *et al.*, 2004). Floods will temporarily be accompanied by increased suspended matter levels and associated contaminants in rivers (Doomen *et al.*, 2007), giving rise to intake problems. For instance, the Dutch water utility Evides had to close down their direct intake of river Meuse surface water for 47 days over the year 2006 due to the presence of high suspended matter levels during floods. In addtion, high rainfall periods frequently lead to overflows of sewage systems negatively affecting the quality of the receiving waters.

Groundwater in the coastal region of The Netherlands is very vulnerable to salinisation, due to the combination of sea level rise, low river flows, and land subsidence. For that reason, surface water is nowadays the main source for drinking water production in the western (low-lying) part of The Netherlands, including the cities of Amsterdam, Rotterdam and The Hague. Different treatment techniques are applied to produce drinking water, in which dune infiltration plays a crucial role. These include abstraction from the rivers Meuse and Rhine, temporary storage in large natural and artificial drinking water reservoirs, extraction after artificial dune infiltration and bank filtration. Dutch drinking water utilities generally apply a range of advanced treatment technologies (e.g. UV disinfection, hydrogen peroxide, ozone, nanofiltration, reversed osmosis) to produce high quality drinking water from surface water.

Recently, changes in salinisation of the Rhine-Meuse estuary in The Netherlands within a time span of 50 years have been analysed (Doomen and Van den Berg, 2005). This study shows that effects of climate change cannot be predicted reliably without a systematic knowledge of the effects of other major developments on salinisation within the delta, such as restoring river deltas, measures to ensure freshwater supply to regional water systems (agricultural function), and dredging to increase the navigation depth. Future salinity gradients in the estuary are mainly determined by autonomous developments related to climate change. An increased salinisation of the Rhine-Meuse estuary is caused by a combination of increased water levels at sea, more frequent and heavier winds from sea, and lower river discharge. This will negatively influence freshwater availability in the area, especially during long drought spells when agricultural and public demand for fresh water is high. Increased salinisation will also threaten the salinity of bank filtrate abstracted in the Rhine-Meuse delta. Locally, the yearly average chloride concentration may increase substantially by salinisation, resulting in a decreased functionality of the abstraction.

NEED FOR ADAPTIVE STRATEGIES

Climate change and other pressures from environmental and spatial developments stress the necessity to develop innovative concepts for drinking water production. It is now generally acknowledged within the drinking water sector that adaptive strategies must be developed and implemented to secure the supply of high quality drinking water over the next decades. We distinguish between two different approaches, which are either demand or supply based. In this paper we describe four types of adaptive strategies for drinking water utitilities. These are summarised in Table 1 and further described in the text.

Table 1. Overview of promising adaptive strategies to cope with the effects of climate change by drinking water utilities.

Potential mitigation measure	Examples
Energy-efficient production technologies	Low-pressure UV Advanced membrane-based systems (low pressure UV, TFC membranes, AiRO process)
Optimised distribution of drinking water	Distribution related to the area and the demand Pressure reduction and tuning
Methane recovery	Methane degassing and re-use
New water cycle concepts and household measures	Source-separated sanitation Re-use of embedded energy of (waste) water Central softening

Demand strategies aim at changing the customer awareness towards the use of drinking water. Measures at the demand side ask for a sectoral approach, preferably at river basin scale or regional scale. Such measures include education and communication to emphasize a reduction in drinking water consumption (especially to lower peak demands), introduction of progressive water pricing or the setting of specific regulations restricting or banning the use of drinking water for less valued purposes (e.g. car washing or lawn watering) during periods of water scarcity.

Stimulating the sustainable (and secure) use of additional water sources, such as filtered rain water, at the household level may also increase customer awareness. Discussion and research focus mainly on the microbiological risks associated with such systems. A study by Van den Berg *et al.* (2008) indicates that when rain water (from the roof) is used for toilet flushing, individual households can save between 10% (during a very dry year, 425 mm/year) to 25% (during a wet year, 1242 mm/year) on their drinking water use.

Supply strategies aim at the development and implementation of adaptive measures to manage the quantity and quality of available water sources for drinking water production.

Alternative sources concept

Worldwide, studies have focussed on the exploitation of sea water, precipitation, brackish ground water and domestic wastewater, using innovative treatment such as membrane technology, as alternatives to the commonly applied water resources (fresh groundwater and surface water). Some concepts are already widely applied in specific regions. Desalination, for example, is a commonly applied technique in densely populated (semi)arid areas, like the Mediterranean countries. Drawbacks for application in other areas are the high investment costs and energy consumption in comparison with presently applied treatment technologies.

The Dutch water cycle company Waternet has investigated the potential use of alternative sources for the production of drinking water from river Rhine (Lek canal) water (Bernhardi and Van den Berg, 2006). These include abstraction of water from large freshwater reservoirs (Lake IJssel and Lake Marken), abstraction of surface water from artificial canals (draining the low-lying polders), use of brackish seepage water, and the use of waste water effluent from a large domestic WWTP (see Figure 2).

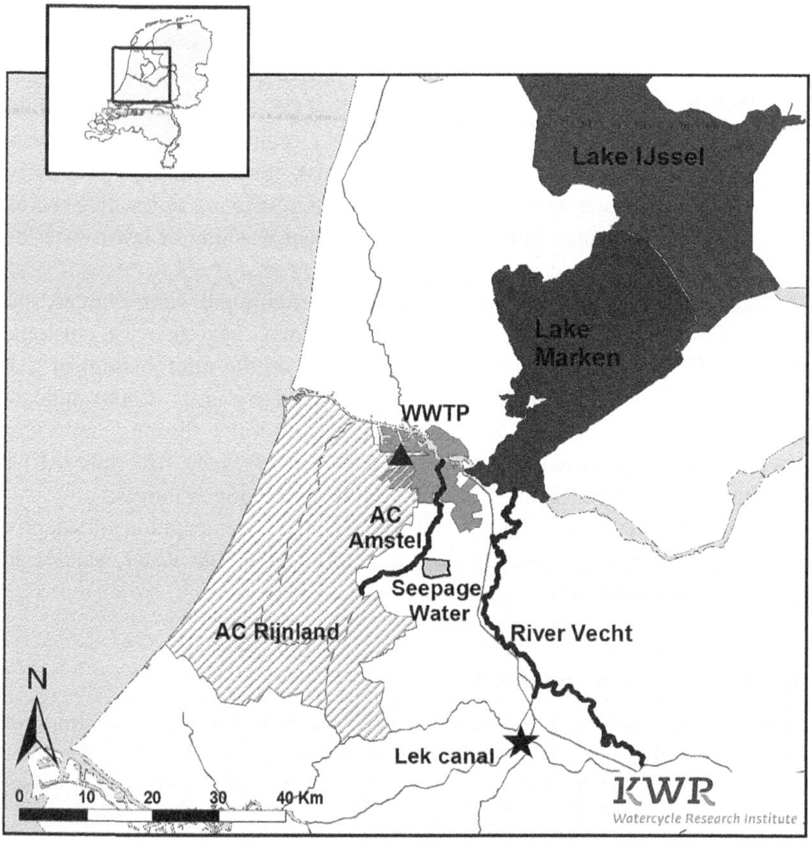

Figure 2. Location of alternative drinking water sources around the city of Amsterdam.

Table 2 shows that each potential alternative source has typical quality aspects, which may require redesign of the present treatment procedure for producing drinking water. Other aspects need to be addressed as well, such as availability, variability and energy demand. The availability of some of the proposed sources is too small to become the sole alternative for the present water source. The energy demand of applying alternative sources will vary widely given differences in infrastructural needs and treatment procedures needed to produce and distribute high quality drinking water.

Table 2. Assesment of the relative suitability of potential sources for the future drinking water supply of Waternet utility (Bernhardi and Van den Berg, 2006).

Source	Salinity	Micro pollutants	Seasonal variability	Availability	Quality aspect	Energy demand
River Rhine*	0	0	High	+	Temperature	0
Lake IJssel	0	+	High	+	Algae	–
Lake Marken	0	+	High	+	SPM	0
AC Rijnland	–	–	High	+	Chloride	–
AC Amstel	–	–	High	–	Chloride	–
River Vecht	–	––	High	–	Effluents	–
Seepage water	––	+	Low	0	Chloride	––
WWTP	0	––	Low	+	Micropollutants	––

*Present source (Lek canal); SPM = suspended matter; AC = artificial canals; WWTP waste water treatment plant; –– very poor; – poor; 0 neutral; + favourable.

In another project, eight possible sources for drinking water production (shallow, deep and hard groundwater, river bank filtrate, brackish groundwater, surface water, WWTP effluent and rain water) have been evaluated for treatment (investment and operational) costs (Ramaker *et al.*, 2005). It was concluded that rain water and brackish ground water could be cost competitive alternatives for current drinking water sources.

Multiple sources concept

The use of multiple sources (e.g. groundwater plus surface water) and the infrastucture to distribute water from these sources within the supply area, decreases the dependency of water utilities upon a single resource. A combination of resources, as shown in Table 2, could be an interesting and sustainable option.

A few years ago the "fresh keeper" concept was developed in order to prevent salinisation of fresh groundwater lenses which are overlying brackish groundwater in the western part of The Netherlands (Kooiman *et al.*, 2004). This concept has advantages over the traditional single-depth extraction of either freshwater or brackish groundwater. Additional deep wells change the flow direction of brackish groundwater protecting existing freshwater wells from upconing of brackish groundwater (see Figure 3).

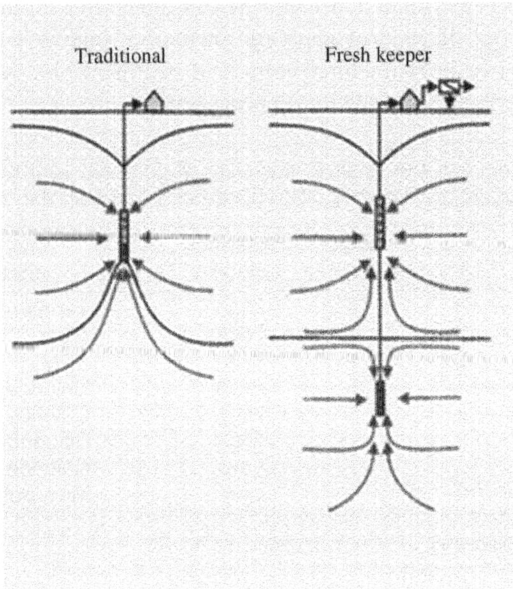

Figure 3. Comparison between traditional single-depth groundwater extraction (with upconing of brackish groundwater (left) and the Fresh keeper concept (right)).

The brackish groundwater extracted can be utilized for drinking water production using e.g. reversed osmosis, thus improving the capacity of well fields. The technique makes optimal use of the existing treatment and distribution infrastructure. Pilot studies are now initiated to test this concept under field conditions. The legal and hydrogeochemical aspects of membrane concentrate disposal and choice of appropriate target aquifers needs special attention (Stuyfzand *et al.*, 2007). In two pilot studies (by the water companies Vitens and Brabant Water) membrane concentrate will be injected in deeper brackish aquifers.

Flexible sources concept

Flexible water resources management and treatment is a promising concept for continuous and sustainable production and supply of high quality drinking water.

Application of the Flex Water concept (Ramaker *et al.*, 2005) is based on large treatment systems, designed for constant production, combined with flexible sources te deal with peak demands. This concept optimizes the use of seasonally available natural resources in combination with temporary storage to match peak demand and extreme events. A major advantage of the Flex Water concept is the possibility to link water sources with limited capacity. This promotes shifts between different sources.

The Flex Water concept is expected to be cost efficient in remote areas and newly developed urban areas. The use of locally available water resources is an important element of this concept. A good example is the closed city concept (De Graaf and Van de Ven, 2005), based on the use of local rainfall for drinking water production in urban areas. Studies are initiated to experiment with such self supporting systems. During dry periods recycled effluent of waste water treatment plants and regional surface water may be used as supplementary sources to create a more flexible system. This can reduce vulnerability of urban areas to changing water dynamics and temporal water stress.

An example of the Flex Water concept is the use of excess rain water or pretreated surface water as additional sources for drinking water in combination with temporary storage in aquifers using Aquifer Storage and Recovery (ASR). Stored water can be extracted during periods of water scarcity. At several locations (e.g. the US, Israel, UK, Canada, and Australia) this concept is already applied widescale. In The Netherlands one succesful pilot has been completed by the drinking water utility WML and more pilots are ongoing. One of these pilots focuses on the application of ASR on the Dutch Wadden Island of Terschelling. These small islands face major problems with drinking water supply during summer months when demands increase exponentially due to visiting tourists. Succesful implementation depends mainly on minimising changes in water quality of the injected (pretreated) water (Prommer and Stuyfzand, 2005). On the Wadden Islands, salinisation of the injected water poses a risk.

POTENTIAL FOR MITIGATION MEASURES

In selecting appropriate adaptive strategies, also the contribution of drinking water utilitues to climate change and possible mitigation measures should be taken into account. The total carbon footprint of the Dutch drinking water sector is 0.44 million tonnes CO_2-eq. per year. This resembles 0.36 kg CO_2-eq. per m^3 drinking water. The Global Warming Potential (GWP) of wastewater treatment is about 2.5 times higher than that of drinking water production (Frijns *et al.*, 2008). Compared to the total yearly GWP of The Netherlands, 212 million tonnes CO_2-eq. (MNP, 2007), the contribution of the drinking water sector is limited (0.2%). Dutch drinking water utilities, however, are committed to reduce their carbon footprint and to implement mitigation measures.

In Figure 4, the GWP distribution of the Dutch drinking water utilities is presented. Energy consumption provides for the largest share of the GWP. The energy use is determined by the required pressure increase with pumps and the specific energy consumption of process units, such as UV lamps, saturation in flotation, and ozonation. On average, the energy use for production and distribution of drinking water is 0.47 kWh/m3 (Vewin, 2006). Distribution contributes to 25% of the energy use. These figures, however, vary substantially between drinking water companies. The energy consumption depends on the scale of the production process, the source for water production, the treatment steps applied, as well as the topography and distribution distances.

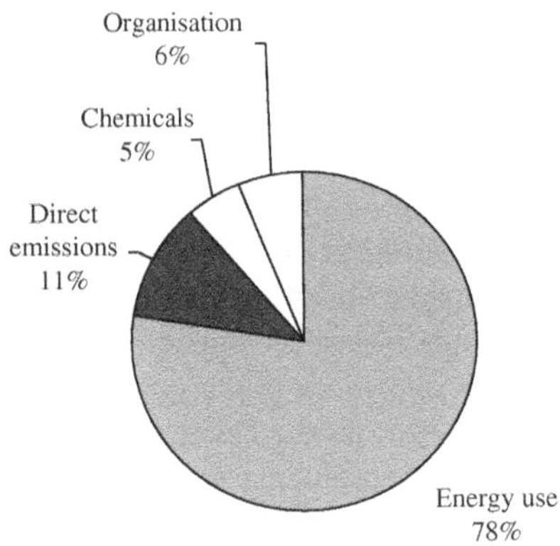

Figure 4. GWP distribution of drinking water production and distribution.

Direct emissions of greenhouse gases contribute to 11% of the GWP. This is mainly caused by methane emissions from degassing groundwater (Drijver *et al.*, 2007). Indirect emissions result from the use of chemicals and the energy use related with the organisation, e.g. office heating and transport. The energy required to produce drinking water production chemicals (e.g. sodium hydroxide and ferrous sulphate) account to the GWP of drinking water. Also here, the chemical use varies substantially between water companies.

Mitigation measures

The energy consumption is mainly related to the applied treatment procedure. As we expect that more advanced treatment will be needed to deal with substances like hormones and medicine residues in the future, the energy use is likely to increase. Next to the application of more advanced and energy-intensive treatment processes, the likely use of alternative sources (e.g. brackish groundwater) to deal with the effects of increasing salinisation of available water resources, will also result in higher energy consumption. The challenge thus is to improve the energy efficiency of drinking water production. Table 3 gives an overview of potential mitigation measures.

Table 3. Overview of potential mitigation measures for drinking water production and distribution.

Potential mitigation measure	Examples
Energy-efficient production technologies	Low-pressure UV Advanced membrane-based systems (low pressure UV, TFC membranes, AiRO process)
Optimised distribution of drinking water	Distribution related to the area and the demand Pressure reduction and tuning
Methane recovery	Methane degassing and re-use Source-separated sanitation
New water cycle concepts and household measures	Re-use of embedded energy of (waste) water Central softening

DISCUSSION AND SYNTHESIS

Knowledge of climate change effects on drinking water supply is increasing. It is now generally accepted by drinking water supply companies in The Netherlands that climate change may affect water resources management in the near future. Measures are needed to cope with the impacts of dynamic natural conditions on water supply, taking into account both quantity and quality aspects. In this paper some innovative adaptive concepts to deal with the anticipated changes are presented. These include the use of new water resources, as well as multiple sources and flexible sources concepts.

An integrated approach needs to be developed that takes into account the specific functions of water systems, such as drinking water, agriculture, cooling and ecology. A successful water management tool to improve the sustainability of resources is water allocation during periods of water shortage. In The Netherlands such a water allocation scheme the priority list water scarcity has recently been developed. It is anticipated that public water supply should prevail over other sectors, given the importance of clean drinking water for public health.

A continuous reduction of chemical pollution of water resources by improving industrial processes and waste water treatment, as well as introduction of cleaner agricultural practices will decrease the contamination of surface waters with pollutants that require intense treatment for drinking water production. Water managers should aim at a controlled permit system based on the maximum acceptable concentrations in the receiving water, taking into account the relevant functions of the water system. Such strategies should preferentially be developed and implemented at a river basin scale, in accordance with the principles of the Water Framework Directive.

Adaptive strategies can not be implemented without considering the energy consumption and costs. Although the carbon footprint of production and distribution of drinking water in The Netherlands is limited compared to the total footprint, ongoing research identifies numerous options to decease the energy use by drinking water utilities and implement promising new concepts, such as a climate neutral water cycle. We suggest that drinking water utilities should incorporate mitigation measures (e.g. energy use) in the selection of climate adaptation strategies. For example, the alternative sources strategy might result in a higher energy use for drinking water production. Demand strategies, on the other hand, might result in combined water and energy savings at households.

Finally, it should be realized that the high uncertainties associated with climate change impacts make efficient planning and investment extremely complicated. Transition to flexible adaptive strategies will probably take decades because investment depreciation periods of current drinking water assets are long. This implies that long-term strategies within an indicated timepath have to be developed. These strategies must be based on thorough knowledge of current assets to seize opportunities for change.

ACKNOWLEDGEMENTS

This study is a product of the Joint Research Programme of the Dutch Water Sector (BTO). It was co-financed by the EU FP6 Integrated Project TECHNEAU. The results on climate mitigation measures are part of a study financed by the Dutch Ministry of Housing, Spatial Planning and the Environment and the Foundation for Applied Water Research (Stowa).

REFERENCES

Bernhardi, L. and Van den Berg, G.A. (2006). Inventarisation opportunities for the use of alternative sources for the drinking water supply of Waternet. Report KWR06.031, Nieuwegein, The Netherlands (in Dutch).

Cirkel, G., van Griensven, E. and Broers E. (2006). Climate change and groundwater use, H_2O, **22**, 39–42 (in Dutch).

De Graaf, R.E. and Van de Ven, F.H.M. (2006). The closed city as a strategy to reduce vulnerability of urban areas for climate change. *Proceedings Innovations in Coping with Water and Climate risks: IWA Conference*, Amsterdam, September 25–27, 2006, pp. 221–227.

Doomen, A. and Van den Berg, G.A. (2005). Quick scan salinisation Rhine-Meuse estuary. Report KWR 05.074, Nieuwegein, The Netherlands (in Dutch).

Doomen, A.M.C., Wijma, E., Zwolsman, J.J.G. and Middelkoop, H. (2007). Predicting suspended sediment concentrations in the Meuse River using a supply-based rating curve. *Hydrological Processes*, **22**, 1846–1856.

Drijver, B., Kappelhof, J., Polman, E. and Van Breukelen, B. (2007). Methane extraction from ground water as a sustainable alternative with economic potential, H_2O, **19**, 52–55 (in Dutch).

EEA, (2001). Sustainable water use in Europe; part 3 extreme hydrological events: floods and droughts. Environmental issue report 21, European Environment Agency.

EEA, (2007). Climate change and water adaptation issues. EEA Technical report No. 2, European Environment Agency.

Frijns, J., Mulder, M. and Roorda, J. (2008). Towards a climate neutral water cycle. Stowa report, Utrecht, The Netherlands (in Dutch).

Hoogenboezem, W., Wagenvoort, A.J. and Blaauboer, K. (2004). The occurrence of toxic cyanobacteria in some Dutch surface waters used for the production of drinking water. Report of the Association of River waterworks RIWA, Nieuwegein, The Netherlands.

Kooiman, J.W., Stuyfzand, P.J., Maas, C. and Kappelhof, J.W.N.M. (2004). Pumping brackish groundwater to prepare drinking water and keep salinizing wells fresh: a feasibility study. In: *Groundwater and Saline Intrusion: Proceedings 18th SWIM, Cartagena*, Araguás, L. (ed.), Instituto Geologico y minero de España, Madrid, pp. 625–635.

MNP (2007). Greenhouse gas emissions in the Netherlands, National Inventory Report.

Prommer, H. and Stuyfzand, P.J. (2005). Identification of temperature-dependent water quality changes during a deep well injection experiment in a pyretic aquifer. *Environmental Science and Technology*, **39**, 2200–2209.

Mooij, W., Hülsmann, S., De Senerpont Domis, L., Nolet, B., Bodelier, P., Boers, P., Pires, L., Gons, H., Ibelings, B., Noordhuis, R., Portielje, R., Wolfstein, K. and Lammens E. (2005). The impact of climate change on lakes in the Netherlands: a review. *Aquatic Ecology*, **39**, 381–400.

Ramaker, T.A.B., Meuleman, A.F.M., Bernhardi, L. and Cirkel, G. (2005). Climate change and drinking water production in the Netherlands: a flexible approach. *Water Science and Technology*, **51**, 37–44.

Segrave, A., Pronk, W., Ramaker, T. and Zuleeg, S. (2007). Global trends affecting the water cycle; winds of change in the world of water. Techneau report.

Stuyfzand, P.J., Kooiman, J.W., Oosterhof, A. and Raat, K.J. (2007) Problems and solutions when recharging brackish aquifers with membrane concentrate from a less brackish groundwater source. In: *Management of Aquifer Recharge for Sustainability, Proceedings of the 6th International Symposium on Managed Artificial Recharge of Groundwater (ISMAR6)*, Fox, P. (ed.), Phoenix, pp. 47–60.

Van den Berg, G.A., de Rijk, S., Mesman, G., Abrahamse, A. and de Graaf, R. (2008). Opportunities for decentralised drinking water supply in The Netherlands. KWR report BTO 2008.018, Nieuwegein, The Netherlands (in Dutch).
Van Vliet, M.T.H. and Zwolsman, J.J.G. (2008). Impact of summer droughts on the water quality of the Meuse River. *Journal of Hydrology*, **353**, 1–17.
Vewin (2006). Water in Beeld. Rijswijk (in Dutch).

Enhancing consumer relations: the role of trust and confidence

C. Fife-Schaw[a], J. Barnett[a], G. Morrison[b], J. Åström[b] and J. Hartung[b]

[a]Dept of Psychology, University of Surrey, Guildford, UK (E-mail: c.fife-schaw@surrey.ac.uk)
[b]Chalmers University of Technology, Göteborg, Sweden

Abstract This paper reflects a recent debate in the risk communication literature on trust that has implications for relationships between the water industry and consumers. This literature makes a clear distinction between trust and confidence: trust being related to beliefs about the motives and values of bodies such as water companies and confidence rather referring to perceptions of their past performance as an indicator of their future performance. A new conceptual model discussing the way in which trust and confidence relate to each other will be outlined. Using an illustration from a recent waterborne outbreak in Sweden we will discuss the relevance of this model for consumer facing practice in the water industry. Finally we will suggest a range of broader implications of the trust/confidence distinction for industry relations with consumers.

Keywords calicivirus, confidence, consumer, trust, waterborne outbreak

INTRODUCTION

This paper discusses the concepts of consumer trust and confidence as they apply to the drinking water industry and draws on work that was conducted as part of the EU funded TECHNEAU project which is concerned to explore technological and regulatory options for securing safe drinking water into the future. Here we discuss some of recent ideas about consumer trust and in particular we focus on the distinction between trust and confidence and their hypothesised roles in fostering consumer acceptance and/or cooperation. A case study of a water contamination incident in Sweden is described and we finish by highlighting some implications for how the industry might better relate to its customers.

It is commonly assumed that public trust is a key factor in encouraging cooperative action on the part of citizens (Siegrist *et al.*, 2003). This assumption is a major component underling recent European legislation requiring public participation in decision making cf. the requirement to involve citizens in river basin management under the EU Water Framework Directive. Turning the focus from citizens to consumers, it is similarly assumed that consumer trust is important in determining the acceptability of any proposed change to the service offered to consumers. Where suppliers have to deal with water-related incidents and accidents again it is often supposed that suppliers and regulators that are

 TECHNEAU: Safe Drinking Water from Source to Tap. Edited by Theo van den Hoven and Christian Kazner. ISBN: 9781843392750. Published by IWA Publishing, London, UK.

trusted will be able to contain negative public reactions more effectively than those that are not and that it is vital to safeguard existing trust in the way that such events are handled. As we shall see however, these assumptions about the role of trust in securing acceptance of the actions of a regulator or supplier are not as straightforward as they may first appear.

THE DISTINCTION BETWEEN TRUST AND CONFIDENCE

The literature on consumer trust and confidence is unfortunately marked by a lot of terminological confusion with the terms *trust* and *confidence* often used interchangeably and imprecisely. This has not fostered clarity or consensus about the role of trust and confidence in consumers' relationships with suppliers and regulators. Additionally, languages other than English provide a different nuance or do not distinguish between these concepts by the allocation of a separate word to denote them and as a consequence the literature contains many apparently inconsistent findings at least some of which can be attributed to this terminological confusion.

Following Siegrist *et al.* (2003) we draw the distinction between *social trust,* which involves some judgement of similarity of values and intentions (so-called 'morality' information) and *confidence* which is a belief based on past experience that events will occur as expected – i.e. somebody is in control. Social trust, in handing over agency to another, is fundamentally a feature of a social relationship where one has to impute openness, fairness and integrity (among many other possible characteristics) to another. Does this agency have your best interests at heart? Confidence that something will happen on the other hand does not necessarily involve trusting the motives or values of the agents involved; it does, however, involve assessments of the competence of that agent.

Although in most cases confidence and social trust will go together they do not have to. On regularly experiencing the delivery of good quality water to their tap a consumer might have developed *confidence* that there will continue to be good quality water coming out of their tap. It may not be necessary to have to trust the motives and values of the water company and to judge whether these are consistent with one's own well-being and interests. In the context of electricity supplies Siegrist *et al.* (2003) show that where past competence has led to high confidence in the supply, trust in the supplier is essentially unimportant. Similarly in England most consumers have confidence in their water companies to provide safe drinking water yet there is often unhappiness at media reports of the private companies' profit taking for the benefit of foreign shareholders suggesting that the companies have a concern with the interests of shareholders that is in some respects in competition with the interests of consumers.

According to the analysis of Siegrist *et al.*, trust comes into play when something has gone wrong and it is no longer possible to be confident on the basis of past performance. Similarly in situations where there is little past experience upon which to base estimates of competence and thus confidence, social trust will become more important and may be used to infer the likely competence of the relevant body. Where a change is proposed, say the introduction of a new billing system, there will be no direct experience for consumers to use as a basis for their confidence estimates and thus social trust based on an assessment of the supplier's motives becomes important.

Earle and Siegrist (2006) have produced a conceptual framework that explicitly acknowledges the distinctions between trust and confidence and brings them together in a single model. Their Trust, Confidence and Cooperation (TCC) framework is a general model intended to be applicable to all aspects of trust between an individual and both known and unknown others including utilities. The framework is presented in Figure 1.

The model is focused on explaining why people might cooperate with another agent. Cooperation implies any compliance or acceptance behaviour made as the result of a request and could include prompt payment of water bills, acceptance of interruptions to supplies, acceptance of price increases, willingness to use less

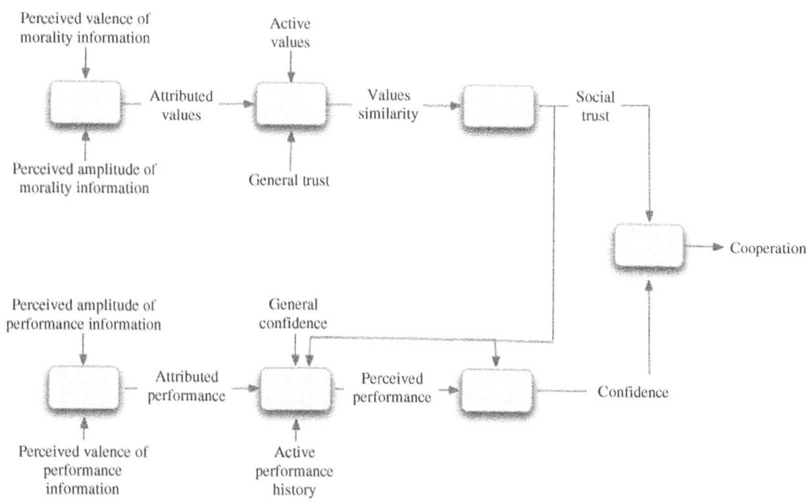

Figure 1. Earle and Siegrist's (2006) trust, confidence and cooperation framework.

water etc. The model suggests that social trust is based on morality-relevant information, while confidence is based on performance–relevant information. In times of low uncertainty, when morality information is less relevant, social trust does not play the main role in cooperation – people will pay their bills, say, without questioning the motives of the company. It also suggests that social trust becomes more important in times of uncertainty, when morality information becomes more relevant and it hypothesises that social trust will affect judgments of confidence both directly and via effects on perceived performance (Earle and Siegrist, 2006, p388, see also Haddad, 2004; Po *et al.*, 2005).

Social trust is more fundamental than confidence and judgements of confidence presume pre-existing relations of trust. It is assumed that where social trust levels are high some performance failings might lower confidence but would not fundamentally undermine a willingness to cooperate. As a corollary, when social trust is low, performance failures should lead to a quick negative response from consumers, such as complaints or a lack of cooperation.

TRUST AND CONFIDENCE IN A WATER CRISIS MANAGEMENT CONTEXT – THE CASE OF LILLA EDET

The TECHNEAU project is in the process of cataloguing successful and unsuccessful approaches to interactions with consumers and the recent case of viral contamination resulting in a waterborne outbreak in Lilla Edet in Sweden is an interesting contemporary example. What follows is a chronology of events with a commentary on how this may have impinged on assessments of trust and confidence.

In the week before the 11th of September 2008 high levels of turbidity were detected in the Göta River which supplies drinking water to some 700,000 people through six drinking water treatment plants. The drinking water supply in the municipalities of Lilla Edet and Göteborg provides for approximately 12,800 and 600,000 people respectively with the ability in Göteborg to close the source water intake at the riverside. Göteborg closed the intake on the 6th of September and drew on its reserve sources, Rådasjön and Delsjön lakes. Consequently, Göteborg was effectively protected from this faecal contamination event. Lilla Edet has no reserve and decided instead to increase the chlorine dose to protect their drinking water from microbial contamination. On the 11th of September 150 cases of gastro-intestinal illness were reported in TTELA (the web version of the local newspaper). Information on the outbreak, including a press release, was continuously updated by the Västra Götaland County Medical Office on their website. As later verified by the epidemiological survey, most of this number fell sick on the evening of the 9–10th of September. Due to several reports of illness during the day of the11th of September the environmental chief of Lilla Edet was informed of the situation by the county doctor for infectious diseases at the regional hospital in Uddevalla. Water samples were taken and nothing was found for the analysis of pathogenic viruses but mapping of cases commenced under the auspices of the Swedish Institute for Infectious Disease Control (SMI).

In the evening of the 11th of September consumers were informed by the municipality through the media that they should not drink tap water unless it was boiled first. The next day the number of notified cases rose to 350 and consumers were told to continue boiling water and raise general hygiene standards (thorough hand washing etc.). However on the same day TTELA reported that the Lilla Edet municipal supply had no alternative reserve water source and had already been aware of indications of faecal contamination (*E.coli*) in the river. They also noted that Göteborg had avoided the problem by taking water from other reserve sources on the 6th of September. Ironically, there was also an article in TTELA on 11th September concerning the problem of sewer overflows in Trollhätten municipality which is upstream the Lilla Edet intake.

On 15th September TTELA reported that the outbreak might be attributed to an illegal discharge into the Göta river. Two people had seen a belt of sewage and toilet paper in the river two kilometres upstream of the Lilla Edet intake on the 6th of September. A police report to this effect had been made by the environmental chief on the 13th of September and she comments in the media on the possibility of a ship, a septic tank or a private discharge as causes. The police also receive reports that a travelling circus had been seen emptying a septic tank into the river though later investigations question the reliability of these reports and by 22nd September it is known that the circus had not been in the area at the time. It is also noted by the police on the 22nd that there is no independent evidence that the belt of sewage and toilet paper reported by the public was in fact sewage.

Records of municipality website hits (which increased from a normal 2000–2500 per day to 3000–6000 during the outbreak) and telephone calls (with extra opening hours on the telephone exchange during the outbreak) demonstrate a marked increase during the outbreak. The municipality were proactive in dealing with the media, particularly through the environment officer) and providing relevant information.

In part of a longer interview with TTELA the political leader of the municipality stated that he was still drinking the tap water. The environmental chief responded the next day (the 17th of September) saying that the advice to the public was still to boil the water.

A week later (the 22nd of September) an expert committee including representatives from SMI, the national food administration, the Lilla Edet municipality, and the national water catastrophe group, put forward their finding that it was most likely that drinking water was the cause of the outbreak. A decision was made to send out a questionnaire to determine the temporal and spatial spread of the gastrointestinal illness. On the 18th of September samples from 11 of 15 individuals with reported gastrointestinal illness in the municipality were positive for *calicivirus*. Preliminary positive results from the analysis of *calicivirus* in the source water however were not confirmed by later analyses. In the source water sampling regime (three times per week for microbial analysis) the highest *E. coli* levels were detected on the 8th of September with 1100 MPN/100 ml (5 km downstream) and 1400 MPN/100 ml (40 km downstream). These values correspond well to the continuously monitored (elevated) turbidity which was first noted on the 6th of September. Drinking water samples in Lilla Edet were taken on several occasions without positive detection of *calicivirus*, although somatic coliphages were detected in samples taken on the 17th of September.

Sixteen days after the notification of the outbreak on the 27th of September the municipality home webpage declared that samples had shown the water was now free from microbial contaminants and it was now possible to drink tap water without boiling it first. This decision is made jointly with the SMI, the national water catastrophe group and the national food authority.

TTELA reports on 26th of November that the questionnaire showed that at least 2000 people had been affected by the outbreak. The county infectious disease doctor at Uddevalla hospital stated that water was the most likely cause of the outbreak as judged from the results of the epidemiological investigation based on the questionnaire. The municipal environment chief also says in the media that tap water is the likely cause and that those who have municipal water were more represented among the affected than those with private wells. Also she notes that the first cases were in Lilla Edet, followed later by Lödöse and then Nygård which is consistent with the nature of the supply infrastructure. None of these statements confirm whether the source of viral contamination of drinking water was from the source water or the pipe infrastructure.

Some six weeks passed without any further clear statements in the media about the source and exact nature of the outbreak until on the 9th of January 2009 when representatives of the water treatment plant present their version to the public via TTELA. They do not believe that the sickness was waterborne as no *calicivirus* was detected in water samples. They claimed that the detection of somatic coliphages reported earlier were due to the use of non sterilised sampling bottles although this possibility had been excluded by the expert committee. They make it clear in the media that feel that they had been blamed unfairly.

At the time of writing in March 2009 some five months after the event the exact source of the outbreak is still the subject of uncertainty. Research by Chalmers University of Technology is currently investigating point discharges of sewage from three upstream locations that preceded the outbreak. These were combined sewer overflows and an emergency discharge. A quantitative microbial risk assessment is presently being undertaken to assess the relative risk due to these discharge scenarios.

In January 2009 we sent out a questionnaire that specifically asked about trust and confidence after the outbreak. The questionnaire (N = 392) is still being analysed, but early findings show that the public relied heavily on media reporting and on the response of politicians, municipality employees and utility workers. The newspaper reporting was thought to be generally accurate and informative.

We asked a series of questions that invited people to reflect on how they felt about their water supply now (in January 2009) compared to how they felt before the outbreak. The pattern of findings suggest that while there are minor shifts in the direction of being less happy with those responsible for the supply generally levels of support are high as are views about the quality of their supply. For example, in response to a question asking people to indicate their level of agreement with the statement "The quality of the tap water is a low priority issue among the municipality's politicians" only 26.8% agreed with this statement after the event. This should be compared to 20.8% who indicated that they would have agreed with it anyway before the event. Similarly people were asked about how safe they felt when they drank the water on a 9-point scale ranging from 1 indicating not at all safe through to 9 indicating they thought it was completely safe. After the event the mean score was 6.51/9 compared with a rating of 7.89/9 when asked to reflect on how they felt about it before the event. While this difference is statistically significant people's perceptions of the safety of their water remain high. We will be following up these individuals later in the year to see what the long term effects of the outbreak were.

COMMENTARY

Prior to the outbreak consumers in Lilla Edet had good reason to have high levels of confidence in the municipality. In 2005 the tap water in Lilla Edet was rated as the tastiest in Sweden by the Swedish Water & Wastewater Association (SWWA) and this was the source of considerable local pride. There had been no similar outbreaks in recent memory. Trust issues were probably not particularly salient as the system was functioning as expected so little thought about the motives of the supplier was needed (cf. Siegrist *et al.*, 2003).

The municipality's handling of this event, while not sufficient to prevent a large proportion of the population becoming ill, does demonstrate a concern for the well-being of the public. Once cases were notified the decision to tell people to boil their water was made quickly through the media. A number of key bodies got together to form a steering committee and its deliberations were explained openly to the public again through the media. The authorities were open in explaining that they were uncertain about the origin of the outbreak. The media appeared to work cooperatively with the municipality.

The event nonetheless presented the public with a number of indicators or potential performance failings of those involved bringing into question their competence and thus challenging confidence. The media made it clear that high levels of contamination were noted on the 6th of September and that Göteborg had been able to draw on alternative supplies while Lilla Edet had none to draw on. Consumers might ask why this safeguard was not in place to protect the public in Lilla Edet and this suggests a lack of resilience in the supply system and adequate risk management.

Although it seems that the municipality acted appropriately in warning consumers, departments within the municipality did not agree on the source or seriousness of the outbreak. The comments of the political leader of the municipality indicating that he would drink tap water thus implicitly not accepting the advice of the municipality's own experts caused some public consternation as indicated by responses to our questionnaires. Similarly the environment chief claiming that the incident was waterborne while the waste water utility representatives challenged this in the media (cf. the use of non-sterilised bottles and feeling blamed) was unlikely to have instilled confidence in the public.

That this dispute was still being played out in public two months after the outbreak may have given the impression that those responsible for identifying the cause of the outbreak had not done their jobs properly. It is likely to be concerning that the utility five months after the event are still in the process of determining the magnitude of sewer discharges (combined sewer overflows) from the sewer systems of Lilla Edet and in upstream municipalities, as previous research has pointed out the potentially high microbial loads from these systems (Åström *et al.*, 2009). The environmental department in Lilla Edet has invited researchers from various fields to contribute in the analysis of the underlying causes for the outbreak, but results from research efforts at this stage have not yet been presented to consumers. In most situations confidence is likely to be enhanced by knowing that the cause of a negative event is clearly identified and understood.

Such threats to confidence levels might be expected to have increased concerns about the safety of the water and to damage the standing of the authorities in the public's eyes. Our initial analyses appear to suggest that this has not happened to any great extent in Lilla Edet and while we are not yet in a position to establish this unequivocally it seems likely that the efforts of an already trusted set of authorities to be open and informative have maintained levels of social trust.

CONCLUSIONS

At first glance the distinction between trust and confidence might seem to be rather subtle with few practical implications. However the kinds of activities needed to establish social trust can be quite different from those that establish confidence. In much the same way as individuals need to adapt their behaviour to get other people to trust them so the same can apply to large agencies. Trust is usually developed through openness, clarity about the scope and limits of one's responsibilities, anticipation of others' concerns and a responsiveness to their needs when appropriate. Confidence on the other hand is promoted by demonstrations of competence, usually over a period of time. Some interactions with consumers require demonstrations of competence over time (e.g. delivering a new billing system) but others such as dealing with situations where confidence is dented by failures to perform competently primarily require agencies to ensure that the public continue to have faith in their moral probity. As we see in the example of Lilla Edet even a relatively serious contamination event coupled with some media reporting suggesting a lack of competence can have a limited impact on consumers' relationships with their suppliers if social trust levels are high. Thus agencies need to be clear about what it is that they are trying to address when they interact with the public.

ACKNOWLEDGEMENT

We would like to thank Annika Ekvall, chief of the municipal environmental department of Lilla Edet for her cooperation with our requests for information about this event.

REFERENCES

Åström, J., Pettersson, T.J., Stenström, T.A. and Bergstedt, O. (2009). Variability analysis of pathogen and indicator loads from urban sewer systems along a river. *Water Science and Technology*, **59**(2), 203–12.

Earle, T.C. and Siegrist, M. (2006). Morality information, performance information, and the distinction between trust and confidence. *Journal of Applied Social Psychology*, **36**, 383–416.

Haddad, B. (2004). WateReuse Foundation summary of research needs assessment workshop: Human reactions to water reuse. Appendix I in Southern California Water Recycling Projects Initiative (2004). Successful Public Information and Education Strategies: Technical Memorandum. http://www.usbr.gov/lc/socal/reports/TM_PublicInfoStrategies.pdf (accessed 25 February 2009).

Po, M., Nancarrow, B., Leviston, Z., Porter, N., Syme, G. and Kaercher, J. (2005). *Predicting Community Behaviour in Relation to Wastewater Reuse: What Drives Decisions to Accept or Reject?* Melbourne: CSIRO.

Siegrist, M., Earle, T.C. and Gutscher, H. (2003). Test of a trust and confidence model in the applied context of electromagnetic field (EMF) risks. *Risk Analysis*, **23**, 705–716.

Nanotechnology applications and prospects in the water sector

T.E. Cloete[a], M. Botes[a], M. De Kwaadsteniet[a] and J. Theron[b]

[a]Department of Microbiology and Plant Pathology, University of Pretoria, South Africa (E-mail: eugenecloete@sun.ac.za)

[b]Department of Microbiology, Stellenbosch University, Stellenbosch, South Africa

INTRODUCTION

Since it would appear that conventional detection methods do not adequately assess the risk of waterborne disease, the need for powerful new tools for the detection of pathogenic microorganisms in water is becoming increasingly more important. Although several different molecular methodologies are available, these have only recently been applied in the field of water science and technology. Moreover, the rapid progress of nanotechnology and advanced nanomaterials production offers significant opportunities not only for the detection and remediation of a broad range of environmental contaminants (Theron and Cloete, 2002; Andreescu *et al.*, 2009), but also for the development of new diagnostic assays that may serve as an appealing alternative to current molecular diagnostic techniques.Controlling bacterial growth in water supplies and industrial water systems pose a unique set of challenges. Nanocides provide novel solutions in this regard. Nanowaste is also an area of great concern, especially that nano-products inevitably end up in water systems. This risks associated with this will be discussed.

NANOTECHNOLOGY DEVICES FOR BIOMOLECULAR DETECTION

Advances in nanotechnology and nanosciences are having a significant impact on the field of diagnostics, where a number of nanoparticle-based assays have been introduced for biomolecular detection (Fortina *et al.*, 2005; Gaspar *et al.*, 2007; Jain, 2007). In this section, the most promising nanomaterials and their use in the detection of nucleic acids and microbial pathogens will be highlighted.

 TECHNEAU: Safe Drinking Water from Source to Tap. Edited by Theo van den Hoven and Christian Kazner. ISBN: 9781843392750. Published by IWA Publishing, London, UK.

Quantum dots

Quantum dots (QDs) are colloidal, luminescent inorganic nanocrystals with unique photochemical properties, which include high quantum yields, large extinction coefficients, pronounced photostability, as well as broad absorption spectra coupled to narrow size-tunable photoluminescent emission spectra (Bruchez *et al.*, 1998; Chan and Nie, 1998). Strategies for attaching biorecognition molecules to QDs comprise of direct chemical coupling to a functional group displayed on the QD surface (Chan and Nie, 1998; Gerion *et al.*, 2003) or by non-covalent self-assembly in which proteins are engineered to express positively charged domains that interact through electrostatic attractions with the negative surface of modified QDs (Ding *et al.*, 2004; Goldman *et al.*, 2005). Although functionalized QDs have been used to detect complementary target nucleic acid sequences in chip-based assays and in fluorescent in situ hybridization (FISH) assays (Pathak *et al.*, 2001; Xiao and Baker, 2004), QDs are increasingly being used in immunoassay detection (Alivisatos *et al.*, 2005; Goldman *et al.*, 2006). QD-based immunological assays have been applied to the detection of different bacterial and protozoan pathogens. For these immunoassays, the QDs are conjugated to organism-specific antibodies, or, alternatively, biotinylated organism-specific antibodies are used that are subsequently detected using a QD-streptavidin bioconjugate.

FloDots

In addition to QDs, luminescent dye-doped silica nanoparticles (FloDots), which consist of luminescent organic or inorganic dye molecules dispersed inside a silica matrix, have also been developed for ultra-sensitive bioanalysis and diagnostics (Tapec *et al.*, 2002; Yao *et al.*, 2006). Owing to the silica matrix shielding effect, the doped dye molecules are protected from environmental oxygen, enabling the fluorescence to be constant and thus providing an accurate measurement for bioanalysis (Tapec *et al.*, 2002). Moreover, the silica matrix provides a versatile substrate for surface immobilization and various biorecognition molecules, including oligonucleotides and antibodies, have been conjugated to FloDots (Santra *et al.*, 2001; Tapec *et al.*, 2002; Hilliard *et al.*, 2002). FloDots have been used successfully for the detection of DNA hybridization and in immunoassays. Using FloDots functionalized with oligonucleotides as labels for chip-based sandwich DNA assays, Zhao *et al.* (2003) reported a detection limit of 1 fM target DNA. The high sensitivity of the assay can be ascribed to the fact that each FloDot has the fluorescence intensity of thousands of dye molecules; therefore, each gene hybridization is reported by thousands of fluorophores.

Gold nanoparticles

Gold nanoparticle-based probes have been used in the identification of pathogenic bacteria in a chip-based system (Taton *et al.*, 2000). The assay consists of a capture DNA strand immobilized on a glass chip that recognizes the DNA of interest. A separate sequence on the captured target strand is then labelled with an oligonucleotide-functionalized gold nanoparticle probe. After catalytic reduction of silver onto the gold nanoparticle surfaces to amplify the target signal, the capture-strand/target/nanoparticle sandwich is visualized with a flatbed scanner. The target DNA was detected at concentrations of 50 fM, which represents a nearly 100-fold increase in sensitivity over traditional fluorescence-based assays. Gold nanoparticles have also been used in the detection of genomic DNA from Staphylococcus aureus, without enzymatic amplification, at concentrations of 33 fM (Storhoff *et al.*, 2004). The assay depends upon the target-selective binding of gold nanoparticle probes to consecutive regions of a target DNA sequence in solution, followed by transfer to an evanescently illuminated glass microscope slide for light scattering measurements.

Magnetic nanoparticles

In contrast to gold nanoparticles and QDs, magnetic nanoparticles have not been used in many biological applications. Bound to biorecognition molecules, magnetic nanoparticles can be used to facilitate the separation, purification and concentration of different biomolecules (Richardson *et al.*, 2001; Yang *et al.*, 2004; Gu *et al.*, 2006). To do so, biorecognition molecules such as antibodies can be immobilized on the surface of magnetic nanoparticles through covalent or electrostatic interactions (Gupta and Gupta, 2005). After reacting these magnetic nanoparticles with sample solutions, targeted molecules can be bound by or captured on the surface of these magnetic nanoparticles. By applying a magnetic field, these nanoparticles can subsequently be concentrated and separated from the bulk solution and identified.

THE APPLICATION OF NANOFIBERS AND NANOBIOCIDES IN WATER PURIFICATION

The formation of harmful disinfection byproducts, biofouling and virus penetration are significant obstacles in current water purification methods. Novel filtration systems may include nanofibrous media which remove unwanted particles smaller than 0.3 μm. The antimicrobial activity of nanofibers is improved by encapsulated nanobiocides such as metal nanoparticles (Barhate & Ramakrishna, 2007). This section will focus on functional nanofibers and nanoparticles and possible applications in water purification.

Fabrication of functional nanofibers

Electrospinning is a simple, rapid, efficient and inexpensive method, run by a high voltage supplier, a capillary tube with a pipette or needle of small diameter (spinneret) and a grounded collecting plate, usually a metal screen, plate, or rotating mandrel. The nanofibers with diameters lower than 100 nm are produced by applying a high voltage to electrically charge liquid or melt out of pipette. The electric force overcomes surface tension of the polymer solution leading to evaporation of the solvent and ejection of a thin polymer drawn toward the collector plate (Huang *et al.*, 2003). Industrially scale electrospinning was recently developed. The "Nanospider" designed by Elmarco synthesize nanofibers weighing between 0.1 and 10 g/m^2 and produces 0.1 to 1 g material in less than one minute, sixty times faster than using traditional electrospinning techniques (www.elmarco.com). Nanofibers possess an extremely high surface-to-volume ratio and a small pore size and may be further functionalized by spinning blends of specific polymers, coating of fibers, or inclusion of functional components such as biocides. The successful preparation of antimicrobial nanofibers depends primarily on the selected surfactants and polymers (Huang *et al.*, 2003). Polymer-surfactant interactions influence parameters of the electrospinning process and may alter the three critical factors of electrospun nanofibers, namely rheology, surface tension and conductivity.

Permanent antimicrobial surfaces may be achieved by the following techniques: covalent coupling of poly(quaternary ammonium) compounds to various surfaces (Yao *et al.*, 2006), attaching alkylated pyridinium and ammonium polymers such as poly(4-vinyl-N-alkylpyridinium bromide to glass slides (Tiller *et al.*, 2002), modification of electrospun poly(vinylidene fluorid-co-hexafluoropropylene fibrous membranes (Yao *et al.*, 2006), addition of N-halamine additives to nylon 6 nanofiber membranes (Tan *et al.*, 2007), combination of polycarbonate/chloroform solution and quaternary ammonium salt (benzyl triethylammonium chloride) (Kim *et al.*, 2007), electrospinning of polyurethane cationomers containing different amounts of quaternary ammonium groups (Jeong *et al.*, 2007), developing a bi-layer composite membrane consisting of 1.0% chitosan and poly(acrylic acid), with a bottom support layer composing of poly(ethylene glycol) diacrylate and poly(acrylic acid) (Don *et al.*, 2005) and replacing flux-limiting asymmetric porous membranes with a three-tier composite structure consisting of chitosan, polyacrylonitrile and terephthalate (Yoon *et al.*, 2006). These electrospun nanofibers show great potential as antibacterial coating for fibrous media.

Nanobiocides

Nanobiocides are divided into three categories, namely, metal and metal oxides, engineered/synthesized nanomaterials and natural antibacterial substances (Li *et al.*, 2008). The antimicrobial activity of embedded metal nanoparticles in nanofibers or polymer matrices will be discussed, followed by the application of carbonaceous nanomaterial, fullerenes and natural substances as biocides.

Nanofibers embedded with silver nanoparticles showing effective antimicrobial activity and good filtration properties are synthesized using the following polymers: polycaprolactone combined with polyacrylonitrile (Jeong *et al.*, 2007), poly(vinyl pyrrolidone) (Bai *et al.*, 2008), polyethane foams, polysulfone (Zodrow *et al.*, 2009) and poly(amidoamine) dendrimers (Balogh *et al.*, 2001). Commercial membranes may also be regenerated and improved by the addition of silver nanoparticles. These water filtration membranes may ensure effective control on a wide variety of bacteria and viruses and have protection against biofouling.

The incorporation of zinc oxide nanoparticles into blend films of chitosan and poly(vinyl alcohol) improved film properties and effectively killed Staphylococcus aureus (Vicentini *et al.*, 2009). Copper nanoparticles were functionalized with an acrylic group to copolymerize with other acrylic monomers. The released particles were very active against marine biofouling strains (Anyaogu *et al.*, 2008).

TiO_2 photocatalysis is currently being used in industrial scale water purification systems by removing contaminants such as Giardia, Cryptosporidium and oocysts (www.purifics.com). Nanocomposite water membranes containing polyamide and 5.0 wt% TiO_2 nanoparticles exibited increased water flux and strong mechanical properties. TiO_2 nanoparticles incorporated into an isotactic polypropylene polymeric matrix also showed high biocidal activity (Lee *et al.*, 2008).

Carbon nanotube filters effectively remove bacteria and viruses from water and allow efficient water flow. Argonide developed NanoCeram® cartridges by combining nano alumina into a microglass fiber matrix to form an electropositive (~+50 mV) filter media and converted into pleated cartridges. All bacteria, viruses and small molecule organic pollutants that might have escaped reverse osmosis are captured (www.argonide.com). Aluminum hydroxide fibers (2 nm in diameter) combined into a composite filter with a high electropositive charge was patented in 2005. Bacteria, viruses, nano sized colloidal particles, sub-micron inorganic and metallic particles are removed (Tepper & Kaledin, 2005). Incorporation of C60 in ceramic membranes successfully reduced the attachment and respiratory activity of Escherichia coli K12 (Chae *et al.*, 2009).

Antimicrobial peptide nanoparticles are designed through self-immobilization mediated by biomineralization reactions and hereby improving their effectiveness through increased stability and protection from inactivation. A cationic decapeptide, KSL (KKVVFKVKFK)10, isolated from a combinatorial peptide array has a wide activity spectrum (Brogden, 2005). Nisin A, a lantibiotic encapsulated into poly-L-lactide nanoparticles exhibited high antimicrobial activity against Lactobacillus delbrueckeii subsp. bulgaricus for up to 45 days (Salmaso *et al.*, 2004). The metal-chelating properties of chitosan nanoparticles have been mainly used in wastewater treatment. The growth of E. coli stopped immediately after exposure to the chitosan–Ag-nanoparticle composite (AgNP) with a high release rate of cellular components. Chitosan triphosphate nanoparticles loaded with Ag+ and Zn_2+ showed an increased biocidal activity (Du *et al.*, 2009). Emulsification of chitosan nanoparticles lead to adhesion onto the surface of E. coli and S. aureus and subsequently cell surface alterations and leakage of intercellular components (Xing *et al.*, 2009). The high availability, low cost and high antimicrobial activity of these nanobiocides make them attractive in the use of water purification.

POTENTIAL RISKS IN NANOTECHNOLOGY

The public's confidence in nanotechnology can easily be broken by a safety incident. In 2006, the international media raised alarms about nanotoxicology after a German bathroom cleaner, Magic Nano, was pulled off the market within three days. This product was responsible for severe respiratory problems in more than one hundred people of which six were hospitalized with pulmonary edema. Subsequent research concluded that the ethanol-water aerosol was responsible for the respiratory problems. This incident clearly illustrates, although the nanoparticles were not the causative agent, that thorough research is needed before a product containing nanoparticles or nanomaterials are commercialized (Stern and McNeil, 2008). A study reporting that inhaled carbon nanotubes cause lesions similar to asbestos, a known carcinogen, has also received a lot of media attention in the science community (Poland *et al.*, 2008).

In vitro toxicity studies

Assessing the safety of nanoparticles and nanomaterials in mammalian cell lines is a simple technique to collect a large amount of data. Numerous in vitro studies have revealed that several nanoparticles and nanomaterials are cytotoxic towards human and animal cell lines (Lewinski *et al.*, 2008). Nanoparticles and nanomaterials also caused DNA damage, and chromosomal and gene mutations in genotoxicity assays (Landsiedel *et al.*, 2009). There is however in nanotoxicology a lack of standardized protocols and a database with reference materials that is reproducible and credible. This makes comparing the toxicity of different nanoparticles and nanomaterials problematic. There are a number of other limitations in in vitro assays such as the ability of nanoparticles to interact with dyes and reagents used cell viability assays and the agglomeration of nanoparticles with proteins in test mediums.

The impact of foreign materials on the immune system in its entirety can only be evaluated in vivo. Macrophage and dendritic cell recruitment are intercellular effects of the immune system that cannot be monitored in vitro. Furthermore, the biodistribution and translocation of nanoparticles and nanomaterials to secondary organs can only be assessed during in vivo studies. Data collected from in vitro and in vivo assays have also been shown not to correlate. In vitro assays must thus be validated with in vivo results to be considered of importance (Fischer and Chan, 2007).

In vivo toxicity studies

Nanoparticles and nanomaterials can enter the human body through the following six routes; intravenously, dermally, subcutaneously, through inhalation, intraperitoneally and orally. They can be absorbed to various biological components, depending on the physical properties of the nanomaterial and the route of administration, before being translocated to secondary organs. Nanoparticles and nanomaterials that have entered the lungs can translocate from the respiratory tract, through the nasal epithelium to the olfactory bulb and has been detected in the brain of laboratory animals after inhalation. The main target organs of nanoparticles and nanomaterials in the body are the liver and spleen, members of the reticuloendothelial system. The kidney is also a common target site and has been shown to be a vital clearance organ of nanoparticles and nanomaterials in the body (Stern and McNeil, 2008).

Toxicity observed during in vivo studies ranged from very low to high, manifesting in the latter as inflammation, granulomas formation, oxidative stress and even necrosis in rare cases. Toxicity depended on the type of nanomaterial, concentration, route of administration and experimental outlay of the study (Stern and McNeil, 2008).

Diseases and clinical disorders

Research has shown that nanoparticles are linked to neurological conditions such as Alzheimer's and Parkinson's diseases. Nanoparticles are associated with elevated levels of autophagic vacuoles in cells. This is probably as a result of a cellular response trying to clear the cells of nanoparticles. Impaired autophagy can result in morphological changes and subsequent cell death and may therefore play a role in neurological and other diseases (Stern and Johnson, 2008). Nanoparticles have also been shown to initiate β-amyloid peptide fibrillation and the formation of senile plagues, a key feature of Alzheimer's pathology (Wu *et al.*, 2008).

β-amyloid peptide is an amyloidogenic protein, a group of proteins that form fibril aggregates under specific conditions that are highly insoluble and precipitates to form plagues. They are part not only characteristic in the pathology of Alzheimer's and Parkinson's disease but also Huntington's, Creutzfeld-Jacob's and prion's disease, to name a few. Nanoparticles are also involved in the fibrillation of other amyloidogenic proteins, such as β-2-microglobulin and myoglobin (Heegard *et al.*, 2007; Linse *et al.*, 2007).

Epidemiology studies have also linked natural nanoparticles with cardiovascular diseases (Chen *et al.*, 2008). The role engineered nanoparticles plays in the inflammation of the endothelial have been investigated in vitro since it is critical for the development of cardiovascular diseases. One example is the inflammation observed in human aortic endothelial cells after exposure to engineered iron oxide and zinc oxide nanoparticles (Gojova *et al.*, 2007).

Mechanisms of nanotoxicity

The physiochemical characteristics of a nanoparticle or nanomaterial influence its toxicity and must therefore be assessed during toxicity assays. Gonzalez *et al.* (2008) proposes that the following information must be documented; the method the nanoparticles are synthesized and prepared, particle size and distribution, the composition and relevant coatings, specific surface area, crystal structure, purity of the sample and the aggregation status in the relevant media.

Nanoparticles and nanomaterials can interact with biological macromolecules because of their small size and unique physiochemical characteristics. These interactions with macromolecules such as DNA, cytoskelet elements, collagen, membrane structure and proteins can adversely affect their original function in the body.

Nanoparticle's and nanomaterial's ability to produce reactive oxygen species (ROS), is an important toxicity mechanism, disrupts a cell's redox equilibrium and so causes oxidative stress. Nanoparticles and nanomaterials can generate

ROS either spontaneously or by targeting subcellular organelles and biological systems. The end result of oxidative stress is excess cell profileration, apoptosis, lipid peroxidation or mutagenesis (Li *et al.*, 2008).

FUTURE PERSPECTIVES

New technologies employing marker-free and live measurements in in vitro assays are currently under evaluation and validation. These in vitro toxicity assays are unaffected by the unique physiochemical characteristics of nanoparticles (Schekenburger *et al.*, 2008). New cell cultures have also been developed that better emulates in vivo toxicity (Lilienblum *et al.*, 2008). Toxicogenomics are new a subdiscipline of toxicology that can provide valuable information about the complex nature of nanoparticle's toxicity. The number of studies involving metabolomics and proteomomics, which falls under this discipline, is limited. Toxicogenomics uses molecular biomarkers that are highly sensitive and can be developed in high through-put automated systems (Fowler *et al.*, 2008).

REFERENCES

Andreescu, S., Njagi, J., Ispas, C. and Ravalli, M.T. (2009) Applications of advanced nanomaterials for environmental monitoring. *J. Environ. Monit.* **11**, 27–40.

Anyaogu, K.C., Fedorov, A.V. and Neckers, D.C. (2008). Synthesis, characterization, and antifouling potential of functionalized copper nanoparticles. *Langmuir*, **24**(8), 4340–4346.

Bai, J., Li, Y., Zhang, C., Liang, X. and Yang, Q. (2008). Preparing AgBr nanoparticles in poly(vinyl pyrrolidone) (PVP) nanofibers. *Colloids and Surfaces A: Physicochemical and Engineering Aspects*, **329**(3), 165–168.

Balogh, L., Swanson, D.R., Tomalia, D.A. Hagnauer, G.L. and McManus, A.T. (2001). Dendrimer silver complexes and nanocomposites as antimicrobial agents. *Nano Letters*, **1**(1), 18–21.

Barhate, R.S. and Ramakrishna, S. (2007). Nanofibrous filtering media: Filtration problems and solution from tiny materials. *Journal of Membrane Science*, **296**(1), 1–8.

Brogden, K.A. (2005). Antimicrobial peptides: pore formers or metabolic inhibitors in bacteria? *Nature Reviews in Microbiology*, **3**(3), 238–50.

Bruchez, M., Moronne, M., Gin, P., Weiss, S. and Alivisatos, A.P. (1998) Semiconductor nanocrystals as fluorescent biological labels. *Science*, **281**, 2013–2016.

Chae, S.-R., Wang, S., Hendren, Z.D., Wiesner, M.R., Watanabe, Y. and Gunsch, C.K. (2009). Effects of fullerene nanoparticles on Escherichia coli K12 respiratory activity in aqueous suspension and potential use for membrane biofouling control. *Journal of Membrane Science*, **329**(1), 68–74.

Chan, W.C. and Nie, S. (1998). Quantum dot bioconjugates for ultrasensitive nonisotopic detection. *Science*, **281**, 2016–2018.

Chen, Z., Meng, H., Xing, G., Yaun, H., Zhao, F., Liu, R., Chang, X., Gao, X., Wang, T., Jia, G., Ye, C., Chai, Z. and Zhao, Y. (2008). Age-related differences in pulmonary and cardiovascular responses to SiO_2 nanoparticle inhalation: nanotoxicity has susceptible population. *Environmental Science and Technology*, **42**(23), 8985–8992.

Ding, S.Y., Rumbles, G., Jones, M., Tucker, M.P., Nedeljkovic, J., Simon, M.N., Wall, J.S. and Himmel, M.E. (2004). Bioconjugation of (CdSe)ZnS quantum dots using a genetically engineered multiple polyhistidine-tagged cohesin/dockerin protein polymer. *Macromol. Mat. Eng.*, **289**, 622–628.

Don, T.M., Chen, C.C., Lee, C.K., Cheng, W.Y. and Cheng, L.P. (2005). Preparation and antibacterial test of chitosan/PAA/PEGDA bilayer composite membranes. *Journal of Biomaterial Science Polymer Edition*, **16**(12), 1503–1519.

Du, W.-L., Niu, S.-S., Xu, Y-L., Xu, Z.-R. and Cheng-Li Fan, C.-L. (2009). Antibacterial activity of chitosan tripolyphosphate nanoparticles loaded with various metal ions. *Carbohydrate Polymers*, **75**(3), 385–389.

Fischer, H.C. and Chan, W.C. (2007). Nanotoxicity: the growing need for in vivo study. *Current Opinion in Biotechnology*, **18**(6), 565–571.

Fortina, P., Kricka, L.J., Surrey, S. and Grodzinski, P. (2005) Nanobiotechnology: The promise and reality of new approaches to molecular recognition. *Trends Biotechnol.*, **23**, 168–173.

Fowler, B.A., Conner, E.A. and Yamauchi, H. (2008). Proteomic and metabolomic biomarkers for III-V semiconducters: And prospects for application to nanomaterials. *Toxicology and Applied Pharmacology*, **233**(1), 110–115.

Gaspar, J.F., Baptista, P.V. and Rueff, J. (2007) Gold nanoparticle-based systems in Genetics. *Curr. Pharmacogen.*, **5**, 39–47.

Gerion, D., Chen, F., Kannan, B., Fu, A., Parak, W.J., Chen, D.J., Majumdar, A. and Alivisatos, A.P. (2003) Room-temperature single-nucleotide polymorphism and multiallele DNA detection using fluorescent nanocrystals and microarrays. *Anal. Chem.*, **75**, 4766–4772.

Gojova, A., Guo, B., Kota, R.S., Rutledge, J.C., Kennedy, I.M. and Barakat, A.I. (2007). Induction of inflammation in vascular endothelial cells by metal oxide nanoparticles: effect of particle composition. *Environmental Health Perspectives*, **115**(3), 403–409.

Goldman, E.R., Mattoussi, H.M., Anderson, G.P., Medintz, I.L. and Mauro, J.M. (2005). Fluoroimmunoassays using antibody-conjugated quantum dots. *Meth. Mol. Biol.*, **303**, 19–34.

Gonzalez, L., Lison, D. and Kirsch-Volders, M. (2008). Genotoxicity of engineered nanomaterials: a critical review. *Nanotoxicology*, **2**(4), 252–273.

Gu, H., Xu, K., Xu, C. and Xu, B. (2006). Biofunctional magnetic nanoparticles for protein separation and pathogen detection. *Chem. Commun.*, **9**, 941–949.

Gupta, A.K. and Gupta, M. (2005). Synthesis and surface engineering of iron oxide nanoparticles for biomedical applications. *Biomaterials*, **26**, 3995–4021.

Heegard, P.M., Boas, U. and Otzen, D.E. (2007). Dendrimer effects on peptide and protein fibrillation. *Macromolecular Bioscience*, **7**(8), 1047–1059.

Hilliard, L.R., Zhao, X. and Tan, W. (2002). Immobilization of oligonucleotides onto silica nanoparticles for DNA hybridization studies. *Anal. Chim. Acta*, **470**, 51–56.

Huang, Z.M., Zhang, Y.Z., Kotaki, M. and Ramakrishna, S. (2003). A review on polymer nanofibers by electrospinning and their application in nanocomposites. *Composites science and technology*, **63**(13), 2223–2253.

Jain, K.K. (2007). Applications of nanobiotechnology in clinical diagnostics. *Clin. Chem.*, **53**, 2002–2009.

Jeong, E.H., Yang, J. and Youk, J.H. (2007). Preparation of polyurethane cationomer nanofiber mats for use in antimicrobial nanofilter applications. *Materials Letters*, **61**(18), 3991–3994.

Kim, S.J, Nam, Y.S., Rhee, D.M., Park, H.-S. and Park, W.H. (2007). Macromolecular Nanotechnology: Preparation and characterization of antimicrobial polycarbonate nanofibrous membrane. *European Polymer Journal*, **43**(8), 3146–3152.

Landsiedel, R., Kapp, M.D., Schulz, M., Wiench, K. and Oesch, F. (2009). Genotoxicity investigations on nanomaterials: Methods, preparation and characterization of test material, potential artifacts and limitations – Many questions, some answers. *Mutatation Research – Reviews in Mutation Research*, **681**(2), 241–258.

Lee, H.S., Im, S.J., Kim, J.H. and Jin, H. (2008). Polyamide thin-film nanofiltration membranes containing TiO2 nanoparticles. *Desalination*, **219**(1), 48–56.

Lewinski, N., Colvin, V. and Drezek, R. (2008). Cytotoxicity of nanoparticles. *Small*, **4**(1), 26–49.

Li, N., Xia, T. and Nel, A.E. (2008). The role of oxidative stress in ambient particulate matter-induced lung diseases and its implications in the toxicity of engineered nanoparticles. *Free Radical Biology and Medicine*, **44**(9), 1689–1699.

Lilienblum, W., Dekant, W., Foth, H., Gebel, T., Hengstler, J.G., Kahl, R., Kramer, P.J., Schweinfurth, H. and Wollin, K.M. (2008). Alternative methods to safety studies in experimental animals: role in the risk assessment of chemicals under the new European Chemicals Legistation (REACH). *Archives of Toxicology*, **82**(4), 211–236.

Linse, S., Cabaleiro-Lago, C., Xue, W.F., Lynch, I., Lindman, S., Thulin, E., Radford, S.E. and Dawson, K.A. (2007). Nucleation of protein fibrillation by nanoparticles. Proceedings of the National Academy of Science of the United States of America, **104**(21), 8691–8696.

Pathak, S., Choi, S.K., Arnheim, N. and Thompson, M.E. (2001). Hydroxylated quantum dots as luminescent probes for in situ hybridization. *J. Am. Chem. Soc.*, **123**, 4103–4104.

Poland, C.A., Duffin, R., Kinloch, I., Maynard, A., Wallace, W.A., Seaton, A., Stone, V., Brown, S., Macnee, W. and Donaldson, K. (2008). Carbon nanotubes introduced into the abdominal cavity of mice show asbestos-like pathogenicity in a pilot study. *Nature Nanotechnology*, **3**(7), 423–428.

Richardson, J., Hawkins, P. and Luxton, R. (2001). The use of coated paramagnetic particles as a physical label in a magnetoimmunoassay. *Biosens. Bioelectron.*, **16**, 989–993.

Salmaso, S., Elvassore, N., Bertucco, A., Lante, A. and Caliceti, P. (2004). Nisin-loaded poly-L-lactide nano-particles by CO_2 anti-solvent precipitation for sustained antimicrobial activity. *International Journal of Pharmaceutics*, **287**(1), 163–173.

Santra, S., Zhang, P., Wang, K., Tapec, R. and Tan, W. (2001). Conjugation of biomolecules with luminophore-doped silica nanoparticles for photostable biomarkers. *Anal. Chem.*, **73**, 4988–4993.

Schnekenburger, J. and Kemper, B. (2008). Digital holography imaging – a 3D view on drug induced apoptosis. *Screening Trends in Drug Discovery*, **9**, 24–26.

Stern, S.T. and Johnson, D.N. (2008). Role for nanomaterial-autophagy interaction in neurodegenerative disease. *Autophagy*, **4**(8), 1097–1100.

Stern, S.T. and McNeil, S.E. (2008). Nanotechnology safety concerns revisited. *Toxicological Sciences*, **101**(1), 4–21.

Storhoff, J.J., Lucas, A.D., Garimella, V., Bao, Y.P. and Müller, U.R. (2004). Homogeneous detection of unamplified genomic DNA sequences based on colorimetric scatter of gold nanoparticle probes. *Nature Biotechnol.*, **22**, 883–887.

Tan, K. and Obendorf, S.K. (2007). Fabrication and evaluation of electrospun nanofibrous antimicrobial nylon 6 membranes. *Journal of Membrane Science*, **305**(1), 287–298.

Tapec, R., Zhao, X.J. and Tan, W. (2002). Development of organic dye-doped silica nanoparticles for bioanalysis and biosensors. *J. Nanosci. Nanotechnol.*, **2**, 405–409.

Taton, T.A., Mirkin, C.A. and Letsinger, R.L. (2000). Scanometric DNA array detection with nanoparticle probes. *Science*, **289**, 1757–1760.

Tepper, F. and Kaledin, L. (2005). Nanosize electropositive cfibrous adsorbent. US Patent 6,838,005.

Tiller, J.C., Lee, S.B., Lewis, K. and Klibanov, A.M. (2002). Polymer surfaces derivatized with poly(vinyl-N-hexylpyrinium) kill airborne and waterborne bacteria Biotechnology and Bioengineering, **79**(4), 465

Theron, J. and Cloete, T.E. (2002). Emerging waterborne infections: Contributing factors, agents, and detection tools. *Crit. Rev. Microbiol.*, **28**, 1–26.

Vicentini, D.S., Smania Jr., A., Laranjeira, M.C.M. (2009). Chitosan/poly (vinyl alcohol) films containing ZnO nanoparticles and plasticizers. Material Science and Engineering C. in press.

Wu, W.H., Sun, X., Yu, Y.P., Hu, J., Zhao, L., Liu, Q., Zhao, Y.F. and Li, Y.M. (2008). TiO_2 nanoparticles promote beta-amyloid fibrillation in vitro. *Biochemical and Biophysical Research Communications*, **373**(2), 315–318.

Xiao, Y. and Barker, P.E. (2004). Semiconductor nanocrystal probes for human metaphase chromosomes. *Nucl. Acids Res.*, **32**, e161.

Xing, K., Chen, X.G., Kong, M., Liu, C.S., Cha, D.S., Park, H.J., (2009). Effect of oleoyl-chitosan nanoparticles as a novel antibacterial dispersion system on viability, membrane permeability and cell morphology of Escherichia coli and Staphylococcus aureus. *Carbohydrate Polymers*, **76**(1), 17–22.

Yang, H.H., Zhang, S.Q., Chen, X.L., Zhuang, Z.X., Xu, J.G. and Wang, X.R. (2004) Magnetite-containing spherical silica nanoparticles for biocatalysis and bio-separations. *Anal. Chem.*, **76**, 1316–1321.

Yao, L., Haas, T.W., Guiseppi-Elie, A., Bowlin, G.L., Simpson, D.G., and Wnek, G.E. (2003). Electrospinning and stabilization of fully hydrolyzed poly(vinyl alcohol) fibers. *Chemical Materials*, **15**(9), 1860–1864.

Yao, G., Wang, L., Wu, Y., Smith, J., Xu, J., Zhao, W., Lee, E. and Tan, W. (2006). FloDots: Luminescent nanoparticles. *Anal. Bioanal. Chem.*, **385**, 518–524.

Yoon, K., Kim, K., Wang, X., Fang, D., Hsiao, B.S. and Chu, B. (2006). High flux ultrafiltration membranes based on electrospun nanofibrous PAN scaffolds and chitosan coating. *Polymer*, **47**(7), 2434–2441.

Zhao, X., Tapec-Dytioco, R., Wang, K. and Tan, W. (2003). Efficient collection of trace amounts of DNA/mRNA molecules using genomagnetic nano-capturers. *Anal. Chem.*, **75**, 3476–3483.

Zodrow, K., Brunet, L., Mahendra, S., Li, D., Zhang, A., Li, Q. and Alvarez, P.J.J. (2009). Polysulfone ultrafiltration membranes impregnated with silver nanoparticles show improved biofouling resistance and virus removal. *Water research*, **43**(3), 715–723.

www.argonide.com accessed 10/03/2009

www.elmarco.com/ accessed 24/02/2009

www.purifics.com/technology/ceramic.html accessed 16/03/2009

Alternative Sources

Recycled water and desalinated seawater replenish and replace drinking water in Australia

S.J. Khan

UNSW Water Research Centre, University of New South Wales, NSW, 2052, Australia
(E-mail: s.khan@unsw.edu.au)

Abstract During the period 2000–2009, provision of urban drinking water supplies has grown to becoming one of the key environmental, social and political issues in Australia. As traditional water sources have begun to reach sustainable limits, alternative supplies from seawater desalination and planned indirect potable reuse (IPR) have begun to be implemented on a large scale. Neither approach had been implemented in Australia prior to 2006. Australia now has one of the largest seawater desalination plants in the world and is currently constructing or planning new plants to supply most of its major cities. Similarly, one of the largest IPR schemes in the world is almost fully constructed and is expected to supplement drinking water supplies when storage levels drop below 40% of capacity. This scheme involves reusing the vast majority of treated municipal effluent produced in South East Queensland for a range of purposes including industrial cooling water and supplementation of Brisbane's drinking water supply in Lake Wivenhoe.

Keywords indirect potable reuse, replacement flows, seawater desalination

INTRODUCTION

As a result of both rapid population growth and recent droughts, most large cities in Australia entered the 21st century with an urgent requirement for new urban water resources. The historical response to drought and population growth in Australia has been to increase storage capacity by building new dams on relatively untapped rivers. However, increased awareness of environmental and economic costs, coupled with the long-term inadequacy of many earmarked future dams, have caused governments to reconsider plans for their construction.

With strong demand management policies already in place in most large cities, Australian State Governments have been left with few options for the increasingly urgent replenishment of drinking water supplies. Consequently, many States, including Western Australia, Queensland, New South Wales, South Australia and Victoria have found the need to begin planning and construction of major water supply projects involving indirect potable water recycling (IPR) and/or seawater desalination. The approach taken has varied between different

 TECHNEAU: Safe Drinking Water from Source to Tap. Edited by Theo van den Hoven and Christian Kazner. ISBN: 9781843392750. Published by IWA Publishing, London, UK.

regions according to existing environmental, water supply, and infrastructure constrains. Some of the major IPR and seawater desalination schemes are summarised in the following sections.

INDIRECT POTABLE WATER RECYCLING

During the last two decades, there have been a number of planned IPR schemes proposed in Australia, which have not eventuated due to strident community opposition. These unsuccessful proposals have generally been in South East Queensland with the cities of Caloundra/Maroochy (planning during 1995–1998), Caboolture (1996–1997) and Toowoomba (2005–2006) being the most prominent examples.

Despite the earlier failures, by 2007 it had become clear that a number of Australia's largest cities would need to adopt varying approaches to water recycling in order to make full use of available water supplies. Major schemes have since been planned or developed in New South Wales, Queensland and Western Australia, as described in this section. A number of smaller proposals have also been mooted around the country, but planning is yet to be finalised.

Australia is the first country to have developed national guidelines for IPR with the release of Phase 2 of the Australian Guidelines for Water Recycling (AGWR) – 'Augmentation of Drinking Water Supplies' in 2008 (NRMMC, EPHC and NHMRC, 2008). The AGWR are notable for the risk management framework that they provide, rather than simply relying on end-product (recycled water) quality testing as the basis for managing water recycling schemes. All of the schemes described in this section are being developed and assessed under the framework of the AGWR.

South East Queensland

In January 2007, The Queensland State Government announced that it would construct a proposed IPR scheme that would supplement the potable water supplies of South East Queensland, including Brisbane. The scheme is known as the Western Corridor Recycled Water Project (WCRWP) and the major components are illustrated in Figure 1. The WCRWP uses the vast majority of treated municipal effluent produced in South East Queensland, which was previously discharged into Moreton Bay via the Brisbane River. Municipal effluent is collected from six wastewater treatment plants at Bundamba, Goodna, Oxley, Wacol, Luggage Point and Gibson Island. This is then subjected to advanced water treatment (AWT) at three new advanced water treatment plants (AWTPs) at Bundamba, Luggage Point and Gibson Island.

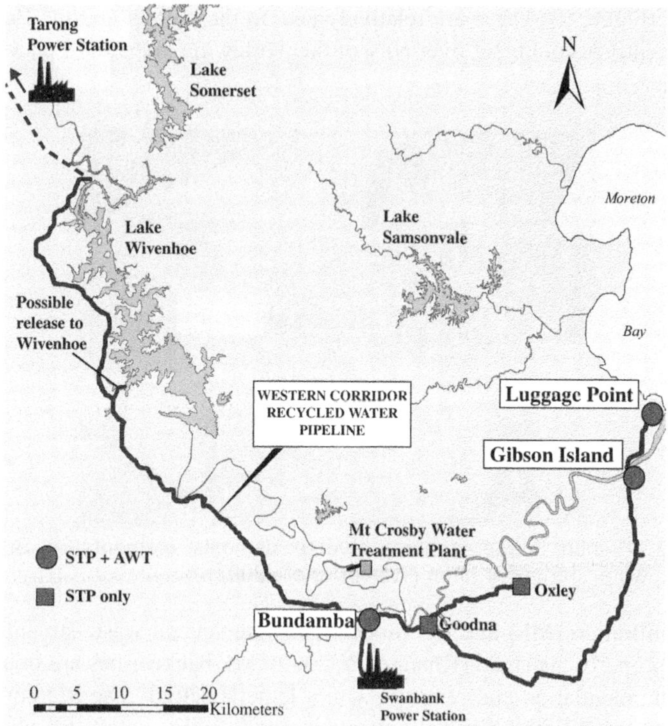

Figure 1. Schematic illustration of the Western Corridor Recycled Water Project (Gardner and Dennien, 2007).

Much of the advanced-treated water is now used to supply cooling water to Swanbank and Tarong coal-fired power stations. Most of the remaining water will be used to recharge Brisbane's main drinking water reservoir, Lake Wivenhoe at Logan Inlet. This IPR aspect is expected to be initiated once the storage levels in Lake Wivenhoe drop below 40% of capacity.

All three AWTPs incorporate treatment by reverse osmosis (RO) and advanced oxidation (UV/H_2O_2).The design philosophy of the scheme is in accordance with the 'multiple barriers' approach, incorporating seven largely independent treatment steps operating under a risk management framework.

The RO units at each plant are comprised of a three-staged array, designed to achieve approximately 85% recovery. The plants at Gibson Island and Luggage Point incorporate standard 8-inch diameter membranes. However, the Bundamba plant has been designed around new generation 18-inch diameter membranes, manufactured by Koch Membrane Systems as 'Megamagnum' elements (Figure 2). These are relatively new to the market and the Bundamba plant currently accounts for over 60% of the worldwide inventory (Walker *et al.*, 2007).

Figure 2. 18-inch 'Megamagnum' reverse osmosis elements at Bundamba advanced water treatment plant (Walker *et al.*, 2007).

Microfiltration (MF) and RO process generate several waste-streams, which must be properly managed (Khan *et al.*, 2009). MF backwashes are treated with ferric salt coagulation and removal with a lamella settler. The RO concentrate stream is treated with chemical coagulation and sedimentation for phosphorus removal and with a denitrification filter with methanol addition for biological nitrogen removal. The treated waste-streams from Luggage Point and Gibson Island are discharged to the Brisbane River, where they are diluted and dispersed into Moreton Bay. The waste-stream from the Bundamba plant is discharged to the Bremmer River.

The WCRWP was initially designed to provide around 80 gigalitres per year (GL/year) of purified water. However, while current strict water restrictions remain in place, the actual water production is expected to be limited to only around 50 GL/year due to limited wastewater production.

Perth (Western Australia)

Perth, the state capital of Western Australia, sits on what is known as the Swan Coastal Plain and approximately 60% of drinking-water supplies are sourced from groundwater. The unconfined aquifer directly below Perth is known as the Superficial aquifer and is on average about 50 metres thick. The Superficial aquifer has two important mounds. These are the Gnangara mound to the north of the city and the smaller Jandakot mound to the south. Below the Superficial aquifer there are a number of confined aquifers, the largest of which is the Leederville, which is typically several hundred metres thick, and the Yarragadee, which in many areas is more than a kilometre thick.

The Leederville aquifer is an important drinking-water source for Perth. The Superficial aquifer has also been an important water source for many decades and it is very common for Perth households to extract water directly from it for garden and lawn watering. The exploitation of this groundwater is rapidly approaching sustainable limits in some areas, causing the water table to drop markedly.

The organisation that manages Perth's water resources, Water Corporation, have signalled that they see managed aquifer recharge as a means of potentially supplying around 27 GL per year of public drinking water injected via the Gnangara mound by 2015. The source water would be advanced-treated recycled water from an AWTP to be located at the site of the Beenyup Wastewater Treatment Plant. It would then be transported via a (roughly 40 km) pipeline to a nature reserve further north of Perth. At various points en route, water would be injected into the confined Leederville aquifer or infiltrated into the Superficial aquifer.

However, before committing to the hundreds of millions of dollars needed to build such a scheme, Water Corporation have elected to begin by the construction of a smaller trial, through which annually 1.5 GL of recycled water is currently injected into the Leederville aquifer at Beenyup. Before injection, the treated effluents are upgraded to IPR standard by AWT including MF, RO, and advanced oxidation. The water is then injected to a depth of around 200 metres. Water Corporation have predicted that this water would then take a minimum of 50 years before reaching drinking-water production bores more than three kilometres away.

Detailed monitoring of water quality is now being undertaken prior to injection into the aquifer and also via a number of extraction wells as the water moves horizontally away from the Gnangara mound. This monitoring program is intended to be undertaken over a period of three years. As well as demonstrating the technical feasibility, an important aspect of the trial project is to enhance community confidence in potable recycling by managed aquifer recharge.

Sydney – Replacement Flows (New South Wales)

Sydney's main drinking water supply is captured by Warragama Dam on Lake Burragorang. Warragamba Dam retains and allows the diversion of water that would otherwise flow into the Hawkesbury-Nepean river system in western

Sydney. However, in order to maintain river health of the Hawkesbury-Nepean, the Sydney Catchment Authority is currently required to release around 12 GL/year from Warragamba Dam.

The NSW Government and Sydney Water Corporation have identified an effective means for preserving much of these environmental releases so that they may be retained as part of Sydney's drinking water supply. The project, known as the 'Replacement Flows Project', involves treating municipal effluent from the St Marys, Penrith and Quakers Hill Sewage Treatment Plants (STPs) in north western Sydney at an AWTP located at the St Marys STP. The AWTP will produce up to 18 GL/year of highly treated recycled water, which will be discharged to the Hawkesbury-Nepean River to replace water that is currently released from Warragamba Dam.

The AWTP for the Replacement Flows project will utilise ultrafiltration and RO membrane treatment followed by chlorination prior to environmental release. The North Richmond Water Filtration Plant is located downstream of the Hawkesbury-Nepean discharge point and supplies potable water for the town of North Richmond and surrounding areas. Accordingly, the AWT processes and associated risk management have been designed and carefully assessed with this subsequent water use in mind.

The AWTP will produce 8 ML/day of RO concentrate, which will be discharged to one of Sydney's major sewers, the Northern Suburbs Ocean Outfall Sewer. The concentrate will be pumped around 20 km to the sewer, from where it will be transported more than of 30 km further to Sydney's North Head Sewage Treatment Plant.

A pilot AWTP was constructed and operated at St Marys STP in 2008. This pilot was used to refine the operational parameters of the AWT process and confirm the production of high water quality suitable for reuse as a supplementation to the downstream drinking water supply. Risk assessment activities involved monitoring for chemical and microbial contaminants, as well as challenge testing the treatment system by spiking in high concentrations of selected potential contaminants. Construction of the full-scale AWTP began in May 2008 and highly treated recycled water is expected to flow to the Hawkesbury-Nepean River from 2010.

SEAWATER DESALINATION

Seawater desalination is an increasingly important water supply strategy for a number of Australia's largest cities. Perth was the first Australian city to have its drinking water partly supplied by desalinated seawater. A 45 GL/year plant was constructed to supply Perth and began operation 2006. The West Australian

Government has announced plans for a second plant to supply Perth, which will be sited near Binningup and produce a further 45 GL/year desalinated water by 2011.

A 45 GL/year plant at Tugan (QLD Gold Coast) began operation in December 2008 and now supplies water to the South East Queensland Water Grid, servicing Brisbane and surrounding cities. A 90 GL/year seawater desalination plant to supply Sydney with drinking water is currently under construction and expected to be operational before the end of 2009.

The South Australian Government has announced plans to construct a 50 GL/year seawater desalination plant (expandable to 100 GL/year) at Port Stanvac to supply Adelaide with drinking water by December 2010. The Victorian State Government is currently planning what will be Australia's largest seawater desalination plant for the near future, to supply Melbourne with 150 GL/year of drinking water by the end of 2011.

These plants will generally provide 15–25% of their respective population's drinking water needs.

Perth (Western Australia)

In December 2006, the Western Australian Government announced the opening of Australia's first seawater desalination plant to supply the city of Perth with up to 17% of its potable water requirements. At a peak capacity of 144 ML/day, the plant is currently the largest seawater desalination plant outside the Middle East (Crisp, 2006).

The Perth desalination plant is located at Kwinana, about 40 kilometres south of Perth. It delivers desalinated water to the city's distribution system via 26 kilometres of large diameter pipelines and a 9 MW pumping station. The plant itself incorporates two separate streams each capable of producing 72 ML/day, via a two-pass RO system. It operates with a first pass recovery of 45% and a second pass recovery of 90%.

The intake to the plant is 300 m offshore and at a depth of 11 m in Cockburn Sound. After screening, two streams of feedwater are pumped through a pre-treatment facility incorporating dual media pressure filters and cartridge filters. The water is then pressurised to over 600 m via six 2.5 MW high-pressure pumps and passed through 12 fist-pass SWRO racks. The permeate from the first pass is then passed through 6 second-pass brackish water RO racks, for a combined 40% recovery. A post-treatment system adds minerals to stabilise the water before being pumped to 12 ML storage tank.

The RO concentrate is discharged to Cockburn Sound, which is a semi-enclosed waterway, with some parts deemed to be of high conservation value.

Accordingly, a particular focus was assigned to the management of the discharge to achieve minimal ecological disruption (Khan *et al.*, 2009). The discharge outlet is 1.2 m in diameter and has a 160 m long, forty port diffuser where the ports are spaced at 5m intervals with a 0.22 m nominal port diameter, located 470 m offshore, at a depth of 10 m, adjacent to the plant in Cockburn Sound (Crisp, 2007). The diffuser is a bifurcated double-T-arrangement and incorporates a discharge angle of 60°. This design was adopted with the expectation that the plume would rise to a height of 8.5 m before beginning to sink due to its elevated density. It was designed to achieve a plume thickness at the edge of the mixing zone of 2.5 m and, in the absence of ambient cross-flow, 40 m laterally from the diffuser to the edge of the mixing zone.

The possibility of reduced seawater oxygen levels in Cockburn Sound as a result of increased stratification was raised as a particular concern during the planning and assessment for the plant. Accordingly, a series of models and tests were used to ensure the plant would meet the required criteria at the edge of the mixing zone. A detailed study concluded that any effect on dissolved oxygen levels would be infrequent and minor (Rhodes, 2006). However, it recommended that because of the uncertainty of predictions for long calm periods, a monitoring program should be implemented as part of an adaptive management plan. Extensive real-time monitoring was undertaken in Cockburn Sound for the first year of operations to ensure the model predictions were correct and that the marine habitat and fauna are protected.

Visual confirmation of the plume dispersion was achieved by the use of 52 L of Rhodamine dye added to the plant discharge (Okely *et al.*, 2007). The expulsion of the Rhodamine dye from one of the plant diffusers is shown in Figure 3 (Water Corporation, 2006). The dye was reported to have billowed to within about 3m of the water surface before falling to the seabed and spilling along a shallow sill of the Sound towards the ocean. The experiment showed that the dye had dispersed beyond what could be visually detected within a distance of around 1.5 km, well short of a protected deeper region of Cockburn Sound about 5 km from the diffuser (Okely *et al.*, 2007). The environmentally benign dye experiment was first commissioned in December 2006 and repeated in April 2007 when conditions were calm.

In May 2007, the Western Australian Government announced plans to construct a second seawater desalination plant, also to supply Perth. This plant will be sited at Binningup, 155 km south of Perth. It will provide 45 GL/year drinking water, the same as the first plant. The second plant, to be completed by 2011, will bring the amount of the state's water supply from desalinated water to one third.

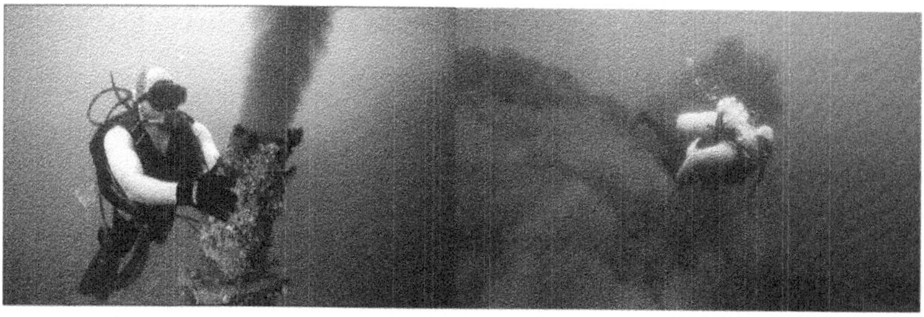

Figure 3. Rhodamine dye tests undertaken at the Perth Seawater Desalination Plant (Water Corporation, 2006).

Sydney (New South Wales)

A 90 GL/year seawater desalination plant to supply Sydney with drinking water is currently under construction and expected to be operational before the end of 2009. This plant has been designed to be expandable to produce up to 180 GL/year in the future. In addition to the plant itself, the Sydney desalination project requires the construction of an 18 km pipeline to deliver the desalinated seawater to a major drinking water distribution connection in Sydney. The construction of this pipeline involves tunnelling both under Botany Bay and a highly developed section of urban Sydney. Plant construction activities started in September 2007 and pipeline work started in February 2008.

Like all proposed seawater desalination plants in Australia, the key treatment technology is RO. Computer modelling has indicated that to match current water quality, the desalinated water would require a two-pass RO arrangement, where up to 100% of the RO permeate would be further purified through a second RO pass (Port and Roddy, 2008).

An initial RO pre-treatment process was developed during the planning stage, which included flash mixing and flocculation followed by two stage filtration. Following the planning process, Sydney Water carried out pilot plant testing of the selected granular media pre-treatment process as well as an ultrafiltration (UF) membrane pre-treatment process. This piloting process is reported to have concluded that a second pre-treatment filtration stage is not required and that flocculation does not provide a benefit and therefore need not be used. Furthermore, it was determined that the filtration rate could be increased thereby reducing the filter area by 25% compared with original concept design. These outcomes from the piloting process resulted in considerable cost savings, in excess of AU$100 million (Port and Roddy, 2008).

Pre-treatment backwash water will be treated on site to remove suspended solids. The solids will be dewatered and transported to an off-site landfill.

The supernatant will be discharged to the ocean along with the seawater concentrate.

A model developed by the University of New South Wales Water Research Laboratory was used to assess suitable locations for both the intake and outlet for the plant. Strings of transducers to measure current, depth and temperature were later installed on the seabed to validate the model.

The seawater intake was constructed on a rocky reef in a relatively high energy zone in water depths of 22–28 m. The location of the intake is designed to minimise entrainment of sand and subsequent deposition in the tunnel to the plant.

The outlet discharge system consists of tunnel risers with caps on the sea floor. To assess the impact of the seawater concentrate discharge on local marine biota, ecotoxicity studies were conducted. Mixtures of RO seawater concentrate with filter backwash supernatant and various membrane cleaning chemicals were made up at various dilution ratios with seawater to simulate the discharge stream. A range of indicator organisms common to the area was then exposed to these mixtures. The results of this testing indicated that the desired 95% species protection is achieved when the targeted salinity dilution rate is within one part per thousand above background (Port and Roddy, 2008).

A marine and estuarine monitoring program was implemented over a two year period prior to construction and will continue through construction and following commissioning for at least three years. This program is monitoring water quality as well as marine biota and planktonic algae species. The data collected prior to construction provides a baseline with which to compare post commissioning data.

CONCLUSIONS

Seawater desalination and IPR have emerged in Australia as very significant contributors to the water supplies of the country's largest cities. Most of these schemes are designed to be expandable and their role in overall water provision is certain to continue to grow. The situation in Australia is notable for the way in which different approaches to IPR and seawater desalination have been adopted depending on locally specific factors. This approach has been necessary to work within existing water supply regimes and infrastructure. Similarly, it is expected that future developments, both in Australia and internationally, will need to continue to work to adapt major water supply infrastructure needs to local conditions.

Planning for major IPR and seawater desalination schemes in Australia has been a hasty, and sometimes rushed, process aimed at meeting deadlines imposed by recently revised water supply modelling. Time will tell whether this rapid decision making process will prove to have been as well conceived as it may have otherwise been, given more time for consideration. Nonetheless, much of the planning has focused on addressing environmental issues, particularly with respect to waste streams such a MF backwashes and RO concentrates that must be disposed of to the environment.

Seawater desalination and IPR schemes must now be considered as important options for enhanced water supply for all water-stressed Australian coastal cities. However, inland regions face obvious limitations with seawater desalination and environmental constraints associated with membrane waste disposal remains a major limitation to IPR.

REFERENCES

Crisp, G. (2006). Australia's first big plant thinks "green". *The International Desalination and Water Reuse Quarterly*, **16**(3), 19–26.

Crisp, G. (2007). Desalination – the focus of Australia's current membrane interest. In: *AWA Membranes Specialty Conference II Australian Water Association*, Melbourne, Vic.

Gardner, T. and Dennien, B. (2007). Why has SEQ decided to drink purified recycled water? In: *Water Reuse and Recycling*, Khan, S.J., Stuetz, R.M. and Anderson, J.M. (eds), UNSW Publishing, Sydney.

Khan, S.J., Murchland, D., Rhodes, M. and Waite, T.D. (2009). Management of concentrated waste streams from high pressure membrane water treatment systems. *Critical Reviews in Environmental Science & Technology*, **39**(5), 367–415.

National Resource Management Ministerial Council, Environment Protection and Heritage Council and National Health and Medical Research Council (2008). National Guidelines for Water Recycling: Managing Health and Environmental Risks (Phase 2) – Augmentation of Drinking Water Supplies.

Okely, P.N., Antenucci, J.P. and Imberger, J. (2007). Field Investigation of the Impact of the Perth Seawater Desalination Plant Discharge on Cockburn Sound During Summer, Centre for Water Research, University of Western Australia, Perth.

Port, C.L. and Roddy, S.J. (2008). Sydney desalination plant: pre-treatment pilot testing. *Water*, **35**(2).

Rhodes, M. (2006) Marine management is high priority. *The International Desalination and Water Reuse Quarterly*, **16**(3), 30–31.

Walker, T., Roux, A. and Owens, E. (2007). Western Corridor Recycled Water Project – The largest recycled water scheme in the southern hemisphere. In: *Water Reuse and Recycling*, Khan, S.J., Stuetz, R.M. and Anderson, J.M. (eds), UNSW Publishing, Sydney, 498–511.

Water Corporation (2006). Photographs taken by the West Australian Newspaper and provided courtesy of Water Corporation, Perth, WA. Perth.

Drinking water safety in Windhoek, Namibia: routine monitoring, trace organics, pathogenic indicators and salinity – comparing different water sources

J.G. Menge[a], P.L. du Pisani[a], E. König[b] and T. Theron-Beukes[b]

[a]City of Windhoek, PO Box 59, Windhoek, Namibia
(E-mail: jgm@windhoekcc.org.na, pdp@windhoekcc.org.na)
[b]WINGOC, PO Box 2103, Windhoek, Namibia
(E-mail: ekonig@wingoc.com.na, truddy@wingoc.com.na)

Abstract This paper deals with drinking water safety in Windhoek, Namibia, comparing the quality of different water sources with regard to standard parameters as well as trace organics, pathogenic indicators and salinity. As capital of a very dry country with low and irregular rainfall and being in a land locked situation without access to sea water, the city of Windhoek developed a comprehensive water supply system based on different conventional and alternative water sources such as surface water and ground water as well as reclaimed water with direct potable reuse and managed aquifer recharge. The 40 years experience at Windhoek demonstrates that a direct potable reuse system can be a practical, responsible way of augmenting potable water supplies in arid regions. It requires, however, comprehensive planning, training and on-going commitment for its continued success.

Keywords alternative water sources, direct potable reuse, water safety, Windhoek

INTRODUCTION

A dry country with low and irregular rainfall, extended periods of drought and mainly dependant on underground resources and surface dams in ephemeral rivers is very vulnerable. When the main perennial rivers are national borders originating in the neighbouring countries, long-term supply agreements cannot be 100% guaranteed. Large supply schemes are very costly and in certain regions not economical to implement for various reasons. In such cases well-planned use and reuse of water can play a vital role to sustainability. National and local policies should support reuse of wastewater, taking the constraints of the region as well as the potential threats of wastewater reuse into consideration. Active participation through educational programs should encourage planners and engineers to design systems that cater for reuse or which can be transformed to a reuse scheme at any stage.

The practical experience at Windhoek demonstrates that a direct potable reuse system can be a practical, responsible way of augmenting potable water supplies in arid regions. It however requires comprehensive planning, training and on-going commitment for its continued success. (Haarhoff and van der Merwe, 1996; du Pisani, 2006; Menge *et al.*, 2007).

Windhoek is situated in the centre of Namibia, about 750 to 900 km away from perennial rivers, in one of the most arid countries south of the Sahara. The average population growth is 5% over the past 18 years. The conventional sources "fed" by ephemeral rivers are not able to sustain the water demand in successive years of severe drought. Direct potable reuse has been practised since January 1969 to augment the drinking water supply.

This paper gives an overview about the quality of drinking water in Windhoek, Namibia, where direct potable water reuse is practiced for the past 40 years. Much about the background and reasons why reclamation is practiced can be read in the following literature (Isaacson, 1987; Haarhoff, 1996; du Pisani, 2006).

A risk based approach was applied to reduce the risk of hazardous substances during the various treatment stages of the reclamation process and to produce a final product water of a quality that complies with national and international drinking water quality guidelines.

During the planning stages of the New Goreangab Water Reclamation Plant (NGWRP), a quantified risk based approach was followed (Haarhoff, 1998). A comprehensive study was conducted to analyse all data of the raw water sources and various pilot and historic full scale plant data. The results of the study guided the configuration of the different barriers for the NGWRP. The guidelines applied for the final water quality were those of the EPA (1996), WHO (1993) and Rand Water (1994), South Africa.

The quality of three water sources, underground, surface and reclaimed water, which supply Windhoek with drinking water, are compared, as well as the quality of the drinking water in the distribution system. The comparison was based on the 95% probability of all the water quality results for the time frames given.

The fact that there are no similar direct reclamation experiences world wide places a very high burden on the City of Windhoek to ensure the safety of the drinking water. As an indispensible part of water supply, loss of consumer faith cannot be contemplated.

This paper will deal with drinking water safety in Windhoek, comparing the quality of different water sources.

MONITORING

Plant operation and maintenance play an important role. Equally important or even more so is the monitoring of substances which are removed by the different process steps. During 1995 to 2000 a comprehensive monitoring program was established to cover the full water cycle in Windhoek (Liputa, 2008). The focus was to detect possible pollutants entering the water cycle or being generated in the water cycle, emanating from catchment activities and geography, treatment processes or during transport and storage. Because domestic waste effluent is reused, pollution monitoring is seen as part of the program. Emphasis was put on the history of the water sources, the reliability of treatment processes and the characteristics of the distribution network (pressure zones, low flow, population densities served).

The monitoring program is divided into routine monitoring and a health research program. The routine monitoring program covers process control and final water compliance monitoring and the health research program includes the development of new test methods or sampling approaches and pilot experimentation to test new process concepts and applications.

State-of-the-art analytical equipment is used and ISO 17025 principles are applied in the laboratory where water samples are analysed. There is very little advanced water research or testing equipment available in Namibia. Such tests are contracted out mainly to South African laboratories or are being done in Europe as part of cooperation programs.

The research and analytical work that has accompanied the Windhoek reclamation process over a period of 40 years is comprehensive. The intention is to present data which compare the different supply sources with the final water in the distribution system.

WATER SOURCES

Below a brief description of the different components covered in this study is given. On average about 14% of the daily consumption is provided from reclaimed drinking water, about 74% from treated surface water and about 22% from boreholes.

Groundwater

There are three distinct borehole groups, totalling about 60 production boreholes, which supply Windhoek with water. The chemical characteristics of the three groups are different. Each group as it enters the distribution system will be compared. The results of all the individual boreholes will however be treated as one. The different groups are: Pahl Quelle, Klein Aub and Kleine Kuppe. Results are presented in Table 1.

Water quality of the Kleine Kuppe borehole group is affected slightly due to a domestic garden refuse site as well as a military army base where some of the boreholes are located. Since 1996 an artificial aquifer injection scheme is being implemented to augment the underground water and to serve as a storage facility to minimise evaporation losses in the surface dams.

The water is not treated separately, but is only chlorinated as it enters the storage reservoir. There are plans to treat the water to reduce the iron and manganese levels.

Surface water

There are several Dams in ephemeral rivers that supply Windhoek with drinking water. The three Dam system operated by NamWater, a parastatal enterprise are the Von Bach Dam, Swakoppoort Dam and Omatako Dam, which are located 60 km, 80 km and 200 km from Windhoek, respectively.

Due to erratic rainfall the Von Bach Dam and Swakoppoort Dam are not overflowing regularly. In fact in 2006, an exceptional good rainy season, the Von Bach has over flown after a period of 30 years. The Swakoppoort Dam only filled up completely in 2009. Therefore there is a large retention time of water in the Dams. Windhoek lies in the catchment of one of the rivers feeding the Swakoppoort Dam. The effects of this pollution can be clearly experienced and the water is more eutrophic than the other Dams. Intensive cattle farming and some small scale mining activities are located in the catchment of the Von Bach Dam. After early rainfalls there is always an increase in ammonia and natural organic matter (NOM) affecting chlorination in the distribution system. Breakthroughs of Giardia and Cryptosporidium have also occurred.

The Goreangab Dam is part of the Windhoek reclamation scheme. Windhoek is located in its catchment of the dam, which affects its water quality severely. In 1991, the water quality deteriorated to such a state, that all recreational activities on the Dam were terminated. Treatment changed from conventional to advanced, and since then the water is treated in the reclamation plant.

Evaporation losses in central Namibia are extremely high. The volume of water that evaporates annually from the dams, is far more than what is consumed in Windhoek. Planned aquifer storage is therefore a viable alternative to ensure adequate water supply.

Table 1. Comparison of different water sources over the period January 2006 to June 2007 (all values are based the 95th Percentile).

Parameters		Units	Borehole KA	Borehole KK	Borehole PQ	Borehole BH	Surface Water	Reclaimed Water	Reservoirs	Consumer	Nam Group A
Aesthetic	pH	-	7.74	7.93	8.01	7.90	8.02	8.4	7.94	8.01	6.0-9.0
	Conductivity	mS/m 25°C	130	68	**165**	108	29	**160**	75	67	150
	TDS calc	mg/l	871	457	**1106**	722	194	**1072**	503	449	1000
	Turbidity	NTU	5.27	4.64	8.08	4.88	1.1	0.082	1.2	1.5	1.0
	Total alkalinity	mg/l CaCO3	513	237	246	334	111	237	283	228	-
	Total hardness	mg/l CaCO3	**350**	252	281	281	110	242	229	225	300
	Calcium hardness	mg/l CaCO3	220	112	93	158	71	153	145	140	375
	Magnesium hardness	mg/l CaCO3	125	69	85	87	29	87	85	83	290
	Chlorophyll A	µg/l	-	-	-	-	1.02	0.95	-	-	-
	Temperature	°C	28	25	58.5	34.1	27	25.8	27	27	-
Bacteria & Viruses	HPC	per 1 ml	**10000**	152	**10000**	**5076**	**10000**	1	232	295	100
	Total coliform	per 100 ml	**10**	0	**45**	**13.8**	0	0	0	0	0
	Faecal coliform	per 100 ml	-	-	-	-	-	0	-	-	0
	E Coli Tryptone	per 100 ml	**57**	0	**48**	**35**	-	0	0	**1.9**	0
	Faecal streptococci	per 100 ml	-	-	-	-	-	0	-	-	-
	Pseudomonas	per 100 ml	-	-	-	-	-	0	-	-	-
	Clostridium spores	cfu/100ml	-	-	-	-	-	0	-	-	-
	Clostridium viable	cfu/100ml	-	-	-	-	-	0	-	-	-
	Som. coliphage 100ml	PFU/100 ml	-	-	-	-	-	0	?	-	-
Protozoa	Giardia		-	-	-	-	0	0	0	-	-
	Cryptosporidium		-	-	-	-	0	0	0	-	-
Organics & Micropollutants	DOC	mg/l	1.39	3.15	1.5	2.29	4.77	2.64	4.2	4.5	-
	CODdis	mg/l	-	-	-	-	-	17.2	-	-	-
	UV 254_dis	abs/cm	0.0591	0.0649	0.0582	0.0618	0.0759	0.0197	0.0707	0.0662	-
Disinfection byproducts	Free chlorine	mg/l	2.2	2.2	-	1.95	1.26	1.72	2.2	2.2	0.1-5.0
	Total chlorine	mg/l	2.2	2.2	-	2.07	1.68	-	2.2	2.2	0.1-5.0
	Total trihalomethane	µg/l	30	147	18	38	96	26	107	107	-
	Chloroform	µg/l	10	73	6	16	54	3	57	57	-
	Dichloromonobromoform	µg/l	9	43	7	13	26	4	27	27	-
	Monochlorodibromoform	µg/l	9	21	4	6	11	9	16	16	-
	Bromoform	µg/l	2	10	1	3	5	10	7	7	-
Stability	CaCO3 precipitation pot	mg/l CaCO3	-	-	-	-	-	-	-	-	-
Natural & Effluent Organic Matter	TKN	mg/l as N	-	-	-	-	-	4.37	-	-	-
	Ammonia (NH3-N)	mg/l as N	0.27	0.16	0.27	0.215	0.27	0.09	0.1595	0.27	1.0
	Ortho phosphate (P)	mg/l	0.9	0.91	0.9	0.9	0.9	0.9	0.9	0.9	-
	Nitrate (NO3-N)	mg/l as N	1	4.75	1.79	3.07	1	**22**	5.84	5.72	10
	Nitrite (NO2-N)	mg/l as N	0.5	0.345	0.5	0.42	0.5	0.5	0.5	0.5	-
Inorganics	K	mg/l	24.7	10.74	36.9	20.8	9.3	39.57	16.7	15.8	200
	Na	mg/l	**101**	57	**158**	93	18.5	**266**	96	97	100
	Cl	mg/l	42	66	71	61	23.5	**277**	118	110	250
	SO4	mg/l	**265**	68	**496**	**224**	20	187	101	85	200
	F	mg/l	**2.10**	0.36	**4.37**	1.79	0.33	0.61	0.45	0.48	1.5
	Br	mg/l	0.20	0.10	0.30	0.11	0.127	0.14	0.15	0.11	1.0
	Mn	mg/l	-	-	-	**0.38**	0.02	0.01	0.03	0.03	0.05
	Fe	mg/l	**0.55**	**0.62**	**0.35**	**0.54**	0.05	0.02	0.08	0.1	0.1

Reclaimed water

Domestic and industrial effluents are collected and treated separately in Windhoek. Two biological nutrient removal plants treat domestic waste water. The Gammams Wastewater Treatment plant (GWTP) receives the bulk of the domestic effluent, about 33 ML/d. The effluent of GWTP is transferred via a 1 km pipeline to the New Goreangab Water Reclamation Plant (NGWRP).

The GWTP process consists of: primary settling and anaerobic digestion with drying beds, then split into stream a: biological trickling filter with secondary settling and stream b: biological nutrient removal activated sludge plant and 8 maturation ponds. To produce sufficient high quality effluent for drinking water reclamation and irrigation, the biological filter system was integrated into the activated sludge system in 2002.

The NGWRP process with a capacity of 21,000 m^3/d consists of the following elements:

- Raw water is supplied from the Goreangab Dam and secondary treated effluent from the Maturation Pond effluent after the GWTP. Both sources can be mixed at any ratio or only one source used at a time. Mixing is determined by best plant performance. Some wastewater streams like the first filtrate of filters and membranes are recycled internal to the raw water sump, while more contaminated wastes are pumped to a nearby waste water treatment plant.
- Powdered activated carbon can be added as a back up for adsorption, should the ozone process fail, or for taste and odour treatment.
- Pre-ozonation: The raw water mix is treated with the off gas and excess ozone.
- Chemical dosing and coagulation: Ferric chloride is added as primary coagulant to achieve enhanced coagulation for maximum organic removal in the first solids separation step. When needed, hydrochloric acid can be added for pH control or a polymer to aid the flocculation.
- Dissolved air flotation (DAF) is used for suspended solids and organics separation. This process is used successfully since 1984 in Windhoek's reclamation and is regarded as the "heart" of the process.
- Chemical dosing: Caustic soda (NaOH) and permanganate (MnO_4) are added to raise the pH and accelerate the precipitation of iron and manganese on the sand filters.
- Rapid sand filtration: Dual media filters with anthracite and graded sand. The filters are equipped with slow start and filter-to-waste facility for maximum cyst/oocyst removal.

- Ozone contact: Oxygen is produced on-site with a Pressure Swing Adsorption (PSA) plant. Ozone is dosed at three dosage points. Dosage is regulated for maximum Dissolved Organic Carbon (DOC) removal (1–1.5 mg O_3/mg DOC) to oxidize high molecular weight compounds and refractory organic matter to produce molecules that are more biodegradable. A C_t value of 15 to 20 minutes is maintained for Cryptosporidium and Virus destruction.
- Chemical dosing: Peroxide (H_2O_2) is dosed to neutralize any ozone residuals to protect the biological activity in the next step.
- Biological activated carbon (BAC) filters remove the biodegradable matter.
- Granular activated carbon filters (GAC) consisting of primary and secondary filters with a contact time of EBCT = 30 minutes at maximum flow to remove organic molecules from the water.
- Ultrafiltration (UF): Chlorinated water is filtered through ultrafiltration membrane modules to remove bacteria, protozoa and viruses. Detection of protozoa in the final water of the Old Goreangab Reclamation Plant prompted the decision to introduce UF into the process train (Menge, 2001).
- Breakpoint chlorination with a free residual of 1 mg/L and contact time of 1 hour.
- Stabilisation by adding caustic soda to raise the pH for a positive precipitation potential of 4 mg/L.
- High lift pumps transfer the final water to New Western Pump Station.
- Blending: At the New Western Pump Station (NWPS) the reclaimed water is blended with surface water from the Von Bach scheme at a ratio of 1:3 and introduces it into the distribution system.

DISTRIBUTION SYSTEM

The distribution system consists of 17 reservoirs and 12 pressure zones. Figure 1 shows the different sources and water mixes in the distribution system.

COMPARISON OF RESULTS

Risk assessment and surveillance (RAS)

The purpose of RAS in Windhoek, and at the NGWRP in particular, is to identify and monitor critical control points for an immediate response, should a deviation from the normal in the treatment and distribution of water occur. A formal Risk Assessment Plan according to HACCP and the Techneau program

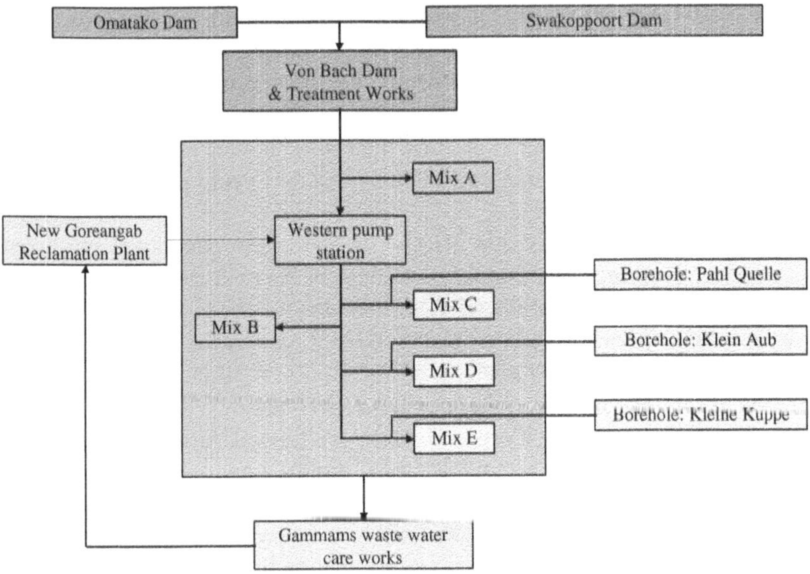

Figure 1. Diagram of Windhoek's Water supply and distribution network.

is being developed, which should be in place by the end of 2009. The plant is ISO 9001 certified and will also be HACCP certified. This area is continuously improved as our knowledge grows and as better tools become available. The following RAS tools are being used at the NGWRP:

1. Plant control

The plant is controlled by a System Control and Data Acquisition (SCADA) system from a central control room. An important aspect is the possibility to put the treatment plant into recycle mode whenever a unit fails to meet a target or when a power breakage occurs. The water from the unit that failed is recycled until it meets the quality target. Only then the water is passed to the next treatment step.

2. Private management agreement

The operation of the NGWRP is contracted out to a private operator, an international consortium called Windhoek Goreangab Operating Company (WINGOC). Du Pisani (2006) gives a detailed overview how this agreement operates in practice. For the purpose of Risk Management, quality targets and absolutes are defined for the raw water, intermediate treatment steps and the final treated water. If these are not met, various penalties apply. In this way it is ensured that quality is not sacrificed and all the barriers are being kept in place.

3. On-line surveillance or monitoring

Besides the analytical program described earlier on, electronic analytical on-line equipment is installed throughout the plant. The data is processed by the SCADA system. Various alarm levels are set, which will notify the operator or automatically put the plant into recycle mode. The following instrumentation is used: Turbidity, pH, conductivity, Total Organic Carbon (TOC), UV_{254} measurement, Dissolved Oxygen (DO) and Particle Counting (PC).

This is an area which still needs to be development to ensure robustness and reliability in a third world country. Turbidity, Conductivity, pH and DO are quite stable, easy to operate and maintain. TOC, UV_{254} and PC however, are sciences that need full time attention for maintenance and processing and at this stage and are not reliable for its intended purpose. Currently a s::scan TOC analyser is being tested under a Techneau cooperation agreement on the raw and final treated water to provide better TOC measurement.

4. Analytical surveillance

Composite and on-line sampling is done at each treatment step or barrier. 24-hour refrigerated composite samples are collected and analysed at the local or external laboratories. On-line samplers are used to collect 1 litre every 15 minutes for 7 days a week to concentrate samples for Giardia, Cryptosporidium and Virus testing.

The analytical data of the routine monitoring program are summarised in Tables 1 and 2. The Namibian Water Quality Guidelines for Drinking Water, Group A for excellent quality, is used to compare the data in Table 1. The parameters that do not comply are highlighted.

RESULTS AND DISCUSSION

In Table 1 the different sources providing Windhoek with drinking water are shown. Four parameters will be discussed. It must be mentioned that the 95% probability was used to evaluate the data. This is an exceptional strict criterion. Most of the parameters which do not comply with the 95% percentile will on average fully comply with the guidelines.

Turbidity is an indicator for the removal of particles or suspended material and also to a large extent dissolved organic and inorganic material. The turbidities

Table 2. Performance of the New Goreangab Water Reclamation Plant over the period January 2006 to June 2007 (all values are based the 95th Percentile).

TEST PARAMETER / PROCESS	Unit	MIX	DAF	DAF %Rem	SF	SF %Rem	SF Log Rem	O3	O3 %Rem	O3 Log Rem	BAC	BAC %Rem	GAC	GAC %Rem	UF	UF %Rem	UF Log Rem	FIN	Perf Gdl	NGnew	FIN %Rem	FIN Log Rem	overall %REM	overall %PROD	overall Log Rem
Chlorophyll A	µg/l	55.93	19.65	**64.9**	8.18	**58.4**												0.95	1.00	1.00	**88.4**		**98.3**		
pH	-	8.25	7.64		8.19			7.99			7.80		7.80		7.72			8.04							
Conductivity	mS/m 25°C	145	150		158			157			155		158		156			160		80				**9.4**	
TDS calc	mg/l	972	1005		1056			1052			1039		1056		1043			1072	1200	500				**9.3**	
Turbidity	NTU	5.454	7.612	**−39.6**	0.403	**94.7**		0.398	**1.1**		0.234	**41.2**	0.300	**−28.2**	0.116	**61.2**		0.082	0.200	0.200	**29.5**		**98.5**		
Colour	mg/l Pt	101.00																		10.00					
Total alkalinity	mg/l CaCO3	250.91	178.99		225.31													237.20					**5.5**		
Total hardness	mg/l CaCO3	238.75	236.50		236.50													242.25		130.00				**1.4**	
Calcium hardness	mg/l CaCO3	156.50	150.00		150.00													152.50		200.00			**2.6**		
Magnesium hardness	mg/l CaCO3	88.50	85.10		84.50													87.40		175.00			**1.2**		
K	mg/l	39.66																39.57	20.00	25.00			**0.2**		
Na	mg/l	219.77																266.04	100.00	100.00				**17.4**	
Cl	mg/l	247.86																277.25	250.00	100.00				**10.6**	
SO4	mg/l	185.50																186.50	200.00	200.00				**0.5**	
F	mg/l	0.61																0.61	1.00	0.70					
Nitrate (NO3-N)	mg/l as N	20.00																22.00	10.00	6.00				**9.1**	
Nitrite (NO2-N)	mg/l as N	0.57																0.50	0.05	0.20					
Br	mg/l	0.39			0.45			0.34			0.32		0.34		0.17			0.14	1.00	1.00			**65.2**		
TTHM/THMFP		182						41	**77**									26	20	40	**37**		**86**		
CHCL3		117						21	**82**									3			**86**		**97**		
CHCL2Br		37						12	**68**									4			**67**		**89**		
CHCLBr2		27						7	**74**									9			**−29**		**67**		
CHBr3		1						1	**0**									10			**−900**		**−900**		
TKN	mg/l as N	5.80																4.37					**24.7**		
Ammonia (NH3-N)	mg/l as N	1.96																0.09	0.10	0.20			**95.3**		
Ortho phosphate (P)	mg/l	11.00																0.90					**91.8**		
TSS:105°GF	mg/l	23.30	14.70	**36.9**																					
FSS:500°GF	mg/l	11.80	9.09	**23.0**																					
VSS:500°GF	mg/l	13.90	10.35	**25.5**																					
DOC	mg/l	10.68	7.22	**32.4**	6.62	**8.3**		6.27	**5.3**		4.80	**23.4**	2.65	**44.7**				2.64	5.00	5.00	**0.6**		**75.3**		
CODdis	mg/l	37.75	30.63	**18.9**	26.50	**13.5**		24.65	**7.0**		23.70	**3.9**	20.50	**13.5**				17.20	15.00		**16.1**		**54.4**		
UV 254_dis	abs/cm	0.2239	0.1419	**36.6**	0.1374	**3.2**		0.0737	**46.4**		0.0658	**10.7**	0.0240	**63.5**	0.0203	**15.4**		0.0197	0.0600		**2.7**		**91.2**		
Temperature	°C	25.07																25.77							
Free chlorine	mg/l																	1.72							
HPC	per 1 ml	520,500			28,550	**94.5**	1.26	376	**98.7**	1.88					6	**98.40**	1.80	1	100	100	**91.7**	1.08	**99.9999**		**6.02**
Total coliform	per 100 ml	18,385			406	**97.8**	1.66	0	**100.0**	3.61					0	**99.98**	2.00	0	0	0	**100.0**	3.61	**99.9995**		**5.26**
Faecal coliform	per 100 ml	2,088			15	**99.3**	2.14	0	**99.3**	2.18					0	**99.34**	2.18	0	0	0	**99.3**	2.18	**99.995**		**4.32**
E Coli Tryptone	per 100 ml	1,865			18	**99.0**	2.01												0	0					
Faecal streptococci	per 100 ml	464			8	**98.3**	1.76	0	**98.8**	1.91					0	**98.77**	1.91	0	0	0	**98.8**	1.91	**99.98**		**3.67**
Pseudomonas	per 100 ml	72			3	**96.2**	1.42								0	**96.36**	1.44	0			**96.4**	1.44	**99.86**		**2.86**
Clostridium spores	cfu/100ml	8,810			63	**99.3**	2.15	1	**98.8**	1.92								0	0		**86.7**	0.88	**99.999**		**4.94**
Clostridium viable	cfu/100ml	25,100			128	**99.5**	2.29	4	**97.1**	1.54								0	0	0	**97.3**	1.57	**99.9996**		**5.40**
Som. coliphage 100ml	PFU/100 ml	4,800			59	**98.8**	1.91	0	**99.8**	2.77					0	**99.83**	2.77	0	0	0	**99.8**	2.77	**99.998**		**4.68**

of the boreholes are very high. Likewise the water received from the surface treatment plant shows elevated turbidities. The high turbidities are reflected in the reservoirs and distribution system. This can be attributed to dissolved iron and manganese which is oxidised by the high chlorine levels, and which precipitate in the reservoirs and distribution system. The lime that is used for stabilisation at the surface water treatment plant contains high levels of iron and manganese. This practise is currently being changed and in future caustic soda will be used. Although the effluent conforms to the guidelines for iron and manganese, there is still an enormous amount of post precipitation in the pipelines and reservoirs taking place. This is confirmed by consumer complaints from time to time. In comparison one can see that the turbidity is effectively removed in the reclamation process.

The bacteriological indicators are an indication of the effectiveness of the treatment processes and in particular disinfection. Chlorine is the main disinfectant used in Namibia. One can see that the levels for free and total chlorine are quite high. It does happen from time to time that low chlorine levels prevail in the distribution system. This is mainly due to a high chlorine demand after the first rains. As mentioned earlier, organic material is washed into the Dams either from human activity or from cattle farming. Sometimes chlorine demands of 23 mg/L treatment are measured. There is no way that the chlorine facilities at the treatment plant can handle this. It puts an enormous burden on the post-chlorination in the network. The temperatures are mostly above 20°C in the distribution system, sometimes being as high as 34°C in certain parts of the system. With the higher levels of Dissolved Organic Carbon (DOC) it is hard to control bacteria growth in the distribution system. This is reflected in the rather high Heterotrophic Plate Counts, sometimes prevailing in the system.

Much is being debated about Natural Organic Matter (NOM), which is indicated by DOC and UV_{254} measurement, and its negative effects on water treatment and distribution. The amount of Trihalomethanes (THMs) formed when high levels of DOC are present during disinfection with chlorine is evident and reflected in the table. One can see that the DOC and THMs from the NGWRP are extremely low.

Salinity, reflected by Conductivity and Total Dissolved Solids (TDS) measurement is a problem in Windhoek. Although the percentage of borehole water introduced is fairly low, the effect of the recycled water is clearly evident. With each cycle the TDS concentration increases. The City of Windhoek is considering to start research on reverse osmosis in the near future at the NGWRP. This will improve the water quality from the NGWRP even more and will lower the TDS.

The health research program covers virus testing for cytopathogenic effect and polymerase chain reaction, toxicity (Waterflea lethality and Urease enzyme) as well as Ames Salomonella mutagenicity testing. These tests are negative for the water sources providing Windhoek with treated drinking water. A special program to trace the removal of endocrine disrupting compounds (EDCs) and medical substances has until now not shown that these compounds are present in the final water of the NGWRP.

CONCLUSIONS

The quality of the drinking water in Windhoek is of an exceptional high standard, which is evident when the 95% percentile is applied to evaluate the data. Water reclamation and reuse cannot be seen in isolation but forms part of the integrated water resource management of the total urban water cycle. When reclaiming wastewater for potable use, there is no room for complacency. As new health concerns or operational challenges appear, relevant ways have to be found to deal with them. In this way the City of Windhoek and its partners are actively involved in applied research and water quality control to improve systems and to safeguard the health of the consumers. The experience over a period of more than 40 years in Windhoek demonstrates that direct reclamation of domestic sewage effluent is a practical way to augment the potable water supply in arid regions. It does however require ongoing commitment, comprehensive planning and training for continued success. In this way potable reclamation in Windhoek is contributing towards a stable economy despite severe droughts and water shortages.

ACKNOWLEDGEMENTS

Over a period of 40 years many diligent people have contributed towards the success of reclamation in Windhoek. Your efforts are appreciated.

REFERENCES

Du Pisani, P.L. (2006). Surviving in an arid land: Direct reclamation of potable water at Windhoek's Goreangab Reclamation Plant. *Desalination,* **188**, 79–88.

Grabow, W.O.K. (2001). New methods for the detection of viruses: Call for review of drinking water quality guidelines. *Water Science and Technology*, **43**(12), 1–8.

Haarhoff, J. and van der Merwe, B. (1996). Twenty-five years of wastewater reclamation in Windhoek, Namibia. *Water Science and Technology*, **33,** 25–35.

Haarhoff, J., van der Walt, C.J. and van der Merwe, B. (1998). Process design considerations for the Windhoek water reclamation plant. *Proceedings of the Water Institute of Southern Africa (WISA)*, held in Cape Town, May 1998, unpublished report.

Isaäcson, M., Sayed, A.R. and Hattingh, W. (1987). *Studies on health aspects of water reclamation during 1974 to 1983 in Windhoek, South West Africa/Namibia.* Report WRC 38/1/87 to the Water Research Commission, Pretoria.

Liputa, G.I., Nikodemus, K. and Menge, J. (2008). Strategic water quality monitoring for drinking water safety in Windhoek, Proceedings of the Water Institute of Southern Africa (WISA), held in Sun City, May 2008, unpublished report.

Menge, J. (2006). Treatment of wastewater for re-use in the drinking water system of Windhoek, *Proceedings of the Water Institute of Southern Africa (WISA)*, held in Durban, May 2006.

Menge, J., du Pisani, P. and van der Merwe, B. (2007). Water quality control in a third world country: challenges to ensure good quality reclaimed water in Windhoek, Namibia. *Proceedings at the 6th International Water Association Specialist Conference on Wastewater Reclamation and Reuse for Sustainability-Guiding the growth of future reuse*, held in Antwerp, Belgium, October 2007.

Menge, J.G., Haarhoff, J., König, E., Mertens, R. and van der Merwe, B. (2001). Occurrence and removal of Giardia and Cryptosporidium at the Goreangab reclamation plant. *Water Science and Technology*, **1**(1), 97–106.

IWA Publishing's authorised EU representative for General Product Safety Regulations is Diane D'Arras, 15 rue Duret, 75116 Paris, France, e-mail: safety@iwap.co.uk.

Printed and bound by CPI Group (UK) Ltd, Croydon, CR0 4YY
12/05/2026
02108880-0004